Local Operators in Integrable Models I

Mathematical
Surveys
and
Monographs

Volume 256

Local Operators in Integrable Models I

Michio Jimbo
Tetsuji Miwa
Fedor Smirnov

AMERICAN
MATHEMATICAL
SOCIETY

Providence, Rhode Island

2020 *Mathematics Subject Classification.* Primary 82B20, 82B23, 16T25, 17B37, 81T40, 81R50.

For additional information and updates on this book, visit
www.ams.org/bookpages/surv-256

Library of Congress Cataloging-in-Publication Data

Names: Jimbo, M. (Michio), author. | Miwa, T. (Tetsuji), author. | Smirnov, F. A., author.
Title: Local operators in integrable models / Michio Jimbo, Tetsuji Miwa, Fedor Smirnov.
Description: Providence, Rhode Island : American Mathematical Society, [2021]- | Series: Mathematical surveys and monographs, 0076-5376 ; volume 256 | Includes bibliographical references and index.
Identifiers: LCCN 2021006693 | ISBN 9781470465520 (paperback) | 9781470465766 (ebook)
Subjects: LCSH: Integral equations. | Operator theory. | Quantum field theory. | Statistical mechanics. | AMS: Statistical mechanics, structure of matter – Equilibrium statistical mechanics – Lattice systems (Ising, dimer, Potts, etc.) and systems on graphs arising in equilibrium statistical mechanics. | Statistical mechanics, structure of matter – Equilibrium statistical mechanics – Exactly solvable models; Bethe ansatz. | Associative rings and algebras – Hopf algebras, quantum groups and related topics – Yang-Baxter equations. | Nonassociative rings and algebras – Lie algebras and Lie superalgebras – Quantum groups (quantized enveloping algebras) and related deformations. | Quantum theory – Quantum field theory; related classical field theories – Two-dimensional field theories, conformal field theories, etc. in quantum mechanics. | Quantum theory – Groups and algebras in quantum theory – Quantum groups and related algebraic methods applied to problems in quantum theory.
Classification: LCC QC20.7.I58 J56 2021 | DDC 530.13–dc23
LC record available at https://lccn.loc.gov/2021006693

∞ The paper used in this book is acid-free and falls within the guidelines
established to ensure permanence and durability.
Visit the AMS home page at `https://www.ams.org/`

10 9 8 7 6 5 4 3 2 1   26 25 24 23 22 21

Contents

Introduction vii

Chapter 1. Formulation of the Problem 1
1.1. Six-vertex Model 1
1.2. Using Tensor Notation 2
1.3. The Main Object of Our Study 6
1.4. Spectral Parameter and R-Matrix 7
1.5. Six-vertex Model on a Plane 9
1.6. XXZ Anti-ferromagnet at Finite Temperature 12
1.7. Density Matrix and Entanglement von Neumann Entropy 16
1.8. Our Strategy in Volume I 17

Chapter 2. Spectral Problem in Matsubara Direction and Quantum Groups 19
2.1. Algebraic Bethe Ansatz 19
2.2. Algebra $U_q(\widehat{\mathfrak{sl}}_2)$ 22
2.3. Bazhanov-Lukyanov-Zamolodchikov Construction 27
2.4. Higher Spins in Matsubara Space 35
2.5. Q-Operators 41
2.6. Destri-deVega Equation 43

Chapter 3. Fermions 47
3.1. Intertwiner and Quasi-intertwiner for Fused Modules 48
3.2. Operators $\mathbf{k}(\zeta,\alpha)$ and $\phi(\mathbf{k})(\zeta,\alpha)$ 50
3.3. Annihilation Operators 59
3.4. Creation Operators 61
3.5. Fermionic Creation Operators 65
3.6. Homogeneous Versus Inhomogeneous Cases: Russian Doll Construction 71
3.7. Commutation Relations Between Creation and Annihilation Operators 76
3.8. Summary 80

Chapter 4. Main Theorem 83
4.1. Fermionic Basis and Difference Equations 83
4.2. Deformed Abelian Differentials 85
4.3. Main Theorem 93
4.4. Completeness in Homogeneous Case 102
4.5. Summary 108

Chapter 5. Applications and Generalisations 111
5.1. Function $\omega(\zeta,\xi|\alpha)$ via Integral Equation 111

5.2. Main Theorem and Inverse Problem 117
5.3. The Case $\alpha = 0$ 126
5.4. XXX Case 149
5.5. Remarks on XYZ Case 154

Appendix A. Quasi-classical Limit and Algebraic Geometry 165
A.1. Algebraic Interpretation of Quantum Results 165
A.2. Canonical Differential in the Classical Case 167
A.3. Riemann Surfaces 171
A.4. Affine Jacobi Variety 174
A.5. Classical Interpretation of Fermionic Basis 178

Notation 181

Bibliography 187

Index 191

Introduction

Integrable models in field theory and statistical mechanics constitute a research area that has made significant advances in the last few decades. Many of the deep results are intimately related in one way or another to the understanding of local operators of the model. The present book is aimed to expound upon the structure of the space of local operators.

We start this introduction by some historical remarks, which are by no means complete. Historical reviews at a more technical level will be given in the first chapters of both Volume I and Volume II.

The theory of quantum integrable models has a dual origin.

The first discovery in the field is due to Bethe [**Be**] who proposed a method for solving a one-dimensional model of isotropic magnet. This method is known now as the Bethe Ansatz. Then comes the famous solution of the two-dimensional Ising model by Onsager [**O1**]. Onsager also pointed out a deep relation between two-dimensional classical statistical mechanics and one-dimensional quantum mechanics [**O2**]. This observation set Bethe's and Onsager's work into a common perspective. The Bethe Ansatz technique was generalised and applied to various other models, including the Bose gas with a δ function interaction. Using this method, Yang and Yang [**YY**] investigated the thermodynamics. Onsager's work was further developed by Lieb [**Li**] and Sutherland [**Su**] who were able to solve the ice model. Finally a giant step was taken by Baxter who solved the eight-vertex and related models [**Ba1**], [**Ba2**], introducing novel machinery and thereby laying the foundation for all further progress in the field [**Ba3**].

The second development came from the theory of soliton equations. The seminal paper by Gardner, Greene, Kruskal, and Miura [**GGKM**] on the Korteweg-de Vries (KdV) equation opened up the whole new subject of integrable non-linear differential equations. Among important contributions are the work of Lax [**La**] who reformulated the equation in terms of Lax pair, and of Faddeev and Zakharov [**ZF**] who reinterpreted the KdV equation as a completely integrable Hamiltonian system. Faddeev and Takhtajan [**FT**] found an integrable model of relativistic classical field theory, known now as the sine-Gordon (sG) model. There arose naturally the issue of quantisation. Faddeev and Korepin [**FK**] successfully addressed the semi-classical quantisation, revealing a striking feature of the model: the spectrum includes solitons. These are excitations which correspond to non-trivial classical solutions and yet do not require the introduction of additional fields, contrary to the naïve particle-field correspondence. Coleman [**C**] and Mandelstam [**Ma**] explained that solitons can be interpreted as fermions in the massive Thirring model. This was the first occurrence of duality in Quantum Field Theory (QFT). Another unusual feature of the model, discovered in [**FK**] by semi-classical quantisation, is that the scattering matrices completely factorise. This property allowed the Zamolodchikov brothers [**ZaS**, **ZZ**] to compute the S-matrices exactly by a bootstrap approach.

The Quantum Inverse Scattering Method (QISM) proposed by Faddeev, Sklyanin, and Takhtajan [**STF**] synthesised these two major streams in the field of integrable models (see also [**KIB**] for a review). The main achievement of QISM is in the formulation of the Algebraic Bethe Ansatz (ABA). This method is based on the notion of R-matrix, a mathematical abstraction of a set of Boltzmann weights which plays a fundamental role in Baxter's work. As it turned out, certain R-matrices coincide with the factorised S-matrices of the Zamolodchikov brothers. The only explanation for this coincidence would be that there are not so many good mathematical objects around. The notion of R-matrix eventually gave birth to the theory of quantum groups. After the first example found by Kulish and Reshetikhin [**KR**], the general formulation was given by Drinfeld [**Dr**] and Jimbo [**Ji1**].

The most complicated part of the theory of integrable models concerns the computation of off-shell objects: form factors, correlation functions, etc. Achievements in this direction include the following. For relativistic models of QFT, Smirnov formulated a system of axioms for form factors, and solved the corresponding equations for a number of models [**Sm**]. Subsequently, correlation functions for the six-vertex model (or equivalently, static correlation functions for the XXZ model) were found by Jimbo, Miwa, and collaborators [**JM**]. The latter work relies heavily on the theory of quantum groups. Surprisingly, form factors in Conformal Field Theory (CFT) and correlation functions on the lattice are described by quite similar (actually dual) equations, though the variables involved have completely different meaning. This is yet another instance which indicates that good mathematical structures are limited in number. Later on, using ABA, Kitanine, Maillet, and Terras [**KMT**] generalised the results of [**JM**] to correlation functions for the XXZ model with a magnetic field. Göhmann, Klümper, and Seel [**GKS**] managed further generalisation to the case of finite temperature. The works [**KMT**, **GKS**] use the determinant formula found much earlier by Slavnov [**Sl**]. All these formulas for correlation functions are given in terms of multiple integrals which are hard to investigate. It is all the more so in the last two cases where the integrand involves certain functions which are not quite explicit.

We finish the historical part of Introduction by mentioning a series of papers by Bazhanov, Lukyanov, and Zamolodchikov [**BLZ1**, **BLZ2**, **BLZ3**]. The construction based on quantum groups proposed there was very important for us.

This monograph summarises results of our long effort on computing expectation values of local operators in integrable models [**BJMST1**, **BJMST2**, **JMS1**, **BJMS**, **JMS2**]. It is not easy to discuss here the details of this work, so we prefer rather to motivate the reader by demonstrating some examples of our results. We shall consider two best known examples, one on the lattice and the other in the continuum.

As an example of a lattice model, we choose the isotropic Heisenberg antiferromagnet. Formally the Hamiltonian is written as

$$H_{\mathrm{XXX}} = \sum_{j=-\infty}^{\infty} \sum_{a=1}^{3} \sigma_j^a \sigma_{j+1}^a \,.$$

To make sense of this formal definition, we start with a finite chain of $2L$ sites $j = -L+1, \cdots, L$. On the space $\left(\mathbb{C}^2\right)^{\otimes 2L}$, we define operators σ_j^a given by conventional Pauli matrices (see (1.2) below) acting non-trivially on the j-th copy of $\mathbb{C}^2$ and as identity elsewhere. We impose the periodic boundary condition and take the limit $L \to \infty$. Denote by $|\mathrm{vac}\rangle$ the ground state, and consider the correlation functions

$$\langle \sigma_1^3 \sigma_n^3 \rangle = \frac{\langle \mathrm{vac}|\sigma_1^3\sigma_n^3|\mathrm{vac}\rangle}{\langle \mathrm{vac}|\mathrm{vac}\rangle}.$$

Long ago, Takahashi [**T**] found a nice formula for $n = 3$:

$$\langle \sigma_1^3 \sigma_3^3 \rangle = \frac{1}{3} - \frac{16}{3}\log 2 + 3\zeta(3),$$

where $\zeta(s)$ is the Riemann zeta function. The appearance of a ζ-value is quite remarkable. Later, Boos and Korepin [**BK**] managed to evaluate the multiple integral formulas, obtaining

$$\langle \sigma_1^3 \sigma_4^3 \rangle = \frac{1}{3} - 12\log 2 + \frac{74}{3}\zeta(3) - \frac{56}{3}\zeta(3)\log 2 - 6\zeta(3)^2 - \frac{125}{6}\zeta(5) + \frac{100}{3}\zeta(5)\log 2.$$

We shall see that this kind of formulas hold in general: $\langle \sigma_1^3 \sigma_n^3 \rangle$ is a polynomial of degree $\left[\frac{n}{2}\right]$ of odd ζ-values with argument up to $2n-3$ (and $\log 2$ as well) with rational coefficients. This was confirmed by [**SST**] where the correlation functions were computed up to $n = 8$. We compute them up to $n = 11$. For this value the exact correlation function is already quite close to Lukyanov's asymptotic prediction [**Lk1**, **LT**]. We are also able to compute the entanglement density matrices, i.e., the expectation values of all local operators, up to 10 sites. In the paper [**BGKS**] evidence was given that the multiple integral formulas factorise even in the case of finite temperature and magnetic field. This work gave an important impetus to our study.

As an example of QFT, we consider the sine-Gordon model. The Hamiltonian is

$$H_{\mathrm{sG}} = \frac{1}{4\pi}\int_{-\infty}^{\infty} \left[\partial_0\varphi(x)^2 + \partial_1\varphi(x)^2 + \mu\big(1-\cos(\beta\varphi(x))\big)\right] dx,$$

where ∂_μ are derivatives with respect to space-time coordinates; we shall also use $\partial_\pm$ for the light-cone derivatives. The commutator of $\varphi(x)$ and $\partial_0\varphi(x)$ is canonical.

The problem which we address here concerns the computation of one-point functions of local operators:

$$\langle \mathcal{O}(0) \rangle = \frac{\langle \mathrm{vac}|\mathcal{O}(0)|\mathrm{vac}\rangle}{\langle \mathrm{vac}|\mathrm{vac}\rangle},$$

where $|\mathrm{vac}\rangle$ is again the ground state. Lukyanov and Zamolodchikov [**LZ**] computed the one-point functions for the operators $e^{ia\varphi(0)}$ (primary fields). Though this is a remarkable result, for application to perturbed CFT one needs to know one-point functions of descendants as well. The simplest non-trivial result was found by Fateev, Fradkin , Lukyanov, Zamolodchikov, and Zamolodchikov [**FFLZZ**]:

$$\langle (\partial_+\varphi)^2(\partial_-\varphi)^2 e^{ia\varphi} \rangle = -m^4 F_1(a)F_1(-a)\langle e^{ia\varphi}\rangle,$$

where we omit the argument 0 writing φ instead of $\varphi(0)$,

$$F_{2j-1}(a)=((j-1)!)^2\gamma\Big(\frac{2a-(2j-1)\beta}{2\eta}\Big)\gamma\Big(\frac{(2j-1)\beta^{-1}-2a}{2\eta}\Big),$$

$\eta=\beta-\beta^{-1}$, and $\gamma(x)=\Gamma(x)/\Gamma(1-x)$. The coupling constant μ has the dimension of mass to the power $2-2\beta^2$ because of the anomalous dimension of $\cos(\beta\varphi)$. The mass scale m is proportional to $\mu^{\frac{1}{2(1-\beta^2)}}$ with the proportionality coefficient known exactly due to Al. Zamolodchikov [**ZaA**]. In general, the levels (number of derivatives) for two light cone components (chiralities) must coincide due to Lorentz invariance. The simplest case above has level 2. The next non-trivial case is one on level 4. We shall see that there are two non-trivial expectation values. To describe them introduce

$$A_\epsilon=\frac{\eta^{-2}}{72}\Big\{\big(3-4a^2(\eta^2+2)\big)(\partial_\epsilon\varphi)^4+12(1-\eta^2)(\partial_\epsilon^2\varphi)^2\Big\},$$

$$B_\epsilon=\frac{\eta^{-2}}{108}(\beta+\beta^{-1})\Big\{(3+4a^2)(\partial_\epsilon\varphi)^4+12(\partial_\epsilon^2\varphi)^2)\Big\},$$

where $\epsilon=\pm$. Then

$$\langle(A_++aB_+)(A_-+aB_-)e^{ia\varphi}\rangle=m^8F_1(a)F_3(-a)\langle e^{ia\varphi}\rangle\,,$$

$$\langle(A_+-aB_+)(A_--aB_-)e^{ia\varphi}\rangle=m^8F_1(-a)F_3(a)\langle e^{ia\varphi}\rangle\,.$$

The second equation follows from the first one by C-symmetry.

We shall see that the general picture is as follows. Given a level L, we take two sets I^+, I^- of distinct odd positive integers, which are equal in size, such that the sum of elements of their union equals L. For such a datum, there is an operator of level L in both chiralities whose expectation value equals

$$m^{2L}\prod_{j\in I^+}F_j(a)\prod_{j\in I^-}F_j(-a)\,.$$

Thus we have a general method for exact computation of expectation values, both on the lattice and in the continuum. The attractive feature is that it is based on the same construction. Moreover our method allows a simple and straightforward generalisation to the case of non-zero temperature, or even to that of the generalised Gibbs ensemble.

Both the isotropic anti-ferromagnet and the sine-Gordon model can be considered as limits of the anisotropic XXZ magnet (inhomogeneous one in the second case). Due to Onsager's observation [**O2**], this magnet is closely related to the six-vertex model. The main tool which we shall use is the fermionic basis. In order to explain its meaning we invoke the analogy with CFT. In CFT, the space of all local operators is organised into a direct sum of irreducible representations of the Virasoro algebra. This structure is decisive for obtaining many exact results. Similarly, for spin chains we consider local observables, i.e., operators with finite supports. The linear span of these operators does not look a very interesting object. However, we manage to introduce on it the structure of a module created by one family of bosonic operators and two families of fermionic operators. The fermionic basis thus constructed has a great advantage that the expectation values for its elements are simple— basically given by determinants. The main goal of Volume I is to explain the construction and applications of the fermionic basis.

We now outline the contents of the present volume.

In Chapter 1, we explain the formulation of the problem and introduce our notation. We consider the six vertex model on an infinitely long cylinder with finite circumference. We call the generatrix the *Space* direction and the circumference the *Matsubara* direction. The main object of our study is the expectation values of quasi-local operators. The latter are operators localised on a finite segment in the *Space* direction having a half-infinite tail of the form $q^{\alpha\sum_{j=-\infty}^{0}\sigma_j^3}$; see (1.9). We explain how the expectation values thus formulated are related in various limits to those at a finite temperature or at zero temperature on the infinite plane, as well as to density matrix and entanglement entroy. We then explain our strategy to evaluate the expectation values by introducing fermionic basis.

In Chapter 2, we prepare algebraic tools which will be extensively used in the later Chapters. We begin with the basics of algebraic Bethe ansatz in the *Matsubara* direction, algebra $U_q(\widehat{\mathfrak{sl}}_2)$ and the universal R-matrix. We then proceed to the Bazhanov-Lukyanov-Zamolodchikov construction based on representations of the q-oscillator algebra Osc. Used originally to construct Q-operators, this method is indispensable for the construction of fermionic basis given in the next Chapter. We also touch upon the Destri-de Vega integral equation for the distribution of Bethe roots, which will become a convenient tool to handle the limit $\mathbf{n} \to \infty$ to infinite *Matsubara* chain.

Chapter 3 is devoted to the construction of fermionic annihilation and creation operators. Starting with the R-matrix whose auxiliary space is the tensor product of oscillator and two-dimensional representations, we introduce certain operator $\mathbf{k}(\zeta,\alpha)$ as a trace of monodromy matrix acting on operators on a finite interval. The fermionic annihilation operators $\mathbf{b}(\zeta,\alpha)$, $\mathbf{c}(\zeta,\alpha)$ are defined from the singular part of $\mathbf{k}(\zeta,\alpha)$. They are shown to anti-commute and are naturally extended to the whole space of quasi-local operators. The bosonic creation operators $\mathbf{t}^*(\zeta,\alpha)$ are simple to construct; they are basically the adjoint action by local integrals of motion of the XXZ chain. In contrast, the construction of fermionic creation operators $\mathbf{b}^*(\zeta,\alpha)$, $\mathbf{c}^*(\zeta,\alpha)$ is more elaborate. Their construction and derivation of their properties, including the (anti-)commutation relations with the annihilation operators, occupy Section 3.5 through Section 3.7. We give in particular the Russian doll construction in Section 3.6 which allows us to express the creation operators in the homogeneous chain compactly in terms of those of finite inhomogeneous chains. Applying the creation operators to the primary field we obtain the fermionic basis of quasi-local operators.

In Chapter 4, we combine the results of the preceding Chapters to prove the main theorem. As a mathematical tool we introduce deformed Abelian differentials and deformed Riemann bilinear identities. We prove the main theorem which states that the expectation values of fermionic basis can be expressed in terms of two functions, $\rho(\zeta|\alpha)$ and $\omega(\zeta,\alpha)$. We show also that the fermionic operators indeed create a basis of quasi-local operators on a homogeneous chain. We end this Chapter explaining the relation of our construction to cohomologies of affine Jacobi varieties.

In Chapter 5, we discuss various applications of the fermionic construction. We first rewrite the function $\omega(\zeta,\alpha)$ into a form suitable for taking the limit $\mathbf{n} \to \infty$. The main theorem allows us to compute expectation values of physically interesting operators, once we know their expression in the fermionic basis. We explain that the main theorem can again be used as a practical tool for finding the relevant coefficients. After examining the limit $\alpha \to 0$, we present numerical computation of the entanglement entropy in the XXZ model. We also consider the RSOS reduction corresponding to the minimal unitary series of CFT. We then present results for the entanglement entropy and correlation functions in the XXX model at zero temperature. Lastly we make a comment on the attempt to extend our approach to the completely anisotropic XYZ model.

In the Appendix we discuss the quasi-classical limit, and explain the meaning of our construction in the light of the theory of Abelian integrals. First we compute the quasi-classical limit of the canonical differential discussed in Chapter 4, expressing it in terms of differentials on a hyper-elliptic Riemann surface. We give a brief summary of basic facts about Riemann surfaces. In the quasi-classical limit, the algebra of local observables turn into the Poisson algebra generated by entries of the classical monodromy matrix. Via separation of variables, the Jacobi inversion formula allows us to express these matrix elements as meromorphic functions on the affine Jacobi varieties. We show that the fermionic creation operators have purely classical origin: on the top cohomology of the affine Jacobi variety, the operators $\mathbf{c}_j^*$ act by removing the first kind differentials and $\mathbf{b}_j^*$ by multiplying by the second kind differentials.

Acknowledgements.

We would like to thank our coauthors Herman Boos and Yoshihiro Takeyama for long and fruitful collaboration.

We are indebted to many colleagues and friends with whom we shared ideas: Vladimir Bazhanov, Vladimir Fateev, Boris Feigin, Frank Göhmann, Nikolai Kitanine, Andreas Klümper, Karol Kozlowski, Sergei Lukyanov, Jean-Michel Maillet, Nikolai Reshetikhin, Nikita Slavnov, Junji Suzuki, Minoru Takahashi, Leon Takhtajan, Veronique Terras, Jörg Teschner, Alexander Zamolodchikov and Alexei Zamolodchikov.

September, 2020

Michio Jimbo

Tetsuji Miwa

Fedor Smirnov

CHAPTER 1

Formulation of the Problem

1.1. Six-vertex Model

Our basic object is the six-vertex model, which is a model of classical statistical mechanics on the square lattice. As mentioned earlier, due to Onsager [**O2**] it is closely related to a quantum mechanical problem on the one-dimensional lattice: the anisotropic Heisenberg XXZ spin chain. We shall comment on this point later (see Section 1.6). We choose to formulate the problem as a model of classical statistical mechanics, because in the scaling limit it leads directly to Euclidean quantum field theory.

Let us describe the model. Consider the square lattice where an arrow is associated to every edge. The ice rule is imposed: at every vertex the number of incoming and outgoing arrows must be the same. There are six possible configurations at every vertex:

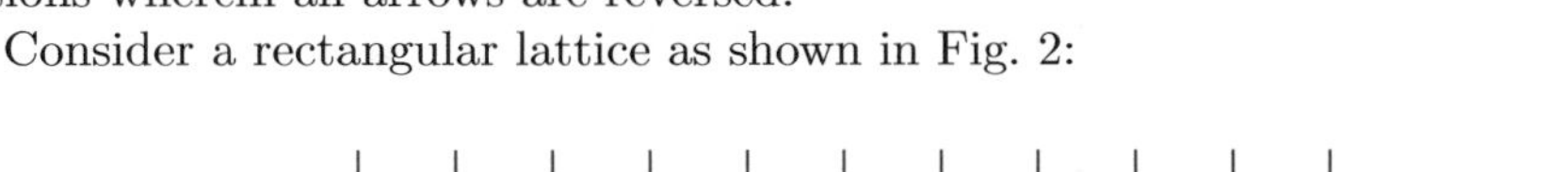

FIG. 1.

Without loss of generality we assign the same Boltzmann weight to two configurations wherein all arrows are reversed.

Consider a rectangular lattice as shown in Fig. 2:

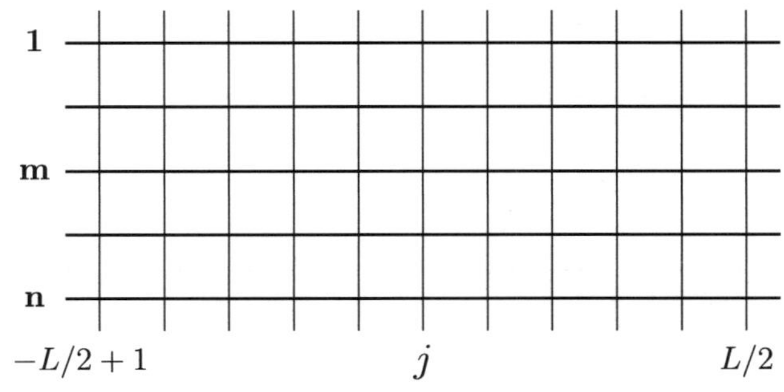

FIG. 2.

Throughout Volume I we shall call the vertical direction the *Matsubara* direction and the horizontal direction the *Space* direction. There are $\mathbf{n}$ rows in the *Matsubara* direction, and L columns in the *Space* direction. Both $\mathbf{n}$ and L are assumed to be even. The rows are counted from top to bottom, and the columns from left to right. As shown in Fig. 2, we use boldface letters such as $\mathbf{m}$ to label rows, and roman letters such as j to label columns. We choose the following limits for them:

$$\mathbf{1} \leq \mathbf{m} \leq \mathbf{n}, \qquad -L/2+1 \leq j \leq L/2\,.$$

It will be clear soon why we choose asymmetric notation for the two directions.

The first problem of statistical mechanics is to compute the partition function $Z_{L,\mathbf{n}}$. We place arrows on the edges of the lattice in all possible ways, compute the product of Boltzmann weights for every such configuration, and sum them up. Then we are interested in taking the limit $\mathbf{n} \to \infty$ and $L \to \infty$ in order to compute the free energy:

$$-\beta f = \lim_{\mathbf{n}\to\infty, L\to\infty} \frac{1}{\mathbf{n}L} \log Z_{L,\mathbf{n}}\,.$$

The limit is finite, and in the normal situation it is supposed to be independent of the boundary condition. Concerning the latter statement, however, it is not the case with the six-vertex model. So we have to explain which boundary condition we are going to deal with.

A conventional choice is to assume the periodic boundary condition in both directions. Geometrically, it corresponds to putting the lattice on a torus. We shall follow this prescription, but we shall not take the limits simultaneously. Rather we shall first take the limit $L \to \infty$, keeping $\mathbf{n}$ finite. Under a mild assumption this already leads to a well-defined object. Then the limit $\mathbf{n} \to \infty$ can be taken if needed. Geometrically, taking the limit $L \to \infty$ corresponds to putting the lattice on an infinite cylinder. To give a more precise definition we need some preparation. This will be done in the next section.

1.2. Using Tensor Notation

Let us return to the Boltzmann weights a, b, c.

For the genuine ice model, the Boltzmann weights are all equal. We shall consider the simplest generalisation of the ice model:

$$a : b : c = 1 : 1 : 2\cos(\pi\nu/2)\,.$$

In general we have more additional parameters (and very many in the inhomogeneous case) as will be clear soon.

The requirement of positivity of the Boltzmann weights leads to two possibilities:

Disordered phase. In this phase ν is real and $0 < \nu < 1$. The model is at criticality. Off criticality it is described by the eight-vertex model. The parameter ν plays the role of a coupling constant.

Ordered phase. In this phase ν is purely imaginary:

$$\nu = i\mu\,, \quad -\infty < \mu \leq 0\,.$$

In this case the model is not critical, and μ is related to the temperature. The point $\mu = -\infty$ corresponds to $T = 0$. The point $\mu = 0$ is critical, the system undergoes a logarithmic (Kosterlitz-Thouless) phase transition at this point.

We shall be mostly interested in the disordered regime. In place of ν we shall use more often

$$q = e^{\pi i \nu}\,. \tag{1.1}$$

It is convenient to use the tensor notation. Consider two spaces

$$\mathfrak{H}_{\mathbf{M}} = \left(\mathbb{C}^2\right)^{\otimes \mathbf{n}}\,, \qquad \mathfrak{H}_{\mathbf{S}} = \left(\mathbb{C}^2\right)^{\otimes L}\,.$$

We shall denote the $\mathbf{m}$-th $\mathbb{C}^2$ of $\mathfrak{H}_{\mathbf{M}}$ by $V_{\mathbf{m}}$, and the j-th $\mathbb{C}^2$ of $\mathfrak{H}_{\mathbf{S}}$ by V_j. We shall use the standard Pauli matrices

$$\sigma^1 = \begin{pmatrix} 0 & 1 \\ 1 & 0 \end{pmatrix}, \quad \sigma^2 = \begin{pmatrix} 0 & -i \\ i & 0 \end{pmatrix}, \quad \sigma^3 = \begin{pmatrix} 1 & 0 \\ 0 & -1 \end{pmatrix}. \tag{1.2}$$

Let $v_+ = {}^t(1,0), v_- = {}^t(0,1)$ be the standard basis of $\mathbb{C}^2$. We shall say that a vector $v_{\epsilon_1} \otimes \cdots \otimes v_{\epsilon_n}$ in the tensor product $\left(\mathbb{C}^2\right)^{\otimes n}$ has weight $\lambda = \epsilon_1 + \cdots + \epsilon_n$ and spin $\lambda/2$, $\epsilon_i = \pm$ being understood as $\epsilon_i = \pm 1$. For any $x \in \mathrm{End}\left(\mathbb{C}^2\right)$ define

$$x_{\mathbf{m}} = I \otimes \cdots \otimes \overset{\mathbf{m}}{\breve{x}} \otimes \cdots \otimes I \quad \in \mathrm{End}(\mathfrak{H}_{\mathbf{M}})$$

with I denoting the unit matrix, and similarly for $\mathfrak{H}_{\mathbf{S}}$.

The permutation operator $P \in \mathrm{End}(\mathbb{C}^2 \otimes \mathbb{C}^2)$ is defined by

$$P(v_1 \otimes v_2) = v_2 \otimes v_1\,.$$

In the matrix form

$$P = \begin{pmatrix} 1 & 0 & 0 & 0 \\ 0 & 0 & 1 & 0 \\ 0 & 1 & 0 & 0 \\ 0 & 0 & 0 & 1 \end{pmatrix}. \tag{1.3}$$

On every edge of the lattice we have an arrow. Let us identify arrows pointing up or to the right with the vector v_+, and arrows pointing down or to the left with v_-. We identify the horizontal or vertical lines with $\mathbb{C}^2$. The Boltzmann weights at the crossing of column j and row $\mathbf{m}$ are combined into the matrix

$$L^{6\mathrm{v}}_{j,\mathbf{m}} = \begin{pmatrix} a & 0 & 0 & 0 \\ 0 & b & c & 0 \\ 0 & c & b & 0 \\ 0 & 0 & 0 & a \end{pmatrix} \qquad \in \mathrm{End}\big(V_j \otimes V_{\mathbf{m}}\big)\,.$$

The summation of the Boltzmann weights over configurations is realised as matrix products of $L^{6\mathrm{v}}_{j,\mathbf{m}}$. Let us illustrate our convention for constructing these products on an example.

In Fig. 2, suppose the arrows on the boundary are fixed, and the summation is performed over the arrows in the bulk. The result is a matrix element of the matrix $T^{6\mathrm{v}}_{\mathbf{S},\mathbf{M}}$ which we define in two steps:

$$T^{6\mathrm{v}}_{j,\mathbf{M}} = L^{6\mathrm{v}}_{j,\mathbf{n}} L^{6\mathrm{v}}_{j,\mathbf{n-1}} \cdots L^{6\mathrm{v}}_{j,\mathbf{1}}\,, \tag{1.4}$$

$$T^{6\mathrm{v}}_{\mathbf{S},\mathbf{M}} = T^{6\mathrm{v}}_{-L/2+1,\mathbf{M}} T^{6\mathrm{v}}_{-L/2+2,\mathbf{M}} \cdots T^{6\mathrm{v}}_{L/2,\mathbf{M}}\,. \tag{1.5}$$

The general rule is that we multiply from left to right with respect to the first index, and from right to left with respect to the second. The reason for that will be clear when we shall explain the relation to the algebra $U_q(\widehat{\mathfrak{sl}}_2)$, see Subsection 2.2.1.

The operator $T^{6\mathrm{v}}_{j,\mathbf{M}}$ is called the *Matsubara* monodromy matrix. The upper and the right boundaries on Fig. 2 correspond to 'incoming' vectors (those on which $T^{6\mathrm{v}}_{\mathbf{S},\mathbf{M}}$ acts), while the lower and the left boundaries correspond to 'outgoing' ones. Several authors draw arrows on pictures like Fig. 2 to indicate these directions. This can be helpful sometimes but can be misleading at other times, so we shall not do so. Notice also that the same $T^{6\mathrm{v}}_{\mathbf{S},\mathbf{M}}$ can be written alternatively as

$$
\begin{aligned}
T^{6\mathrm{v}}_{\mathbf{S},\mathbf{m}} &= L^{6\mathrm{v}}_{-L/2+1,\mathbf{m}} L^{6\mathrm{v}}_{-L/2+2,\mathbf{m}} \cdots L^{6\mathrm{v}}_{L/2,\mathbf{m}}\,, \\
T^{6\mathrm{v}}_{\mathbf{S},\mathbf{M}} &= T^{6\mathrm{v}}_{\mathbf{S},\mathbf{n}} T^{6\mathrm{v}}_{\mathbf{S},\mathbf{n-1}} \cdots T^{6\mathrm{v}}_{\mathbf{S},\mathbf{1}}\,.
\end{aligned} \tag{1.6}
$$

Formulas (1.4) and (1.6) are equivalent simply because operators acting on different tensor components commute.

We shall write (1.4),(1.6) concisely as

$$
\begin{aligned}
T^{6\mathrm{v}}_{j,\mathbf{M}} &= \prod_{\mathbf{m}=\mathbf{1}}^{\overset{\curvearrowleft}{\mathbf{n}}} L^{6\mathrm{v}}_{j,\mathbf{m}}\,, \qquad T^{6\mathrm{v}}_{\mathbf{S},\mathbf{m}} = \prod_{j=-L/2+1}^{\overset{\curvearrowright}{L/2}} L^{6\mathrm{v}}_{j,\mathbf{m}}\,, \\
T^{6\mathrm{v}}_{\mathbf{S},\mathbf{M}} &= \prod_{\mathbf{m}=\mathbf{1}}^{\overset{\curvearrowleft}{\mathbf{n}}} T^{6\mathrm{v}}_{\mathbf{S},\mathbf{m}} = \prod_{j=-L/2+1}^{\overset{\curvearrowright}{L/2}} T^{6\mathrm{v}}_{j,\mathbf{M}}\,.
\end{aligned} \tag{1.7}
$$

These symbols for ordered products originate from the quantum inverse scattering method (QISM) [**STF**],[**KIB**].

It is clear that the partition function on the torus is written as a trace over both the *Space* and the *Matsubara* spaces:

$$
Z^{6\mathrm{v}}_{L,\mathbf{n}} = \mathrm{Tr}_{\mathbf{S}} \mathrm{Tr}_{\mathbf{M}} \left(T^{6\mathrm{v}}_{\mathbf{S},\mathbf{M}} \right)\,.
$$

Now we introduce an important object, the *Matsubara* transfer matrix

$$
T^{6\mathrm{v}}_{\mathbf{M}} = \mathrm{Tr}_j \left(T^{6\mathrm{v}}_{j,\mathbf{M}} \right)\,.
$$

Clearly, we can rewrite the partition function in terms of the *Matsubara* transfer matrix:

$$
Z^{6\mathrm{v}}_{L,\mathbf{n}} = \mathrm{Tr}_{\mathbf{M}} \left(\left(T^{6\mathrm{v}}_{\mathbf{M}} \right)^L \right)\,.
$$

We are interested in the limit $L \to \infty$.

The *Matsubara* transfer matrix is symmetric with non-negative matrix elements, so the Perron-Frobenius theorem applies. Namely the eigenvector $|0\rangle$ with maximal eigenvalue T is unique up to a scalar multiple. Obviously, for $L \to \infty$

$$
Z^{6\mathrm{v}}_{L,\mathbf{n}} = T^L \left(1 + O\Big(\big(\frac{T_{\mathrm{next}}}{T} \big)^L \Big) \right)\,,
$$

where T_{next} is the sub-leading eigenvalue. So with exponential precision the partition function for large L is defined by T.

As far as the limit $L \to \infty$ is concerned, the periodic boundary condition in the *Space* direction is not very important; in the limit $L \to \infty$ we can replace $Z^{6\mathrm{v}}_{L,\mathbf{n}}$ by

$$
\mathrm{Tr}_{\mathbf{S}} \mathrm{Tr}_{\mathbf{M}} \left(X_{\mathbf{M}} T^{6\mathrm{v}}_{\mathbf{S},\mathbf{M}} \right)\,,
$$

where $X_{\mathbf{M}}$ is any matrix acting on $\mathfrak{H}_{\mathbf{M}}$, provided $\langle 0|X_{\mathbf{M}}|0\rangle \neq 0$. Inserting $X_{\mathbf{M}}$ amounts to changing the periodic boundary condition to a more general one.

In fact we have been slightly loose in the foregoing discussion. The ice rule implies that the transfer matrix decomposes into a block diagonal form, each block acting on a subspace of fixed weight. The Perron-Frobenius theorem applies to each of these blocks. It is known that the maximal eigenvector belongs to the subspace of weight 0. If we choose an operator $X_{\mathbf{M}}$ such that $X_{\mathbf{M}} = P_s X_{\mathbf{M}} P_s$, where P_s is the projector onto the subspace of weight s, then the partition function is described by the maximal eigenvalue in the subspace of weight s. Hence the partition function is sensitive to the choice of boundary condition. This is a peculiarity of the six-vertex model. In Volume II we shall take advantage of this opportunity to change the boundary condition. In the present volume, however, we shall always work with an infinite cylinder, implying that the limit has been taken from a finite lattice with the periodic boundary condition.

We finish this section with a generalisation of the partition function introducing one more parameter κ:

$$Z_{L,\mathbf{n}}^{6\mathrm{v},\kappa} = \mathrm{Tr}_{\mathbf{S}}\mathrm{Tr}_{\mathbf{M}}\left(q^{2\kappa S}T_{\mathbf{S},\mathbf{M}}^{6\mathrm{v}}\right),$$

where S is the total spin in the *Space* direction:

$$S = \frac{1}{2}\sum_{j=-L/2+1}^{L/2}\sigma_j^3\,.$$

Graphically, in the limit $L \to \infty$ discussed above, this partition function is presented as follows:

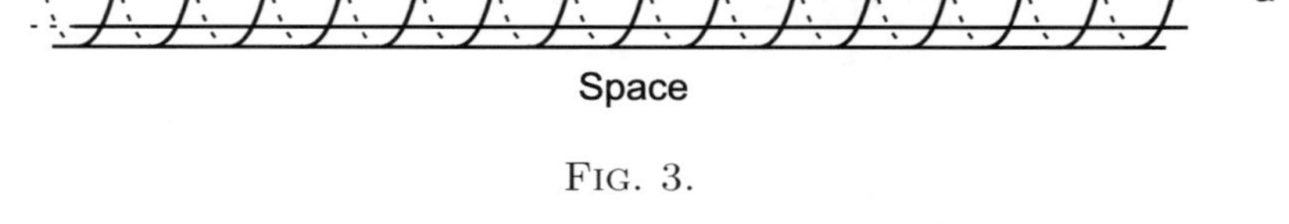

FIG. 3.

Here the vertices represent L-operators, summation over the arrows is implied, the black circles on an edge correspond to insertion of $q^{\kappa\sigma^3}$; in other words, it corresponds to multiplication by q^{κ} if the arrow on this edge points up, and by $q^{-\kappa}$ if it points down. The introduction of κ entails that the *Matsubara* transfer matrix is changed to

$$T_{\mathbf{M}}^{6\mathrm{v},\kappa} = \mathrm{Tr}_j\left(q^{\kappa\sigma_j^3}T_{j,\mathbf{M}}^{6\mathrm{v}}\right). \tag{1.8}$$

For κ purely imaginary, the matrix elements of $T_{\mathbf{M}}^{6\mathrm{v},\kappa}$ are real positive. Using the invariance

$$\left[\sigma_j^3 + \sigma_{\mathbf{m}}^3, L_{j,\mathbf{m}}^{6\mathrm{v}}\right] = 0\,,$$

we can move the line of vertices on Fig. 3 without changing $Z_{L,\mathbf{n}}^{6\mathrm{v},\kappa}$.

In general, in the presence of κ we cannot invoke the Perron-Frobenius theorem to ensure the uniqueness of the eigenvector with the eigenvalue of maximal absolute value.

Still, the uniqueness must hold for small κ by continuity.

1.3. The Main Object of Our Study

We shall use a notation which goes along with the spirit of Quantum Field Theory. Suppose that an element $X \in \mathrm{End}(\mathfrak{H}_{\mathrm{S}})$ acts nontrivially only on a part $V_k \otimes V_{k+1} \otimes \cdots \otimes V_l$ of the tensor product and as identity on the rest of the tensor components. We shall say that X is localised on the interval $[k,l]$, and write $X_{[k,l]}$ to indicate this fact. Eventually we shall be interested in the limit $L \to \infty$. Having this in mind, we use the letter $\mathcal{O}$ to denote a local operator, i.e., one localised on some finite interval.

Further, we define

$$S(k) = \frac{1}{2} \sum_{j=-L/2+1}^{k} \sigma_j^3 .$$

Throughout Volume I we shall deal with operators of the form

$$q^{2\alpha S(0)}\mathcal{O}\,, \tag{1.9}$$

which we call *quasi-local*. In other words, quasi-local operators are those which stabilise as $q^{\alpha\sigma_j^3}$ for $j \ll 0$ and as I_j for $j \gg 0$. In particular, when $\mathcal{O} = I$, we call $q^{2\alpha S(0)}$ primary field, borrowing the term from CFT. On these operators we define the linear functional

$$\mathcal{Z}_{L,\mathbf{n}}^{6\mathrm{v},\kappa}\left\{q^{2\alpha S(0)}\mathcal{O}\right\} = \frac{\mathrm{Tr}_{\mathbf{S}}\mathrm{Tr}_{\mathbf{M}}\left(T_{\mathbf{S},\mathbf{M}}^{6\mathrm{v}} q^{2\kappa S+2\alpha S(0)}\mathcal{O}\right)}{\mathrm{Tr}_{\mathbf{S}}\mathrm{Tr}_{\mathbf{M}}\left(T_{\mathbf{S},\mathbf{M}}^{6\mathrm{v}} q^{2\kappa S+2\alpha S(0)}\right)} . \tag{1.10}$$

Graphically the numerator of (1.10) is given by the partition function:

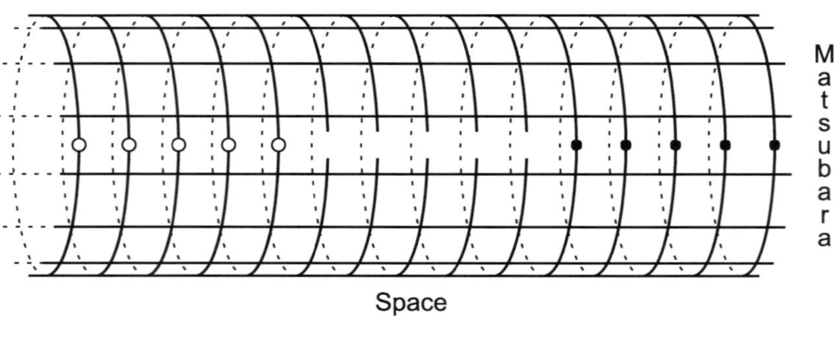

FIG. 4.

Here the summation is performed over the arrows on interior edges, while keeping fixed the arrows on broken edges. The arrows on the lower bank of the cut correspond to the row indices of the matrix $\mathcal{O}$, and those on the upper bank to the column indices. A black circle (resp. an open circle) on an edge indicates multiplication by q^{κ} (resp. $q^{\kappa+\alpha}$) if the arrow points up, and by $q^{-\kappa}$ (resp. $q^{-(\kappa+\alpha)}$) if it points down.

Define the adjoint action of S:

$$\mathbb{S}(\mathcal{O})=[S,\mathcal{O}]\,.$$

The ice rule implies that (1.10) does not vanish identically only if

$$\mathbb{S}(\mathcal{O})=0\,.$$

The definition of $\mathcal{Z}^{6\mathrm{v},\kappa}_{L,\mathbf{n}}$ includes normalisation in order that the limit

$$\mathcal{Z}^{6\mathrm{v},\kappa}_{\mathbf{n}}\left\{q^{2\alpha S(0)}\mathcal{O}\right\}=\lim_{L\to\infty}\mathcal{Z}^{6\mathrm{v},\kappa}_{L,\mathbf{n}}\left\{q^{2\alpha S(0)}\mathcal{O}\right\} \tag{1.11}$$

is well-defined. The main goal of the present volume is to compute $\mathcal{Z}^{6\mathrm{v},\kappa}_{\mathbf{n}}$ for the six-vertex model and its generalisations.

1.4. Spectral Parameter and R-Matrix

The solution of the six-vertex model hinges on introducing the spectral parameter and constructing commuting transfer matrices [**Ba3**]. Using a modern mathematical language, the integrability is due to its relation with the quantum affine algebra $U_q(\widehat{\mathfrak{sl}}_2)$. The meaning of this phrase will be explained later. We shall use extensively this mathematical interpretation. In order to conform with the standard notation for quantum affine algebras, we need to change our L-operator slightly. Namely we introduce

$$L_{j,\mathbf{m}}=\sigma^3_{\mathbf{m}}L^{6\mathrm{v}}_{j,\mathbf{m}}\sigma^3_j\,,$$

or in matrix form

$$L_{j,\mathbf{m}}=\begin{pmatrix}a&0&0&0\\0&-b&c&0\\0&c&-b&0\\0&0&0&a\end{pmatrix}\in\mathrm{End}\big(V_j\otimes V_{\mathbf{m}}\big)\,. \tag{1.12}$$

Let us show that using these Boltzmann weights (one of which is strangely negative) we do not change the partition function on the torus, and change the functional $\mathcal{Z}^{6\mathrm{v},\kappa}_{L,\mathbf{n}}$ only slightly. Using

$$\sigma^3_j\sigma^3_{\mathbf{m}}L^{6\mathrm{v}}_{j,\mathbf{m}}=L^{6\mathrm{v}}_{j,\mathbf{m}}\sigma^3_j\sigma^3_{\mathbf{m}}\,,$$

we obtain

$$\begin{aligned}T_{j,\mathbf{M}}&=\prod_{\mathbf{m}=\mathbf{1}}^{\overset{\curvearrowleft}{\mathbf{n}}}L_{j,\mathbf{m}}=\prod_{\mathbf{m}=\mathrm{even}}^{\overset{\curvearrowleft}{\mathbf{n}}}\sigma^3_{\mathbf{m}}L^{6\mathrm{v}}_{j,\mathbf{m}}\sigma^3_j\cdot\sigma^3_{\mathbf{m}-\mathbf{1}}L^{6\mathrm{v}}_{j,\mathbf{m}-\mathbf{1}}\sigma^3_j\\&=\prod_{\mathbf{m}=\mathrm{even}}^{\overset{\curvearrowleft}{\mathbf{n}}}\sigma^3_{\mathbf{m}}L^{6\mathrm{v}}_{j,\mathbf{m}}L^{6\mathrm{v}}_{j,\mathbf{m}-\mathbf{1}}\sigma^3_{\mathbf{m}-\mathbf{1}}=\prod_{\mathbf{m}=\mathrm{even}}^{\mathbf{n}}\sigma^3_{\mathbf{m}}\cdot T^{6\mathrm{v}}_{j,\mathbf{M}}\cdot\prod_{\mathbf{m}=\mathrm{odd}}^{\mathbf{n}}\sigma^3_{\mathbf{m}}\,.\end{aligned}$$

Set

$$T_{\mathbf{S},\mathbf{M}}=\prod_{j=-L/2+1}^{\overset{\curvearrowright}{L/2}}T_{j,\mathbf{M}}\,.$$

From

$$\prod_{\mathbf{m}=1}^{\mathbf{n}}\sigma^3_{\mathbf{m}}\cdot T^{6\mathrm{v}}_{j,\mathbf{M}}\cdot\prod_{\mathbf{m}=1}^{\mathbf{n}}\sigma^3_{\mathbf{m}}=\sigma^3_j\cdot T^{6\mathrm{v}}_{j,\mathbf{M}}\cdot\sigma^3_j\,,$$

it follows that

$$Z_{L,\mathbf{n}} = \mathrm{Tr}_S\mathrm{Tr}_{\mathbf{M}}\left(T_{\mathbf{S},\mathbf{M}}\right) = Z^{6\mathrm{v}}_{L,\mathbf{n}}\,,$$

$$\mathcal{Z}^{\kappa}_{L,\mathbf{n}}\left\{q^{2\alpha S(0)}\mathcal{O}\right\} = \frac{\mathrm{Tr}_{\mathbf{S}}\mathrm{Tr}_{\mathbf{M}}\left(T_{\mathbf{S},\mathbf{M}}q^{2\kappa S+2\alpha S(0)}\mathcal{O}\right)}{\mathrm{Tr}_{\mathbf{S}}\mathrm{Tr}_{\mathbf{M}}\left(T_{\mathbf{S},\mathbf{M}}q^{2\kappa S+2\alpha S(0)}\right)} = \mathcal{Z}^{6\mathrm{v},\kappa}_{L,\mathbf{n}}\left\{q^{2\alpha S(0)}V\mathcal{O}V\right\},$$

where we set $V = \prod_{j=1}^{L/2}\sigma^3_{-L/2+2j-1}$ and use

$$VT_{\mathbf{S},\mathbf{M}}V = \prod_{\mathbf{m}=\mathrm{odd}}^{\mathbf{n}}\sigma^3_{\mathbf{m}}\cdot T^{6\mathrm{v}}_{\mathbf{S},\mathbf{M}}\cdot\prod_{\mathbf{m}=\mathrm{odd}}^{\mathbf{n}}\sigma^3_{\mathbf{m}}.$$

From now on we shall use $L_{j,\mathbf{m}}$, recalling $L^{6\mathrm{v}}_{j,\mathbf{m}}$ only a couple of times. We use the analogs of (1.4) and derived objects associated with the new L-operator, omitting the superscript '6v'.

Consider the R-matrix

$$R(\zeta) = q^{1/2}\begin{pmatrix} q^{-1}-q\zeta^2 & 0 & 0 & 0\\ 0 & 1-\zeta^2 & -\zeta(q-q^{-1}) & 0\\ 0 & -\zeta(q-q^{-1}) & 1-\zeta^2 & 0\\ 0 & 0 & 0 & q^{-1}-q\zeta^2\end{pmatrix}. \tag{1.13}$$

The normalisation here may look unusual, but it is natural for us from the general point of view explained in Chapter 2. We shall refer to ζ as the spectral parameter. We have the inversion relation:

$$R(\zeta)^{-1} = -\frac{q^{-1}\zeta^2}{(1-q^2\zeta^2)(1-q^{-2}\zeta^2)}R(\zeta^{-1})\,. \tag{1.14}$$

Clearly (choosing $a = q^{-1/2}-q^{1/2}$)

$$L = L(1)\,,\quad L(\zeta) := R(\zeta q^{-1/2})\,. \tag{1.15}$$

The main property of the R-matrix is the Yang-Baxter equation

$$R_{1,2}(\zeta_1/\zeta_2)R_{1,3}(\zeta_1)R_{2,3}(\zeta_2) = R_{2,3}(\zeta_2)R_{1,3}(\zeta_1)R_{1,2}(\zeta_1/\zeta_2)\,. \tag{1.16}$$

We have already emphasised the importance of the maximal eigenvalue of the *Matsubara* transfer matrix for the computation of the partition function. It is equally important for the computation of $\mathcal{Z}^{\kappa}_{\mathbf{n}}$ to know maximal eigenvalues and eigenvectors of $T^{\kappa}_{\mathbf{M}}$ and $T^{\kappa+\alpha}_{\mathbf{M}}$.

These computations are based on the equation (1.16). The main point is that $T^{\kappa}_{\mathbf{M}}$ can be embedded into a one-parametric family of operators. Define

$$T_{j,\mathbf{M}}(\xi) = \overset{\curvearrowleft}{\prod_{\mathbf{m}=1}^{\mathbf{n}}} L_{j,\mathbf{m}}(\xi)\,, \tag{1.17}$$

then

$$T_{\mathbf{M}}(\xi,\kappa) = \mathrm{Tr}_j\left(T_{j,\mathbf{M}}(\xi)q^{\kappa\sigma^3_j}\right),\quad T^{\kappa}_{\mathbf{M}} = T_{\mathbf{M}}(1,\kappa)\,. \tag{1.18}$$

The Yang-Baxter equation implies the commutativity

$$[T_{\mathbf{M}}(\xi_1,\kappa), T_{\mathbf{M}}(\xi_2,\kappa)] = 0 \tag{1.19}$$

valid for all ξ_1, ξ_2. In Chapter 2 we shall see how this relation helps in finding the spectrum of $T_{\mathbf{M}}(\xi, \kappa)$.

Another use of the spectral parameter is that it allows a considerable generalisation of our construction. Namely we may consider a completely inhomogeneous lattice where a parameter ξ_j is attached to every *Space* site j and a parameter $\tau_{\mathbf{m}}$ to every *Matsubara* site $\mathbf{m}$. This amounts to replacing the L-operators as follows:

$$L_{j,\mathbf{m}} \longrightarrow L_{j,\mathbf{m}}(\xi_j/\tau_{\mathbf{m}})\,.$$

The point is that the transfer matrices constructed from

$$T_{j,\mathbf{M}}(\xi|\tau_{\mathbf{1}}, \dots, \tau_{\mathbf{n}}) = \overset{\curvearrowleft}{\prod_{\mathbf{m}=\mathbf{1}}^{\mathbf{n}}} L_{j,\mathbf{m}}(\xi/\tau_{\mathbf{m}})$$

still commute for different values of ξ_j with fixed $\tau_{\mathbf{1}}, \dots, \tau_{\mathbf{n}}$. We then define

$$T_{\mathbf{S},\mathbf{M}}(\{\xi_j\}|\tau_{\mathbf{1}}, \dots, \tau_{\mathbf{n}}) = \overset{\curvearrowright}{\prod_{j=-L/2+1}^{L/2}} T_{j,\mathbf{M}}(\xi_j|\tau_{\mathbf{1}}, \dots, \tau_{\mathbf{n}})\,.$$

Since we want to take the limit $L \to \infty$, it is natural to assume that the ξ_j's obey some fixed pattern for sufficiently large $|j|$. In Volume I we deal mostly with a homogeneous *Space* lattice, i.e., $\xi_j = 1$ for sufficiently large $|j|$. On the other hand, the inhomogeneities in the *Matsubara* direction may be very useful as we shall see in the next section.

1.5. Six-vertex Model on a Plane

In the present volume we shall consider various ways of taking the limit $\mathbf{n} \to \infty$. The most straightforward one for the homogeneous *Matsubara* chain brings us to the six-vertex model on the entire plane. In this case we shall set $\kappa = 0$, which is not actually a restriction, at least for real κ: the maximal eigenvalue in the *Matsubara* direction for $\mathbf{n} = \infty$ is independent of κ. On the other hand, it is useful to keep the *Space* direction partly inhomogeneous. So, we consider the partition function described by Fig. 5.

In principle one can choose independent inhomogeneity parameters above and below the cut. But such a generalisation would contradict to taking our limit from the periodic boundary condition, so we shall not consider it here.

For the ordered regime, the computation of such a partition function is possible by the method of vertex operators. We shall not describe it here in detail, referring the reader to the monograph [**JM**]. The same method suggested also a very similar result in the disordered regime, conjectured in [**JM2**], and proved subsequently in [**KMT**]. However, we would like to cite one result of this method which was an important source of inspiration for us.

We have operators $X_{[1,l]}$ localised on the interval $[1, l]$, and Fig. 5 defines the following linear functional on $\mathrm{End}\big((\mathbb{C}^2)^{\otimes l}\big)$:

$$h(\xi_1, \dots \xi_l, \alpha)(X_{[1,l]}) = \lim_{\mathbf{n}\to\infty} \lim_{L\to\infty} \mathcal{Z}^0_{L,\mathbf{n}}[\xi_1, \dots, \xi_l]\Big\{q^{2\alpha S(0)} X_{[1,l]}\Big\}\,.$$

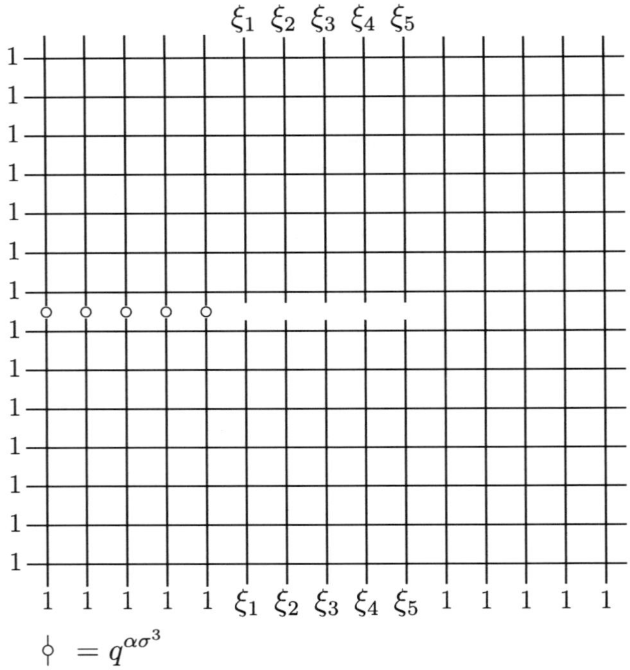

FIG. 5.

Here $\mathcal{Z}^0_{L,\mathbf{n}}[\xi_1,\dots,\xi_l]\Big\{q^{2\alpha S(0)}X_{[1,l]}\Big\}$ is given by (1.10) with $\kappa=0$, $\mathcal{O}=X_{[1,l]}$, and $T^{6\mathrm{v}}_{\mathbf{S},\mathbf{M}}$ replaced by

$$\prod_{j=-L/2+1}^{0} T_{j,\mathbf{M}}\cdot T_{1,\mathbf{M}}(\xi_1)\cdots T_{l,\mathbf{M}}(\xi_l)\cdot\prod_{j=l+1}^{L/2} T_{j,\mathbf{M}}.$$

This functional satisfies a linear q-difference equation

$$h(\xi_1,\dots,\xi_{l-1},\xi_l q^{-1},\alpha)=h(\xi_1,\dots,\xi_{l-1},\xi_l,\alpha)\circ\mathbf{A}(\xi_1,\dots,\xi_l,\alpha)\,, \tag{1.20}$$

which we call the reduced qKZ equation. The notation is as follows. The symbol $\circ$ means composition of operators. Operator

$$\mathbf{A}(\xi_1,\dots,\xi_l,\alpha):\mathrm{End}\big((\mathbb{C}^2)^{\otimes l}\big)\to\mathrm{End}\big((\mathbb{C}^2)^{\otimes l}\big)$$

is defined by

$$\mathbf{A}(\xi_1,\dots,\xi_l,\alpha)(X_{[1,l]})=T_{l,[1,l-1]}(\xi_l)q^{\alpha\sigma^3_l}\theta_l\Big(X_{[1,l]}\ \theta_l(T_{l,[1,l-1]}(\xi_l)^{-1})\Big)\,, \tag{1.21}$$

$$T_{l,[1,l-1]}(\xi_l)=\prod_{p=1}^{\overset{\curvearrowleft}{l-1}} R_{l,p}(\xi_l/\xi_p)\,. \tag{1.22}$$

Here and after, we use the anti-automorphism of $\mathrm{End}(\mathbb{C}^2)$

$$\theta(x)=\sigma^2x^t\sigma^2\,. \tag{1.23}$$

The following is called the crossing symmetry relation.

$$\theta_l\left(R_{l,p}(\zeta)\right) = (1-\zeta^2)(1-q^4\zeta^2)R_{l,p}(q\zeta)^{-1}. \tag{1.24}$$

Note that (1.22) involves only the *Space* sites and is defined via R-matrices, as opposed to the ones used before which mix the *Space* and the *Matsubara* sites and were defined via the L-operators. There is only a minor difference between the R-matrices and L-operators, but we prefer to keep it.

Introduce the adjoint action

$$\begin{aligned} \mathbb{R}_{j,j+1}(\xi_j/\xi_{j+1})(X) &= R_{j,j+1}(\xi_j/\xi_{j+1})\cdot X\cdot R_{j,j+1}(\xi_j/\xi_{j+1})^{-1}\,, \\ \mathbb{P}_{j,j+1}(X) &= P_{j,j+1}XP_{j,j+1}\,. \end{aligned} \tag{1.25}$$

We have then the symmetry relation

(1.26)

$$h(\xi_1,\dots,\xi_j,\xi_{j+1},\dots\xi_l,\alpha)\mathbb{R}_{j,j+1}(\xi_j/\xi_{j+1})^{-1} = h(\xi_1,\dots,\xi_{j+1},\xi_j,\dots,\xi_l,\alpha)\mathbb{P}_{j,j+1}\,,$$

reflecting the same symmetry of $\mathcal{Z}^0_{L,\mathbf{n}}$.

Eq. (1.26) allows us to write difference equations with respect to all ξ_j. We do not write these equations, because the notation becomes cumbersome and they will not be used in the present volume.

The solution to (1.20) which describes $h(\xi_1,\dots,\xi_l,\alpha)$ satisfies three additional conditions 0)–2) below:

0) $h(\xi_1\dots,\xi_l,\alpha)$ is a meromorphic function of $\xi_1,\dots,\xi_l$.

1) In the limit $\xi_l^{\pm1}\to\infty$, $h(\xi_1,\dots,\xi_l,\alpha)$ reduces to $h(\xi_1,\dots,\xi_{l-1},\alpha)$:

$$\begin{aligned} &h(\xi_1,\dots,\xi_l,\alpha)(X_{[1,l]}) \\ &\quad\to \frac{1}{\mathrm{Tr}_l\left(q^{-\alpha\sigma_l^3/2}\right)}h(\xi_1,\dots,\xi_{l-1},\alpha)\Big(\mathrm{Tr}_l\left(q^{-\alpha\sigma_l^3/2}X_{[1,l]}\right)\Big)\,. \end{aligned} \tag{1.27}$$

2) $h(\xi_1,\dots,\xi_l,\alpha)$ is regular on the diagonals $\xi_i=\xi_j$, $i\neq j$.

Together with these requirements, the equations (1.20), (1.26) uniquely define $h(\xi_1,\dots,\xi_l,\alpha)$.

In [**JM2**], there is given a multi-dimensional integral representation of the function $h(\xi_1,\dots,\xi_l,\alpha)$. One of the motivations for our work was dissatisfaction with this formula. We shall see later that actually the multiple integrals can be carried through, and that the function $h(\xi_1,\dots,\xi_l,\alpha)$ is defined in terms of only *one* transcendental function. But there is another, more important, reason to talk about the equation (1.20).

Consider the equation dual to (1.20). Namely, take a function $g(\xi_1,\dots,\xi_l)$ with values in $\mathrm{End}\big((\mathbb{C}^2)^{\otimes l}\big)$, which is meromorphic in all its variables and satisfies

$$g(\xi_1,\dots,\xi_{l-1},\xi_l,\alpha) = \mathbf{A}(\xi_1,\dots,\xi_l,\alpha)g(\xi_1,\dots,\xi_{l-1},\dots\xi_lq^{-1},\alpha)\,. \tag{1.28}$$

We shall refer to this as the dual reduced qKZ equation. From (1.20) and (1.28) it is clear that the scalar function

$$h(\xi_1,\dots,\xi_l,\alpha)g(\xi_1,\dots,\xi_l,\alpha)$$

is invariant under $\xi_j\to\xi_jq$ for all j, i.e., it is a quasi-constant. In practice this quasi-constant can be easily found, which allows an inductive computation (see Section 4.1).

There is an obvious difference between the way we treat (1.20) and (1.28); we are interested in one special solution to the former, whereas we are interested in all solutions to the latter. In the present volume the solutions in both cases will be described exhaustively. An interesting circumstance is that, while the solution to (1.20) is specific to the case of infinite *Matsubara* chain, the solutions to (1.28) have much wider applications as explained in Chapter 4.

1.6. XXZ Anti-ferromagnet at Finite Temperature

Let us discuss the relation to the XXZ spin chain. Consider the six-vertex model with a homogeneous spectral parameter in the *Space* direction, and fully non-homogeneous ones in the *Matsubara* direction.

We have a one-parametric commuting family of operators

$$T_{\mathbf{S}}(\zeta) = \mathrm{Tr}_{\mathbf{m}}\Bigl(\overset{\curvearrowright}{\prod_{j=-L/2+1}^{L/2}} R_{j,\mathbf{m}}(\zeta)\Bigr) . \tag{1.29}$$

The space transfer matrix corresponding to the $\mathbf{m}$-th row is $T_{\mathbf{S}}(q^{-1/2}/\tau_{\mathbf{m}})$.

We have

$$T_{\mathbf{S}}(1)/\left(q^{1/2}(q^{-1}-q)\right)^L = \mathrm{Tr}_{\mathbf{m}}\Bigl(\overset{\curvearrowright}{\prod_{j=-L/2+1}^{L/2}} P_{j,\mathbf{m}}\Bigr) = U ,$$

where

$$U = \overset{\curvearrowright}{\prod_{j=-L/2+1}^{L/2-1}} P_{j,L/2} , \tag{1.30}$$

is the operator of cyclic shift by one site to the left. Periodicity in the *Space* direction is implied: $L/2+k$ is identified with $-L/2+k$. The operator $T_{\mathbf{S}}(\zeta)$ is cyclic-invariant, i.e., commuting with U. Set

$$\check{t}_{\mathbf{S}}(\zeta) = T_{\mathbf{S}}(\zeta)U^{-1}/\left(q^{1/2}(q^{-1}-q\zeta^2)\right)^L . \tag{1.31}$$

We have $\check{t}_{\mathbf{S}}(1) = I$.

Clearly $\check{t}_{\mathbf{S}}(\zeta)$ does not change if we replace each $R_{j,\mathbf{m}}(\zeta)$ by

$$\widetilde{R}_{j,\mathbf{m}}(\zeta) = \zeta^{-\sigma^3_{\mathbf{m}}/2} R_{j,\mathbf{m}}(\zeta)\zeta^{\sigma^3_{\mathbf{m}}/2} . \tag{1.32}$$

Since (1.32) depends only on ζ^2, so does $\check{t}_{\mathbf{S}}(\zeta)$.

Let us study the behaviour of $\check{t}_{\mathbf{S}}(\zeta)$ as $\zeta^2 \to 1$. To this aim we set

$$\check{R}_{j,\mathbf{m}}(\zeta) = \widetilde{R}_{j,\mathbf{m}}(\zeta)P_{j,\mathbf{m}} , \tag{1.33}$$

where P is the permutation operator (1.3). Explicitly we have

$$\check{R}_{j+1,j}(\zeta) = q^{1/2}(q^{-1}-q\zeta^2)\Bigl(I + \frac{1-\zeta^2}{q^{-1}-q\zeta^2}h_{j,j+1}\Bigr) , \tag{1.34}$$

where

$$h = \sigma^+\otimes\sigma^- + \sigma^-\otimes\sigma^+ - q\tau^+\otimes\tau^- - q^{-1}\tau^-\otimes\tau^+ , \tag{1.35}$$

and $\tau^{\pm} = \frac{1}{2}(1 \pm \sigma^3)$. Noting that

$$h^2 = -(q+q^{-1})h\,,$$

we rewrite (1.34) further and define

$$\check{r}_{j,j+1}(u) = \check{R}_{j+1,j}(\zeta)/q^{1/2}(q^{-1} - q\zeta^2) = \exp\left(u h_{j,j+1}\right)\,,$$

where

$$u = -\frac{1}{q+q^{-1}} \log\left(\frac{q^{-1}\zeta^2 - q}{q^{-1} - q\zeta^2}\right)\,. \tag{1.36}$$

We will compute

$$\log \check{t}_{\mathbf{S}}(\zeta) = \sum_{p=1}^{L-1} \mathbf{I}_p u^p + O(u^L). \tag{1.37}$$

For this purpose, introduce a set of parameters $\mathbf{u} = \left(u_{-L/2+1}, u_{-L/2+2}, \dots, u_{L/2}\right)$ and define

$$\check{t}_{\mathbf{S}}[\mathbf{u}] = \mathrm{Tr}_{\mathbf{m}}\Big(\prod_{j=-L/2+1}^{\overset{\curvearrowright}{L/2}} \check{r}_{\mathbf{m},j}(u_j) P_{j,\mathbf{m}}\Big) U^{-1}\,. \tag{1.38}$$

We have $\check{t}_{\mathbf{S}}(\zeta) = \check{t}_{\mathbf{S}}[\mathbf{u}]|_{u_{-L/2+1}=\cdots=u_{L/2}=u}$. The cyclic symmetry takes the form of $\check{t}_{\mathbf{S}}[\mathbf{u}] \cdot U = U \cdot \check{t}_{\mathbf{S}}[\tilde{\mathbf{u}}]$ where $\tilde{\mathbf{u}} = (u_{L/2}, u_{-L/2+1}, \dots, u_{L/2-1})$. In particular, the equality

$$\check{t}_{\mathbf{S}}[u_{L/2}, 0, u_{-L/2+2}, \dots, u_{L/2-1}] = U^{-1} \cdot \check{t}_{\mathbf{S}}[0, u_{-L/2+2}, \dots, u_{L/2}] \cdot U. \tag{1.39}$$

holds. We proceed to

(1.40)

$$\check{t}_{\mathbf{S}}[\mathbf{u}] = \mathrm{Tr}_{\mathbf{m}}\Big(\check{r}_{\mathbf{m},-L/2+1}(u_{-L/2+1}) \prod_{j=-L/2+1}^{\overset{\curvearrowright}{L/2-1}} \check{r}_{j,j+1}(u_{j+1}) \prod_{j=-L/2+1}^{\overset{\curvearrowright}{L/2}} P_{j,\mathbf{m}}\Big) U^{-1}$$

$$= \mathrm{Tr}_{\mathbf{m}}\Big(\exp\left(u_{-L/2+1} h_{\mathbf{m},-L/2+1}\right) \prod_{j=-L/2+1}^{\overset{\curvearrowright}{L/2-1}} \exp\left(u_{j+1} h_{j,j+1}\right) \cdot P_{L/2,\mathbf{m}}\Big)\,.$$

Setting $u_{-L/2+1} = 0$ we can remove the trace:

$$\check{t}_{\mathbf{S}}[0, u_{-L/2+2}, \dots, u_{L/2}] = \prod_{j=-L/2+1}^{\overset{\curvearrowright}{L/2-1}} \exp\left(u_{j+1} h_{j,j+1}\right)\,.$$

It is obvious that terms in $\log \check{t}_{\mathbf{S}}[\mathbf{u}]$ of homogeneous degree less than L are all present in the union of terms appearing in $\log \check{t}_{\mathbf{S}}[0, u_{-\frac{L}{2}+2}, \dots, u_{\frac{L}{2}}]$, $\log \check{t}_{\mathbf{S}}[u_{-\frac{L}{2}+1}, 0, \dots, u_{\frac{L}{2}}]$, $\cdots$, $\log \check{t}_{\mathbf{S}}[u_{-\frac{L}{2}+1}, \dots, u_{\frac{L}{2}-1}, 0]$. Therefore, our task is to compute these terms.

Using the Campbell-Hausdorff formula, we define operators

$$h^{(p,m)} \in \mathrm{End}\left(\left(\mathbb{C}^2\right)^{\otimes(m+1)}\right)\,, \quad (2 \le m \le p)\,,$$

so that the following operator identity holds:

$$\log \check{t}_{\mathbf{S}}[0,u,\dots,u] = h^{(L-1)}_{[-L/2+1,L/2]} + O(u^L), \tag{1.41}$$

$$h^{(L-1)}_{[-L/2+1,L/2]} = u \sum_{j=-L/2+1}^{L/2-1} h_{j,j+1} + \sum_{p=2}^{L-1} u^p \sum_{m=2}^{p} \sum_{j=-L/2+1}^{L/2-m} h^{(p,m)}_{[j,j+m]}. \tag{1.42}$$

Here we denote by $h^{(p,m)}_{[j,j+m]}$ the operator $h^{(p,m)}$ acting on the interval $[j,j+m]$. It is a linear combination of iterated commutators of $h_{j,j+1}, h_{j+1,j+2}, \dots, h_{j+m-1,j+m}$, and its commutator degree is p. For example,

$$h^{(3,2)}_{j,j+2} = \frac{1}{12}\big([X,[X,Y]] + [Y,[Y,X]]\big) \text{ where } X = h_{j,j+1}, Y = h_{j+1,j+2}.$$

Collecting the terms in (1.42) and its cyclic shifts taking care not to double count, we reach the formula for $\mathbf{I}_p$ in (1.37):

$$\mathbf{I}_p = \begin{cases} \sum_{j=-L/2+1}^{L/2} h_{j,j+1} & (p=1) \\ \sum_{m=2}^{p} \sum_{j=-L/2+1}^{L/2} h^{(p,m)}_{[j,j+m]} & (p \geq 2) \end{cases}$$

We cannot compute the coefficients of u^p with $p \geq L$ in this way, but our goal is the limit of infinite L in any case. So, we will write formulas admitting incorrectness in those terms, e.g.,

$$T_{\mathbf{S}}(\zeta)/(q^{1/2}(q^{-1} - q\zeta^2))^L = \exp\left\{\sum_{p=1}^{\infty} \mathbf{I}_p u^p\right\} U.$$

Let us rewrite

$$h_{j,j+1} = \sigma^+_{j+1}\sigma^-_j + \sigma^-_{j+1}\sigma^+_j + \frac{q+q^{-1}}{4}(\sigma^3_{j+1}\sigma^3_j - 1) + \frac{q-q^{-1}}{4}(\sigma^3_{j+1} - \sigma^3_j)\,.$$

Summing up $h_{j,j+1}$ we observe that the last term containing a total difference cancels due to the periodic boundary condition. So we obtain

$$\mathbf{I}_1 = \sum_{j=-L/2+1}^{L/2} \left(\sigma^+_{j+1}\sigma^-_j + \sigma^-_{j+1}\sigma^+_j + \frac{q+q^{-1}}{4}(\sigma^3_{j+1}\sigma^3_j - 1)\right) = H_{\mathrm{XXZ}}\,,$$

which is the Hamiltonian of the XXZ spin chain.

This form uses the generators of the Lie algebra $\mathfrak{sl}_2$, and is convenient for applying the Campbell-Hausdorff formula which involves only commutators. For example we compute:

(1.43)

$$\begin{aligned} h^{(2,2)}_{[j,j+2]} &= \frac{1}{2}\big[h_{j,j+1}, h_{j+1,j+2}\big] \\ &\equiv \frac{1}{2}\Big\{\sigma^+_{j+2}\sigma^3_{j+1}\sigma^-_j - \sigma^-_{j+2}\sigma^3_{j+1}\sigma^+_j \\ &+ \frac{q+q^{-1}}{2}\Big(\sigma^-_{j+2}\sigma^+_{j+1}\sigma^3_j - \sigma^+_{j+2}\sigma^-_{j+1}\sigma^3_j - \sigma^3_{j+2}\sigma^+_{j+1}\sigma^-_j + \sigma^3_{j+2}\sigma^-_{j+1}\sigma^+_j\Big)\Big\}, \end{aligned}$$

where $\equiv$ means modulo total differences which cancel in the sum from $j = -L/2+1$ to $j = L/2$.

The crossing symmetry relation (1.24) can be rewritten as

$$R_{j,\mathbf{m}}(q^{-1}\zeta^{-1}) = -q^{-1}\zeta^{-2}\theta_j\big(R_{j,\mathbf{m}}(\zeta)\big)\,,$$

where the anti-automorphism θ is defined by (1.23). So, we have

$$\begin{aligned} T_{\mathbf{S}}(q^{-1}\zeta^{-1}) &= (-q^{-1}\zeta^{-2})^L\theta_{\mathbf{S}}\big(T_{\mathbf{S}}(\zeta)\big) \\ &= (q^{-1/2}(q-q^{-1}\zeta^{-2}))^L\,U^{-1}\exp\left\{\sum_{p=1}^{\infty}(-1)^{p-1}\mathbf{I}_p u^p\right\}. \end{aligned} \tag{1.44}$$

The second equality follows from

$$\theta_{\mathbf{S}}(\mathbf{I}_p) = (-1)^{p-1}\mathbf{I}_p\,. \tag{1.45}$$

We shall not prove this property in general, but rather consider two examples $\mathbf{I}_1$ and $\mathbf{I}_2$ at hand. The density $h_{j,j+1}$ of $\mathbf{I}_1$ comprises quadratic and linear terms in the σ-matrices, which are respectively even and odd with respect to $\theta_{\mathbf{S}}$. Since the odd part is a total difference, it does not contribute to the Hamiltonian. Similarly, the density $h_{[j,j+2]}$ of $\mathbf{I}_2$ contains odd and even terms, but the latter one is a total difference, hence $\mathbf{I}_2$ is odd.

Now we proceed to the XXZ chain at finite temperature following [**Kl**, **Su**]. Let us return to $T_{\mathbf{S},\mathbf{M}}$. In the space transfer matrix $T_{\mathbf{S}}(q^{-1/2}/\tau_{\mathbf{m}})$, we choose $\tau_{\mathbf{m}} = \tau^{-1}q^{-1/2}$ for even $\mathbf{m}$ and $\tau_{\mathbf{m}} = \tau q^{1/2}$ for odd $\mathbf{m}$. Then from (1.38) and (1.44) we have

$$\begin{aligned} \mathrm{Tr}_{\mathbf{M}}\left(T_{\mathbf{S},\mathbf{M}}\right) &= \big(T_{\mathbf{S}}(\tau)T_{\mathbf{S}}(\tau^{-1}q^{-1})\big)^{\mathbf{n}/2} \\ &= C^{\mathbf{n}L}\exp\left\{\mathbf{n}\sum_{p=1}^{\infty}\mathbf{I}_{2p-1}u^{2p-1}\right\}, \end{aligned} \tag{1.46}$$

where now $u = -(q+q^{-1})^{-1}\log\big((q^{-1}\tau^2-q)/(q^{-1}-q\tau^2)\big)$ and C is an irrelevant constant.

Now set $u = -\beta/\mathbf{n}$ and take the limit $\mathbf{n}\to\infty$ for fixed β. In the present volume we shall use both T and $\beta = 1/T$ when we work with the Gibbs ensemble. Clearly only the term with $p=1$ counts in (1.46), and we obtain

$$\lim_{\mathbf{n}\to\infty} C^{-\mathbf{n}L}\,\mathrm{Tr}_{\mathbf{M}}\left(T_{\mathbf{S},\mathbf{M}}\right) = e^{-\beta H_{\mathrm{XXZ}}}\,.$$

In the present context, it is natural to use the magnetic field h defined by

$$2i\nu\kappa = -h\beta$$

in place of κ. We conclude that

$$\lim_{\mathbf{n}\to\infty} Z^{\kappa}_{\mathbf{n}}\left\{q^{2\alpha S(0)}\mathcal{O}\right\} = \frac{\mathrm{Tr}_{\mathbf{S}}\left(e^{-\beta(H_{\mathrm{XXZ}}+hS)}q^{2\alpha S(0)}\mathcal{O}\right)}{\mathrm{Tr}_{\mathbf{S}}\left(e^{-\beta(H_{\mathrm{XXZ}}+hS)}q^{2\alpha S(0)}\right)}\,. \tag{1.47}$$

So, in this limit we obtain expectation values of quasi-local operators at finite temperature normalised by the expectation value of the 'primary field' $q^{2\alpha S(0)}$. In particular, for $\alpha = 0$ we obtain the expectation values of local operators normalised by the partition function.

This construction suggests the following generalisation. Adjusting properly the *Matsubara* inhomogeneities and taking the limit $\mathbf{n}\to\infty$, one can obtain the expectation values over the generalised Gibbs ensemble

$$\lim_{\mathbf{n}\to\infty} \mathcal{Z}^{\kappa}_{\mathbf{n}}\left\{q^{2\alpha S(0)}\mathcal{O}\right\} = \frac{\mathrm{Tr}_{\mathbf{S}}\left(e^{-\sum_{p=0}^{\infty}\beta_p \mathbf{I}_p} q^{2\alpha S(0)}\mathcal{O}\right)}{\mathrm{Tr}_{\mathbf{S}}\left(e^{-\sum_{p=0}^{\infty}\beta_p \mathbf{I}_p} q^{2\alpha S(0)}\right)}, \tag{1.48}$$

where we set $\mathbf{I}_0 = S$.

Certainly, the actual computation of (1.47), (1.48) depends on how explicitly one can do the formulas for $\mathcal{Z}^{\kappa}_{\mathbf{n}}\left\{q^{2\alpha S(0)}\mathcal{O}\right\}$. We shall see that they can be done very explicitly.

1.7. Density Matrix and Entanglement von Neumann Entropy

The notions introduced in this section are quite general, but for the sake of definiteness let us consider the XXZ anti-ferromagnet at finite temperature. For simplicity we set also $h = 0$. Consider a finite interval $[1, n]$ on the lattice. Let $\{O_J\}$ be a basis of operators localised on $[1, n]$, and $\{O'_J\}$ be the dual basis with respect to the pairing given by trace. Set $\alpha = 0$ and define

$$\langle \mathcal{O}\rangle_T = \frac{\mathrm{Tr}_{\mathbf{S}}\left(e^{-\beta H_{\mathrm{XXZ}}}\mathcal{O}\right)}{\mathrm{Tr}_{\mathbf{S}}\left(e^{-\beta H_{\mathrm{XXZ}}}\right)},$$

recall that $\beta = 1/T$. For all J compute the expectation values $\langle O_J\rangle_T$.

Then we can define formally the density matrix

$$\rho(n, T) = \sum_J \langle O_J\rangle_T \, O'_J\,, \tag{1.49}$$

so that for any $\mathcal{O}$ localised on the same interval we have

$$\langle \mathcal{O}\rangle_T = \mathrm{Tr}_{[1,n]}(\rho(n,T)\ \mathcal{O})\,.$$

The density matrix is normalised:

$$\mathrm{Tr}_{[1,n]}\left(\rho(n,T)\right) = 1\,. \tag{1.50}$$

The physical meaning of the matrix ρ is the density matrix of a subsystem in the presence of the rest of the system considered as the environment. Even at zero temperature when the entire system is in the pure ground state, the subsystem has the density matrix. The number of degrees of freedom is measured by the entanglement von Neumann entropy:

$$s(n,T) = -\mathrm{Tr}_{[1,n]}\left(\rho(n,T)\log\rho(n,T)\right)\,. \tag{1.51}$$

Let us consider the case $|q| = 1$ which corresponds to the critical XXZ-model. The long distance behaviour of such models is described by Conformal Field Theory (CFT). Here we shall not discuss CFT in detail, this will be done in the Volume II, but we would like to mention that the most important characteristics of a given CFT is its central charge c. Roughly speaking the central charge measures the number of degrees of freedom, in that sense it is close to the entropy. So, it is not very surprising that for large n the two are related. Namely, for zero temperature and $n \gg 1$ we have [**HLW**]

$$s(n,0) \simeq \frac{c}{3}\log n + A\,, \tag{1.52}$$

where A is some non-universal constant. For $n \gg 1$ and nT being kept finite we have [**Ko, CC**]

$$s(n,T) - s(n,0) \simeq \frac{c}{3}\log\left(\frac{\sinh(nT)}{nT}\right). \tag{1.53}$$

In Chapter 5 we shall see that the asymptotics (1.52), (1.53) agree with exact computations already for rather small n ($n = 8, 9, 10$). So, the entanglement entropy is a very robust measure of the long distance CFT. For other physical quantities, such as correlation functions for instance, one has to consider longer distances in order to have good agreement with CFT.

1.8. Our Strategy in Volume I

Let us return to the functional $\mathcal{Z}^{\kappa}_{\mathbf{n}}$ as it was originally defined, and evaluate it on $\mathcal{O} = X_{[1,l]}$. Start with finite L, and evaluate the numerator taking a part of traces over V_j explicitly:

$$\begin{aligned}&\mathrm{Tr}_{\mathbf{M}}\mathrm{Tr}_{\mathbf{S}}\left(T_{\mathbf{S},\mathbf{M}}q^{2\kappa S+2\alpha S(0)}X_{[1,l]}\right)\\ &\quad=\mathrm{Tr}_{\mathbf{M}}\Big(\big(T^{\kappa+\alpha}_{\mathbf{M}}\big)^{L/2}\,\mathrm{Tr}_{[1,l]}\left(T_{[1,l],\mathbf{M}}q^{2\kappa S_{[1,l]}}X_{[1,l]}\right)\big(T^{\kappa}_{\mathbf{M}}\big)^{L/2-l}\Big),\end{aligned}$$

where $\mathrm{Tr}_{[1,l]}$ stands for the trace on $V_1 \otimes \cdots \otimes V_l$. Suppose the *Matsubara* transfer matrices $T^{\kappa}_{\mathbf{M}}$ and $T^{\kappa+\alpha}_{\mathbf{M}}$ have eigenvectors $|\kappa\rangle$ and $|\kappa+\alpha\rangle$ corresponding to maximal eigenvalues. We denote the corresponding eigenvalues by $T(\kappa)$ and $T(\kappa+\alpha)$. In the limit $L \to \infty$ we can replace

$$\begin{aligned}&T^{\kappa+\alpha}_{\mathbf{M}} \to |\kappa+\alpha\rangle\frac{T(\kappa+\alpha)}{\langle\kappa+\alpha|\kappa+\alpha\rangle}\langle\kappa+\alpha|\,,\\ &T^{\kappa}_{\mathbf{M}} \to |\kappa\rangle\frac{T(\kappa)}{\langle\kappa|\kappa\rangle}\langle\kappa|\,.\end{aligned}$$

Doing the same in the numerator we obtain

$$\mathcal{Z}^{\kappa}_{\mathbf{n}}\Big\{q^{2\alpha S(0)}\mathcal{O}\Big\} = \frac{\langle\kappa+\alpha|\mathrm{Tr}_{[1,l]}\left(T_{[1,l],\mathbf{M}}q^{2\kappa S_{[1,l]}}X_{[1,l]}\right)|\kappa\rangle}{T(\kappa)^l\langle\kappa+\alpha|\kappa\rangle}\,, \tag{1.54}$$

where we assume the genericity condition

$$\langle\kappa+\alpha|\kappa\rangle \neq 0\,. \tag{1.55}$$

So, what is needed for successful computation of $\mathcal{Z}^{\kappa}_{\mathbf{n}}$? One thing is clear: we have to diagonalise the *Matsubara* transfer matrix. The simplest way to do it is to use the algebraic Bethe Ansatz. However, for our goals a more sophisticated procedure will be needed, which is the Bazhanov-Lukyanov-Zamolodchikov construction of Q-operators. This procedure requires some knowledge of the quantum algebra $U_q(\widehat{\mathfrak{sl}}_2)$. All that will be discussed in the next Chapter. Then we have $\mathrm{Tr}_{[1,l]}\left(T_{[1,l],\mathbf{M}}q^{2\kappa S_{[1,l]}}X_{[1,l]}\right)$. How to deal with it?

The first idea would be the following. Introduce operators $A(\xi), B(\xi), C(\xi), D(\xi)$ which act on $\mathfrak{H}_{\mathbf{M}}$ by

$$T_{j,\mathbf{M}}(\xi) = \begin{pmatrix}A(\xi) & B(\xi)\\ C(\xi) & D(\xi)\end{pmatrix}_j\,.$$

The operators $\mathrm{Tr}_{[1,l]}\left(T_{[1,l],\mathbf{M}}q^{2\kappa S_{[1,l]}}X_{[1,l]}\right)$ are sums of products of $A(\xi)$, $B(\xi)$, $C(\xi)$, $D(\xi)$. So one could try to compute (1.54) using the algebraic Bethe Ansatz, namely the commutation relations between $A(\xi), B(\xi), C(\xi), D(\xi)$ and formulas for scalar products of Slavnov type [**Sl**]. Unfortunately, this procedure leads finally to the same kind of integral representations as the ones discussed in Section 1.5 [**KMT**]. In the case of finite temperature, for example, the corresponding formulas become especially complicated [**GKS**]: they are given by multiple integrals, whose integrands contain a function for which there is an additional non-linear integral equation.

We shall take a completely different approach which can be explained in analogy with quantum field theory. Consider the well known example of CFT in two dimensions. Local operators are described by representations of the Virasoro algebra. One can compute their one-point functions on a plane, on a cylinder, on a torus or even on an arbitrary Riemann surface. In the first two cases the computation is simple, while in the last two cases it is far more complicated. However, the set of local operators does not depend on the global geometry. This example teaches us that one should distinguish between the local, or ultra-violet (UV), description, versus the global, or infra-red (IR), one; and that one has to settle the local description before dwelling upon such complicated matters as switching on the environment.

Motivated by the above analogy we ask the question: is it possible to have a 'good' description of quasi-local operators $q^{2\alpha S(0)}\mathcal{O}$? The collection of such operators looks rather structureless. Is it possible to put some order in it the same way as does the Virasoro algebra in CFT? The answer to these questions is positive. It is given by the fermionic construction which we describe in Chapter 3. For that construction, we shall need the theory of $U_q(\widehat{\mathfrak{sl}}_2)$ described in Chapter 2.

CHAPTER 2

Spectral Problem in Matsubara Direction and Quantum Groups

2.1. Algebraic Bethe Ansatz

In this chapter we shall diagonalise the transfer matrix in the *Matsubara* direction $T_{\mathbf{M}}(\xi)$. We shall assume that the *Matsubara* chain is inhomogeneous with inhomogeneity parameters $\tau_{\mathbf{m}}$. As before, consider the block decomposition:

$$T_{j,\mathbf{M}}(\xi) = \begin{pmatrix} A(\xi) & B(\xi) \\ C(\xi) & D(\xi) \end{pmatrix}_j . \tag{2.1}$$

The entries $A(\xi), B(\xi), C(\xi), D(\xi)$ are operators on $\mathfrak{H}_{\mathbf{M}}$; for economy of symbols we do not supply them with index $\mathbf{M}$. The Yang-Baxter equation entails

$$R_{j,k}(\xi_1/\xi_2)T_{j,\mathbf{M}}(\xi_1)T_{k,\mathbf{M}}(\xi_2) = T_{k,\mathbf{M}}(\xi_2)T_{j,\mathbf{M}}(\xi_1)R_{j,k}(\xi_1/\xi_2) . \tag{2.2}$$

This implies, in particular, that

$$[C(\xi), C(\lambda)] = 0 , \tag{2.3}$$

$$A(\xi)C(\lambda) = u(\xi/\lambda)C(\lambda)A(\xi) - v(\xi/\lambda)C(\xi)A(\lambda) , \tag{2.4}$$

$$D(\xi)C(\lambda) = u(\lambda/\xi)C(\lambda)D(\xi) + v(\xi/\lambda)C(\xi)D(\lambda) , \tag{2.5}$$

$$[B(\xi), C(\lambda)] = v(\xi/\lambda)(A(\xi)D(\lambda) - A(\lambda)D(\xi)) , \tag{2.6}$$

where

$$u(\zeta) = \frac{\zeta q - \zeta^{-1}q^{-1}}{\zeta - \zeta^{-1}} , \quad v(\zeta) = \frac{q - q^{-1}}{\zeta - \zeta^{-1}} .$$

We shall not need the last relation (2.6) until Section 5.2. Denote by $|\downarrow\rangle$ the vector $v_- \otimes \cdots \otimes v_-$ in which 'all spins are down'. We have $B(\xi)|\downarrow\rangle = 0$, and

$$A(\xi)|\downarrow\rangle = q^{\frac{\mathbf{n}}{2}}a(\xi)|\downarrow\rangle , \quad D(\xi)|\downarrow\rangle = q^{-\frac{\mathbf{n}}{2}}d(\xi)|\downarrow\rangle ,$$

with[1]

$$a(\xi) = \prod_{\mathbf{m}=1}^{\mathbf{n}} (1 - q^{-1}(\xi/\tau_{\mathbf{m}})^2) , \quad d(\xi) = \prod_{\mathbf{m}=1}^{\mathbf{n}} (1 - q(\xi/\tau_{\mathbf{m}})^2) . \tag{2.7}$$

In Algebraic Bethe Ansatz (ABA), we look for eigenvectors in the form

$$|\lambda_1^+, \ldots, \lambda_k^+\rangle = C(\lambda_1^+) \cdots C(\lambda_k^+)|\downarrow\rangle . \tag{2.8}$$

The reason for putting the superfix '+' will be explained soon. Notice that $s = k - \mathbf{n}/2$ is the spin of the vector (2.8).

The procedure for finding eigenvectors is as follows. We apply $T_{\mathbf{M}}(\xi, \kappa) = q^{\kappa}A(\xi) + q^{-\kappa}D(\xi)$ to this vector, and move $A(\xi)$ and $D(\xi)$ to the right using (2.4), (2.5). We create 2×2^k terms along the way. Among them there are exactly

[1] a and d are opposite to those of [**JMS1**].

two terms which contain only $u(\zeta)$ and no $v(\zeta)$ in the coefficients. These are the 'wanted terms' in Baxter's terminology. All other terms containing at least one $v(\zeta)$ are 'unwanted terms'.

On the other hand, due to (2.3) the vector (2.8) is symmetric with respect to the λ_i^+'s. Then by symmetry one immediately concludes that

$$\begin{aligned} &T_{\mathbf{M}}(\xi,\kappa)|\lambda_1^+,\dots,\lambda_k^+\rangle \\ &= \Bigl(q^{\kappa+\frac{\mathbf{n}}{2}}a(\xi)\prod_{j=1}^{k}u(\xi/\lambda_j^+) + q^{-\kappa-\frac{\mathbf{n}}{2}}d(\xi)\prod_{j=1}^{k}u(\lambda_j^+/\xi)\Bigr)|\lambda_1^+,\dots,\lambda_k^+\rangle \\ &\quad -\sum_{j=1}^{k}v(\xi/\lambda_j^+)\Bigl(q^{\kappa+\frac{\mathbf{n}}{2}}a(\lambda_j^+)\prod_{p\neq j}^{k}u(\lambda_j^+/\lambda_p^+) - q^{-\kappa-\frac{\mathbf{n}}{2}}d(\lambda_j^+)\prod_{p\neq j}^{k}u(\lambda_p^+/\lambda_j^+)\Bigr) \\ &\qquad \times|\lambda_1^+,\dots,\overset{j\text{-th}}{\xi},\cdots,\lambda_k^+\rangle\,. \end{aligned} \tag{2.9}$$

The unwanted terms vanish if the following Bethe equations are satisfied:

$$q^{2\kappa-2s+2}\frac{a(\lambda_j^+)}{d(\lambda_j^+)}\prod_{p\neq j}^{k}\frac{1-q^2(\lambda_j^+/\lambda_p^+)^2}{1-q^{-2}(\lambda_j^+/\lambda_p^+)^2}=1\,,\quad j=1,\dots,k\,, \tag{2.10}$$

where $s=k/2-\mathbf{n}$ is the spin of the vector. If this is the case, and if (2.8) is a non-zero vector, then we obtain an eigenvector with the eigenvalue given in the first line of (2.9).

Let us choose and fix a solution $\{\lambda_i^+\}$ to the Bethe equation (2.10), and continue it analytically with respect to κ. The requirements for the choice of the relevant solution will be discussed later. To simplify the notation we suppress the dependence on λ_j^+ and denote the corresponding eigenvector by $|\kappa\rangle$. The corresponding eigenvalue of $T_{\mathbf{M}}(\xi,\kappa)$ is denoted by $T(\xi,\kappa)$.

Introduce further a definition

$$Q^+(\xi,\kappa)=\xi^{\kappa-s}\prod_{j=1}^{k}(1-(\xi/\lambda_j^+)^2)\,. \tag{2.11}$$

Then the equations (2.10) and (2.9) (in which the sum disappears due to (2.10)) are equivalent to the following Baxter's TQ equation:

$$T(\xi,\kappa)Q^+(\xi,\kappa)=a(\xi)Q^+(\xi q,\kappa)+d(\xi)Q^+(\xi q^{-1},\kappa)\,, \tag{2.12}$$

where $Q^+(\xi,\kappa)$ is of the form (2.11), and $T(\xi,\kappa)$ is a polynomial of ξ^2 of degree $\mathbf{n}$. Later we shall introduce operators $Q^+_{\mathbf{M}}(\xi,\kappa)$ whose eigenvalues are the functions (2.11).

Let us discuss the bra-vectors. Usually one would write them as $\langle\downarrow|\prod_{j=1}^{k}B(\lambda_j^+)$. Then the formulas for the norms are rather complicated [**G**]. What is worse is that in the problem formulated in Chapter 1 we need inner products of Bethe vectors corresponding to different parameters, κ and $\kappa+\alpha$. Then we shall need to use rather complicated formulas from [**Sl**]. We do not want to have complicated formulas at a simple place, so, we shall proceed differently. Namely, let us look for the bra eigenvectors in the form

$$\langle\lambda_1^-,\dots\lambda_{\mathbf{n}-k}^-|=\langle\uparrow|C(\lambda_1^-)\cdots C(\lambda_{\mathbf{n}-k}^-)\,. \tag{2.13}$$

To get a formula of the type (2.9) we need to move $A(\xi)$ and $D(\xi)$ to the left. For this we have the formulas

$$C(\lambda)A(\xi)=u(\lambda/\xi)A(\xi)C(\lambda)-v(\lambda/\xi)A(\lambda)C(\xi)\,,\qquad (2.14)$$
$$C(\lambda)D(\xi)=u(\xi/\lambda)D(\xi)C(\lambda)-v(\xi/\lambda)D(\lambda)C(\xi)\,.$$

Repeating the computation, we find that the bra vector (2.13) is an eigenvector with the eigenvalue $T(\xi,\kappa)$ if there exists a function of the form

$$Q^-(\xi,\kappa)=\xi^{-\kappa+s}\prod_{j=1}^{\mathbf{n}-k}\left(1-(\xi/\lambda_j^-)^2\right)\qquad (2.15)$$

satisfying the same equation

$$T(\xi,\kappa)Q^-(\xi,\kappa)=a(\xi)Q^-(\xi q,\kappa)+d(\xi)Q^-(\xi q^{-1},\kappa)\,.\qquad (2.16)$$

We need to know the scalar products $\langle\kappa+\alpha|\kappa\rangle$ for eigenvectors with different parameters. Let them have spin s. Making the dependence of Bethe numbers on κ and $\kappa+\alpha$ explicit, we have:

$$\begin{aligned}\langle\kappa+\alpha|\kappa\rangle&=\langle\uparrow|\prod_{j=1}^{\frac{\mathbf{n}}{2}-s}C\big(\lambda_j^-(\kappa+\alpha)\big)\prod_{j=1}^{\frac{\mathbf{n}}{2}+s}C\big(\lambda_j^+(\kappa)\big)|\downarrow\rangle\\&=\langle\uparrow|\prod_{j=1}^{\mathbf{n}}C(\lambda_j)|\downarrow\rangle\,,\end{aligned}\qquad (2.17)$$

where we introduced the union

$$\{\lambda_j\}_{j=1}^{\mathbf{n}}=\{\lambda_j^-(\kappa+\alpha)\}_{j=1}^{\frac{\mathbf{n}}{2}-s}\cup\{\lambda_j^+(\kappa)\}_{j=1}^{\frac{\mathbf{n}}{2}+s}\,.$$

Notice that the total number of λ's is independent of s. The expression in the right hand side of (2.17) is well-known; this is the partition function of the six-vertex model with the domain wall boundary condition. Graphically it is presented as

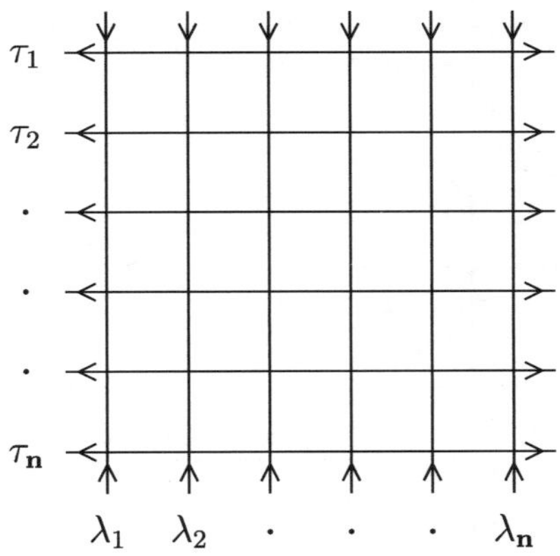

FIG. 6.

Introduce the notation

$$M_{\mathbf{n}}(\lambda_1,\dots\lambda_{\mathbf{n}}|\tau_1,\dots,\tau_{\mathbf{n}})=\langle\uparrow|\prod_{j=1}^{\mathbf{n}}C(\lambda_j)|\downarrow\rangle\,.\qquad (2.18)$$

This is a symmetric odd polynomial of λ_j of degree $2\mathbf{n}-1$ in every variable. It is easy to prove the recurrence relation:

(2.19)

$$M_{\mathbf{n}}(q^{\frac{1}{2}}\tau_1,\lambda_2,\dots\lambda_{\mathbf{n}}|\tau_1,\tau_2,\dots,\tau_{\mathbf{n}})$$
$$=q^{-\mathbf{n}+\frac{3}{2}}(q^{-1}-q)\prod_{j=2}^{\mathbf{n}}(1-q(\lambda_j/\tau_1)^2)(1-q^2(\tau_1/\tau_j)^2)\cdot M_{\mathbf{n}-1}(\lambda_2,\dots\lambda_{\mathbf{n}}|\tau_2,\dots,\tau_{\mathbf{n}})\,.$$

Due to symmetry, this relation along with the initial condition

$$M_{\mathbf{1}}(\lambda|\tau)=(q^{-1}-q)\,\frac{\lambda}{\tau}$$

is sufficient to specify $M_{\mathbf{n}}$. It is possible to write down the explicit solution [**I**] which we shall use in Section 4.4:

(2.20)

$$M_{\mathbf{n}}(\lambda_1,\lambda_2,\dots\lambda_{\mathbf{n}}|\tau_1,\tau_2,\dots,\tau_{\mathbf{n}})$$
$$=\frac{\prod_{j,k=1}^{\mathbf{n}}(\tau_k^2-q\lambda_j^2)(\tau_k^2-q^{-1}\lambda_j^2)\tau_k^{-2}}{\prod_{1\le i<j\le\mathbf{n}}(\lambda_i^2-\lambda_j^2)(\tau_j^2-\tau_i^2)}\det\left(\frac{(q^{-1}-q)\lambda_j\tau_k}{(\tau_k^2-q\lambda_j^2)(\tau_k^2-q^{-1}\lambda_j^2)}\right)_{j,k=1,\dots,\mathbf{n}}.$$

Later we shall see that there is another determinant formula solving (2.19).

2.2. Algebra $U_q(\widehat{\mathfrak{sl}}_2)$

It has been said that the equations (2.12), (2.16) allow an independent definition. This definition is based on the quantum affine algebra $U_q(\widehat{\mathfrak{sl}}_2)$ [**BLZ3**]. One may say that using complicated mathematics for solving the spectral problem in the *Matsubara* direction is like firing a cannon at sparrows; Bethe Ansatz does the job perfectly well. Nevertheless, the algebra $U_q(\widehat{\mathfrak{sl}}_2)$ will be used in Chapter 3 for describing quasi-local operators in the *Space* direction as well. There it will be shown to be an indispensable tool.

2.2.1. General Definitions. We begin with some terminology. In the following, an algebra $\mathcal{A}$ will always mean an associative algebra over $\mathbb{C}$ with unit. Tensor product of algebras $\mathcal{A}\otimes\mathcal{B}$ is given an algebra structure by component-wise multiplication: $(a\otimes b)(a'\otimes b')=(aa')\otimes(bb')$.

A Hopf algebra is a quadruple $(\mathcal{A},\epsilon,\Delta,S)$, consisting of an algebra $\mathcal{A}$ and linear maps called the counit $\epsilon:\ \mathcal{A}\to\mathbb{C}$, the coproduct $\Delta:\ \mathcal{A}\to\mathcal{A}\otimes\mathcal{A}$ and the antipode $S:\ \mathcal{A}\to\mathcal{A}$. The maps ϵ,Δ are homomorphisms of algebras, while S is an anti-homomorphism, i.e., $S(ab)=S(b)S(a)$. In addition we assume the coassociativity

$$(\Delta\otimes\mathrm{id})\circ\Delta=(\mathrm{id}\otimes\Delta)\circ\Delta\,,$$

as well as the properties

$$(\epsilon\otimes\mathrm{id})\circ\Delta=(\mathrm{id}\otimes\epsilon)\circ\Delta=\mathrm{id}\,,$$
$$m\circ(S\otimes\mathrm{id})\circ\Delta=m\circ(\mathrm{id}\otimes S)\circ\Delta=\iota\circ\epsilon\,,$$

where $m:\mathcal{A}\otimes\mathcal{A}\to\mathcal{A}$ is the multiplication map and $\iota:\mathbb{C}\to\mathcal{A}$ is the unit map. We assume also that S is invertible.

Let $\Delta' = \sigma \circ \Delta$ be the opposite coproduct, where $\sigma(a \otimes b) = b \otimes a$. A Hopf algebra $(\mathcal{A}, \epsilon, \Delta, S)$ is said to be quasi-triangular if there exists an invertible element $\mathcal{R} \in \mathcal{A} \otimes \mathcal{A}$ satisfying the following conditions:

$$\mathcal{R}\,\Delta(a)\,\mathcal{R}^{-1} = \Delta'(a) \quad (a \in \mathcal{A}), \tag{2.21}$$

$$(\Delta \otimes \mathrm{id})\mathcal{R} = \mathcal{R}_{1,3}\mathcal{R}_{2,3}\,, \quad (\mathrm{id} \otimes \Delta)\mathcal{R} = \mathcal{R}_{1,3}\mathcal{R}_{1,2}\,, \tag{2.22}$$

$$(S \otimes \mathrm{id})\mathcal{R} = (\mathrm{id} \otimes S^{-1})\mathcal{R} = \mathcal{R}^{-1}\,. \tag{2.23}$$

In the above, we write $\mathcal{R}_{1,2} = \mathcal{R} \otimes 1 \in \mathcal{A}^{\otimes 3}$, $\mathcal{R}_{2,3} = 1 \otimes \mathcal{R} \in \mathcal{A}^{\otimes 3}$, and so forth. Note that the relations (2.22) conform with our previous rule (1.7) of multiplying L-operators : multiplication is performed in the direct order with respect to the first component, and in the inverse order with respect to the second.

The equations (2.21) and (2.22) imply the Yang-Baxter equation

$$\mathcal{R}_{1,2}\mathcal{R}_{1,3}\mathcal{R}_{2,3} = \mathcal{R}_{2,3}\mathcal{R}_{1,3}\mathcal{R}_{1,2}\,. \tag{2.24}$$

For that reason $\mathcal{R}$ is called the universal R-matrix. (Actually, in almost all cases of interest, $\mathcal{R}$ is not an element of $\mathcal{A} \otimes \mathcal{A}$ but rather of its completion. In the present volume we do not discuss this point, leaving the details to standard textbooks [**Ja**], [**Lu**].)

Condition (2.21) has the following meaning on representations. Quite generally, an intertwiner $M \to N$ between two $\mathcal{A}$ modules M, N is a linear map which commutes with the action of $\mathcal{A}$. Given two $\mathcal{A}$ modules V, W, let $R_{V,W} : V \otimes W \to V \otimes W$ denote the action of $\mathcal{R}$ on their tensor product, and let $P : V \otimes W \to W \otimes V$ be the permutation map $P(v \otimes w) = w \otimes v$. Then (2.21) implies that $PR_{V,W} : V \otimes W \to W \otimes V$ is an intertwiner of $\mathcal{A}$ modules.

Somewhat loosely, we shall use the word 'quantum group' as a synonym for 'quasi-triangular Hopf algebra'.

2.2.2. Algebra $U_q(\mathfrak{sl}_2)$. In this and the next subsections, we fix $q \in \mathbb{C}^\times$ with $q^2 \neq 1$ and use the symbol

$$[x]_q = \frac{q^x - q^{-x}}{q - q^{-1}}\,.$$

The simplest (and historically the first) example of a quantum group is $U_q(\mathfrak{sl}_2)$ [**KR**]. Algebra $U_q(\mathfrak{sl}_2)$ is generated by elements E, F and an invertible element K, under the defining relations

$$KE = q^2 EK\,, \quad KF = q^{-2} FK\,, \quad [E, F] = \frac{K - K^{-1}}{q - q^{-1}}\,. \tag{2.25}$$

The coproduct, counit and antipode are given by

$$\begin{aligned} &\Delta(E) = E \otimes 1 + K \otimes E\,, \quad \Delta(F) = F \otimes K^{-1} + 1 \otimes F\,, \quad \Delta(K) = K \otimes K\,, \\ &\epsilon(E) = \epsilon(F) = 0\,, \quad \epsilon(K) = 1\,, \\ &S(E) = -K^{-1}E\,, \quad S(F) = -FK\,, \quad S(K) = K^{-1}\,. \end{aligned} \tag{2.26}$$

The universal R-matrix is not complicated in this case [**Dr**], but we shall not need it. We shall rather need the Casimir element which generates the centre of $U_q(\mathfrak{sl}_2)$,

$$C = \frac{Kq^{-1} + K^{-1}q}{(q - q^{-1})^2} + EF\,. \tag{2.27}$$

We shall consider only those representations on which K is diagonalisable and the operator H given by

$$K = q^H$$

has a well defined action. We say that a vector v has weight $\lambda \in \mathbb{C}$ if $Hv = \lambda v$, i.e., $Kv = q^\lambda v$. An element x in $U_q(\mathfrak{sl}_2)$ has weight λ if $KxK^{-1} = q^\lambda x$.

We shall use the following representations of $U_q(\mathfrak{sl}_2)$.

1. *Verma module with lowest weight.*

Let $\Lambda \in \mathbb{C}$. The Verma module[2] with lowest weight $-\Lambda$ is defined on the space $V^\Lambda = \bigoplus_{j=0}^\infty \mathbb{C}v_j$, with the action

$$Ev_j = v_{j+1}\,, \quad Fv_j = [j]_q[\Lambda - j + 1]_q v_{j-1}\,, \quad Kv_j = q^{2j-\Lambda}v_j\,. \tag{2.28}$$

This representation will be denoted by π^Λ.

We shall use the trace over the Verma module $\mathrm{Tr}^\Lambda(q^{\kappa H}x)$, $x \in U_q(\mathfrak{sl}_2)$. It is a linear functional on $U_q(\mathfrak{sl}_2)$ uniquely defined by the following requirements.

- Cyclicity

$$\mathrm{Tr}^\Lambda(q^{\kappa H}xy) = \mathrm{Tr}^\Lambda(q^{\kappa H}(q^{-\kappa H}yq^{\kappa H})x)\,.$$

- Casimir

$$\mathrm{Tr}^\Lambda(q^{\kappa H}Cx) = \frac{q^{\Lambda+1} + q^{-\Lambda-1}}{(q-q^{-1})^2}\mathrm{Tr}^\Lambda(q^{\kappa H}x)\,.$$

- Values on K^m, $m \in \mathbb{Z}$

$$\mathrm{Tr}^\Lambda(q^{\kappa H}K^m) = \frac{q^{-(\kappa+m)\Lambda}}{1 - q^{2(\kappa+m)}}\,. \tag{2.29}$$

By cyclicity, the trace of an element with non-zero weight vanishes. For the trace of an element of weight zero, we use

$$EF = C - \frac{Kq^{-1} + K^{-1}q}{(q-q^{-1})^2}\,, \quad FE = C - \frac{Kq + K^{-1}q^{-1}}{(q-q^{-1})^2}\,, \tag{2.30}$$

to eliminate E and F, and then apply (2.29).

We are primarily interested in the case $|q| = 1$. In this case, formula (2.29) says that the functional Tr^Λ coincides with the analytic continuation of the usual trace defined in the region $|q^\kappa| < 1$. Poles occur at $\kappa \in \mathbb{Z}$ and these points require special attention. For the moment let us consider the case of generic κ.

2. *Finite dimensional representations.* When $\Lambda = 2s$ is a non-negative integer, the Verma module contains a submodule isomorphic to the Verma module $V^{-\Lambda-2}$. The corresponding quotient is a finite-dimensional representation $V^{(2s)} = \bigoplus_{j=0}^{2s} \mathbb{C}v_j$ of dimension $2s+1$. The action of the generators is defined by the same formulas (2.28) taking $\Lambda = 2s$. We shall denote this representation by $\pi^{(2s)}$. Obviously,

$$\mathrm{Tr}_{V^{(2s)}}(q^{\kappa H}x) = \Big(\mathrm{Tr}^\Lambda(q^{\kappa H}x) - \mathrm{Tr}^{-\Lambda-2}(q^{\kappa H}x)\Big)\Big|_{\Lambda=2s}\,. \tag{2.31}$$

We retain the notation $V = V^{(1)}$ for the two-dimensional representation used most often.

[2]The sign of Λ is changed from the one in [**BJMST2**].

2.2.3. Algebra $U_q(\widehat{\mathfrak{sl}}_2)$. Now we proceed to the definition of $U_q(\widehat{\mathfrak{sl}}_2)$. We begin with the Cartan matrix

$$(a_{i,j})_{i,j=0,1} = \begin{pmatrix} 2 & -2 \\ -2 & 2 \end{pmatrix}.$$

The algebra $U_q(\widehat{\mathfrak{sl}}_2)$ is generated by e_i, f_i, t_i and its inverse t_i^{-1}, $i = 0, 1$, with the defining relations

$$t_i e_j = q^{a_{i,j}} e_j t_i\,, \quad t_i f_j = q^{-a_{i,j}} f_j t_i\,, \tag{2.32}$$

$$[e_i, f_j] = \delta_{i,j} \frac{t_i - t_i^{-1}}{q - q^{-1}}\,, \tag{2.33}$$

$$[e_i, [e_i, [e_i, e_j]_{q^2}]_{q^0}]_{q^{-2}} = 0 \quad (i \neq j)\,, \tag{2.34}$$

$$[f_i, [f_i, [f_i, f_j]_{q^2}]_{q^0}]_{q^{-2}} = 0 \quad (i \neq j)\,, \tag{2.35}$$

where $[x, y]_p = xy - pyx$.

We shall use h and c defined by

$$t_0 = q^{c-h}\,, \quad t_1 = q^h\,. \tag{2.36}$$

The element c is central.

The coproduct, counit and antipode are given by

$$\begin{aligned}
&\Delta(e_i) = e_i \otimes 1 + t_i \otimes e_i\,, \quad \Delta(f_i) = f_i \otimes t_i^{-1} + 1 \otimes f_i\,, \quad \Delta(t_i) = t_i \otimes t_i\,, \\
&\epsilon(e_i) = \epsilon(f_i) = 0\,, \quad \epsilon(t_i) = 1\,, \\
&S(e_i) = -t_i^{-1} e_i\,, \quad S(f_i) = -f_i t_i\,, \quad S(t_i) = t_i^{-1}\,.
\end{aligned}$$

The algebra $U_q(\widehat{\mathfrak{sl}}_2)$ has the Borel subalgebra $U_q\mathfrak{b}^+$ generated by $e_i, t_i^{\pm 1}$ $(i = 0, 1)$, and the opposite Borel subalgebra $U_q\mathfrak{b}^-$ generated by $f_i, t_i^{\pm 1}$ $(i = 0, 1)$. The universal R-matrix of $U_q(\widehat{\mathfrak{sl}}_2)$ has an important feature that it is of the form

$$\mathcal{R} = \overline{\mathcal{R}}\, q^{-h \otimes h/2}\,, \quad \overline{\mathcal{R}} \in U_q\mathfrak{b}^+ \otimes U_q\mathfrak{b}^-\,. \tag{2.37}$$

For completeness of exposition we present the Khoroshkin-Tolstoy formula [**KT**] for the universal R-matrix. *From this point on, we restrict our considerations to the case* $c = 0$. Readers who are not interested in mathematical details can skip this and proceed to specific R-matrices given in (2.46) and after.

To write the formula we need the Drinfeld generators. We set

$$\begin{aligned}
&x_0^+ = e_1\,, \quad x_0^- = f_1\,, \quad x_1^- = e_0 t_1\,, \quad x_{-1}^+ = t_1^{-1} f_0\,, \\
&h_1 = [e_1, e_0]_{q^2}\,, \quad h_{-1} = [f_0, f_1]_{q^{-2}}\,,
\end{aligned}$$

and define $x_m^\pm, h_r$ for general $m \in \mathbb{Z}, r \in \mathbb{Z}\backslash\{0\}$ inductively by

$$\begin{aligned}
&[h_r, x_m^\pm] = \pm \frac{[2r]_q}{r} x_{m+r}^\pm \quad (r = \pm 1)\,, \\
&[x_m^+, x_n^-] = \frac{1}{q - q^{-1}} \Big(\psi_{m+n}^+ - \psi_{m+n}^- \Big)\,,
\end{aligned} \tag{2.38}$$

where $\psi_m^\pm$'s are related to h_r's by

$$\psi^\pm(z) = \sum_{\pm m \geq 0} \psi_m^\pm z^{-m} = t_1^{\pm 1} \exp\Big(\pm (q - q^{-1}) \sum_{\pm r > 0} h_r z^{-r} \Big)\,.$$

Then the h_r's are mutually commutative, and (2.38) holds true for all $r \neq 0$.

The universal R matrix is given by

$$\mathcal{R}=\mathcal{R}^{+}\mathcal{R}^{0}\mathcal{R}^{-}q^{-h\otimes h/2}\,, \tag{2.39}$$

where

$$\mathcal{R}^{+}=\prod_{m\geq 0}^{\curvearrowright}\exp_{q^2}\Bigl(-(q-q^{-1})x^{+}_{m}\otimes x^{-}_{-m}\Bigr)\,, \tag{2.40}$$

$$\mathcal{R}^{0}=\exp\Bigl(-(q-q^{-1})\sum_{r>0}\frac{r}{[2r]_q}h_r\otimes h_{-r}\Bigr)\,, \tag{2.41}$$

$$\mathcal{R}^{-}=\prod_{m>0}^{\curvearrowleft}\exp_{q^2}\Bigl(-(q-q^{-1})t_1^{-1}x^{-}_{m}\otimes x^{+}_{-m}t_1\Bigr)\,, \tag{2.42}$$

and $\exp_p(x)=\sum_{n\geq 0}\frac{x^n}{(n)_p!}$, $(n)_p!=\prod_{j=1}^{n}\frac{1-p^j}{1-p}$.

We have the evaluation homomorphism $ev_\zeta:\ U_q(\widehat{\mathfrak{sl}}_2)\to U_q(\mathfrak{sl}_2)$ given by

$$\begin{aligned} &ev_\zeta(e_0)=\zeta F\,,\quad ev_\zeta(f_0)=\zeta^{-1}E\,,\quad ev_\zeta(t_0)=K^{-1}\,,\\ &ev_\zeta(e_1)=\zeta E\,,\quad ev_\zeta(f_1)=\zeta^{-1}F\,,\quad ev_\zeta(t_1)=K\,. \end{aligned} \tag{2.43}$$

In terms of the Drinfeld generators we have

$$\begin{aligned} &ev_\zeta(x^{+}_{m})=\zeta^{2m+1}K^mE,\quad ev_\zeta(x^{-}_{m})=\zeta^{2m-1}FK^m,\quad (m\in\mathbb{Z})\\ &ev_\zeta(\psi^{+}_{m})=\zeta^{2m}\varphi_m(K)\quad ev_\zeta(\psi^{-}_{-m})=(q\zeta)^{-2m}\varphi_m(K^{-1}),\quad (m\geq 1)\,, \end{aligned} \tag{2.44}$$

where

$$\varphi_m(x)=x^m\left(\frac{q^2(1-q^{2m-2})}{1-q^2}x^{-1}-\frac{C(1-q^2)(1-q^{2m})}{q}+\frac{1-q^{2m+2}}{1-q^2}x\right).$$

Using this we define the representations of $U_q(\widehat{\mathfrak{sl}}_2)$:

$$\pi^{\Lambda}_{\zeta}=\pi^{\Lambda}\circ ev_\zeta\,,\quad \pi^{(2s)}_{\zeta}=\pi^{(2s)}\circ ev_\zeta\,. \tag{2.45}$$

The corresponding modules are denoted by V^{Λ}_{ζ} and $V^{(2s)}_{\zeta}$, respectively.

Now we compute R-matrices by evaluating the universal R-matrix on representations. The simplest case is

$$\begin{aligned} &(ev_{\zeta_1}\otimes\pi^{(1)}_{\zeta_2})\mathcal{R}=\mathsf{r}(\zeta,C)R^{(ev,1)}(\zeta)\,,\\ &R^{(ev,1)}(\zeta)=\begin{pmatrix}1-\zeta^2q^{H+1} & -(q-q^{-1})\zeta F\\ -(q-q^{-1})\zeta E & 1-\zeta^2q^{-H+1}\end{pmatrix}q^{-H\sigma^3/2}\,, \end{aligned} \tag{2.46}$$

where $\zeta=\zeta_1/\zeta_2$, $\mathsf{r}(\zeta,C)$ is a formal power series in ζ^2 satisfying $\mathsf{r}(0,C)=1$ and

$$\mathsf{r}(\zeta,C)\mathsf{r}(\zeta q^{-1},C)\left(1-\zeta^2(q-q^{-1})^2C+\zeta^4\right)=1\,, \tag{2.47}$$

which follows from (2.23).

Notice that in the case $|q|=1$ the power series in question is divergent. This formal series is not to be confused with an analytic solution to (2.47), which has an essential singularity at $\zeta=0$.

In particular,

$$(\pi^{(1)}_{\zeta_1}\otimes\pi^{(1)}_{\zeta_2})\mathcal{R}=\mathsf{r}(\zeta,\pi^{(1)}(C))R(\zeta)\,,\qquad \pi^{(1)}(C)=\frac{q^2+q^{-2}}{(q-q^{-1})^2}\,, \tag{2.48}$$

where $R(\zeta)$ coincides with (1.13). This is the reason for the normalisation chosen in the latter formula. Similar formulas for $(\pi^{(s_1)}_{\zeta_1} \otimes \pi^{(s_2)}_{\zeta_2})\mathcal{R}$ with higher s_1, s_2 will be discussed in Section 2.4.

2.3. Bazhanov-Lukyanov-Zamolodchikov Construction

As we noted in (2.37), the universal R-matrix belongs to $U_q\mathfrak{b}^+ \otimes U_q\mathfrak{b}^-$. Bazhanov, Lukyanov, and Zamolodchikov **[BLZ3]** observed that one can consider its image on representations of these Borel subalgebras rather than those of the full algebra $U_q(\widehat{\mathfrak{sl}}_2)$. We shall do so for the first tensor component of the R-matrix. Following the tradition of QISM, we call the first tensor component the *auxiliary space* and the second component the *quantum space*. In Subsection 2.3.1 we introduce representations $W^\pm$ of $U_q\mathfrak{b}^+$. Since we make several choices for the auxiliary space, we shall use indices to distinguish them. They will be upper case letters $A, B, \dots$ for $W^\pm$ and lower case letters $a, b, \dots$ for $\mathbb{C}^2$. The general rule is that they are taken from the beginning of the alphabet (close to a for 'auxiliary'). We choose $\pi^{(1)}_1$ to be the quantum space in most cases. Especially, Theorem 2.6 is formulated for such quantum spaces. Application of the evaluation map (2.43) to the first component gives rise to a matrix acting on the quantum space, whose entries belong to the algebra $U_q(\mathfrak{sl}_2)$. In this situation we shall use the symbol $*$ to indicate that the 'auxiliary space' is $U_q(\mathfrak{sl}_2)$. Readers who are not interested in technical details can read the definitions of the q-oscillator algebra and its representations, and then go directly to Theorem 2.6.

2.3.1. q-Oscillator Representation of $U_q\mathfrak{b}^+$. By definition, the q-oscillator algebra Osc is the $\mathbb{C}$-algebra generated by $\mathbf{a}, \mathbf{a}^*, q^{\pm D}$ under the relations

(2.49)
$$q^D\mathbf{a}\, q^{-D} = q^{-1}\mathbf{a}, \quad q^D\mathbf{a}^* q^{-D} = q\, \mathbf{a}^*\,, \quad \mathbf{a}\,\mathbf{a}^* = 1 - q^{2D+2}, \quad \mathbf{a}^*\mathbf{a} = 1 - q^{2D}\,.$$

The infinite-dimensional module $W = \oplus_{k\in\mathbb{Z}}\mathbb{C}|k\rangle$ given by

$$q^D|k\rangle = q^k|k\rangle,\ \mathbf{a}|k\rangle = (1-q^{2k})|k-1\rangle,\ \mathbf{a}^*|k\rangle = |k+1\rangle \tag{2.50}$$

has a submodule $W^+ = \oplus_{k\geq 0}\mathbb{C}|k\rangle$ and a quotient module $W^- = W/W^+$. The corresponding representations $\pi^\pm$ are both irreducible. We identifty W^- with $\oplus_{k<0}\mathbb{C}|k\rangle$ as linear space. The representation π^+ is generated by $|0\rangle$, and π^- is generated by $|-1\rangle$.

As in the case of $U_q(\mathfrak{sl}_2)$, there is a linear functional $\mathrm{Tr}(q^{2\kappa D}\bullet)$ on Osc uniquely defined by the following two requirements:

- Cyclicity
$$\mathrm{Tr}(q^{2\kappa D}xy) = \mathrm{Tr}(q^{2\kappa D}(q^{-2\kappa D}yq^{2\kappa D})x)\,,$$
- Value on q^{lD}, $l \in \mathbb{Z}$:
$$\mathrm{Tr}(q^{(2\kappa+l)D}) = \frac{1}{1-q^{2\kappa+l}}\,.$$

In the case $|q| = 1$ and $|q^\kappa| < 1$, the usual trace $\mathrm{Tr}_{W^\pm}$ is related to Tr as
$$\mathrm{Tr}_{W^+}(q^{2\kappa D}x) = \mathrm{Tr}(q^{2\kappa D}x), \quad \mathrm{Tr}_{W^-}(q^{-2\kappa D}x) = -\mathrm{Tr}(q^{-2\kappa D}x)\,,$$

for $x \in Osc$. We extend the linear functional to the one on the tensor product $Osc \otimes Osc$:

$$\mathrm{Tr}(q^{2\kappa(D\otimes 1-1\otimes D)}x)=\sum_m \mathrm{Tr}(q^{2\kappa D}x_m^{(1)})\cdot \mathrm{Tr}(q^{-2\kappa D}x_m^{(2)})$$

for $x=\sum_m x_m^{(1)}\otimes x_m^{(2)}\in Osc\otimes Osc$. Sometimes we write this as $\mathrm{Tr}_{A,B}(q^{2\kappa(D_A-D_B)}x)$. If $|q|=1$ and $|q^\kappa|<1$, we have

$$\mathrm{Tr}_{W^+\otimes W^-}(q^{2\kappa(D\otimes 1-1\otimes D)}x)=-\mathrm{Tr}(q^{2\kappa(D\otimes 1-1\otimes D)}x) \tag{2.51}$$

for $x \in Osc \otimes Osc$.

We have a homomorphism $o_\zeta : U_q\mathfrak{b}^+ \to Osc$ given by

$$o_\zeta(e_0)=\frac{\zeta}{q-q^{-1}}\mathbf{a},\ o_\zeta(e_1)=\frac{\zeta}{q-q^{-1}}\mathbf{a}^*,\ o_\zeta(t_0)=q^{-2D},\ o_\zeta(t_1)=q^{2D}.$$

In terms of the Drinfeld generators it reads

$$o_\zeta(x_m^+)=\frac{\zeta^{2m+1}q^{m(2D+1)}\mathbf{a}^*}{q-q^{-1}},\quad (m\geq 0),$$

$$o_\zeta(x_m^-)=\frac{\zeta^{2m-1}\mathbf{a}\,q^{m(2D+1)-1}}{q-q^{-1}},\quad (m\geq 1),$$

$$o_\zeta(h_r)=\frac{-q^r\zeta^{2r}}{r(q-q^{-1})}\left(1-(1+q^{2r})q^{2rD}\right),\quad (r\geq 1).$$

Copies of 'auxiliary' q-oscillator algebra will be labeled by indices such as A, B.

We define a $U_q\mathfrak{b}^+$-module W_ζ^+ by using

$$\pi_\zeta^+=\pi^+\circ o_\zeta .$$

Using (2.39) we find

$$\left(\pi_\zeta^+\otimes\pi_1^{(1)}\right)\mathcal{R}=\mathsf{s}(\zeta)R_A^+(\zeta)\qquad\in \mathrm{End}(W_\zeta^+)\otimes \mathrm{End}(\mathbb{C}^2), \tag{2.52}$$

with

$$R_A^+(\zeta)=\begin{pmatrix}1-\zeta^2q^{2D_A+2} & -\zeta\mathbf{a}_A\\ -\zeta\mathbf{a}_A^* & 1\end{pmatrix}q^{-D_A\sigma^3}, \tag{2.53}$$

$$R_A^+(\zeta)^{-1}=\frac{1}{1-\zeta^2}q^{D_A\sigma^3}\begin{pmatrix}1 & \zeta\mathbf{a}_A\\ \zeta\mathbf{a}_A^* & 1-\zeta^2q^{2D_A}\end{pmatrix}.$$

In (2.53) we abbreviated $\pi^+(\mathbf{a}_A),\pi^+(\mathbf{a}_A^*),\pi^+(D_A)$ to $\mathbf{a}_A,\mathbf{a}_A^*,D_A$. The equation (2.23) is satisfied if

$$\mathsf{s}(\zeta)\mathsf{s}(\zeta q^{-1})(1-\zeta^2)=1 .$$

The superscript $^+$ in $R_A^+(\zeta)$ indicates that the auxiliary space is π_ζ^+.

We have the crossing symmetry relation

$$\theta(R_A^+(\zeta))=(1-q^2\zeta^2)R_A^+(q\zeta)^{-1}, \tag{2.54}$$

where θ is given in (1.23).

If the auxiliary space is $\pi_\zeta^{(1)}$, the R matrix carries a lower case letter. We rewrite (1.13) as

$$R_a(\zeta)=\begin{pmatrix}1-\zeta^2q^{\sigma_a^3+1} & -(q-q^{-1})\zeta\sigma_a^-\\ -(q-q^{-1})\zeta\sigma_a^+ & 1-\zeta^2q^{-\sigma_a^3+1}\end{pmatrix}q^{-\sigma_a^3\sigma^3/2}. \tag{2.55}$$

The crossing symmetry relation (1.24) can also be written in the form

$$\theta\left(R_a(\zeta)\right) = (1-\zeta^2)(1-q^4\zeta^2)R_a(q\zeta)^{-1}. \tag{2.56}$$

The Borel subalgebra $U_q\mathfrak{b}^+$ allows the involution ι: $e_i \to e_{1-i}$, $t_i \to t_{1-i}$, $i=0,1$. Define

$$\pi_\zeta^- = \pi^- \circ o_\zeta \circ \iota$$

and the corresponding $U_q\mathfrak{b}^+$-module W_ζ^-. Then

$$\left(\pi_\zeta^- \otimes \pi_1^{(1)}\right)\mathcal{R} = \mathsf{s}(\zeta)R_A^-(\zeta) \quad \in \operatorname{End}(W_\zeta^-)\otimes\operatorname{End}(\mathbb{C}^2),$$

where

$$R_A^-(\zeta) = \begin{pmatrix} 1 & -\zeta\mathbf{a}_A^* \\ -\zeta\mathbf{a}_A & 1-\zeta^2q^{2D_A+2}\end{pmatrix} q^{D_A\sigma^3}, \tag{2.57}$$

where $\pi^-(\mathbf{a}_A), \pi^-(\mathbf{a}_A^*), \pi^-(D_A)$ are abbreviated to $\mathbf{a}_A, \mathbf{a}_A^*, D_A$. In addition, for $x\in U_q\mathfrak{b}^+$ the intertwining property certainly holds:

$$R_A^\epsilon(\zeta)(\pi_\zeta^\epsilon\otimes\pi_1^{(1)})\Delta(x) = (\pi_\zeta^\epsilon\otimes\pi_1^{(1)})\Delta'(x)R_A^\epsilon(\zeta) \quad \text{where } \epsilon=\pm.$$

The definitions (2.53), (2.57) go along with our usual prescription: the $(2,1)$-matrix element is a creation operator, while the $(1,2)$-matrix element is an annihilation operator.

REMARK 2.1. Viewed as matrices with values in the algebra Osc, the matrices (2.53), (2.57) are formally related by

$$R_A^-(\zeta) = \sigma^1 R_A^+(\zeta)\sigma^1.$$

The two sides being evaluated with different representations $\pi^\pm$, this equation has to be taken with care.

2.3.2. Intertwiner for $W_{\zeta_1}^\pm \otimes W_{\zeta_2}^\pm$. The q-oscillator representation π_ζ^+ has a peculiar property that an arbitrary tensor product $\pi_{\zeta_1}^+\otimes\cdots\otimes\pi_{\zeta_n}^+$ is always irreducible [**FH**]. Moreover, changing the order of the tensor product gives rise to isomorphic representations. The same is true for π_ζ^-'s. Consequently, there exist intertwiners

$$\check{R}(\zeta_1,\zeta_2): W_{\zeta_1}^\pm\otimes W_{\zeta_2}^\pm \xrightarrow{\sim} W_{\zeta_2}^\pm\otimes W_{\zeta_1}^\pm$$

which satisfy the Yang-Baxter equation

$$\begin{aligned}&\left(I\otimes\check{R}(\zeta_2,\zeta_3)\right)\left(\check{R}(\zeta_1,\zeta_3)\otimes I\right)\left(I\otimes\check{R}(\zeta_1,\zeta_2)\right)\\ &=\left(\check{R}(\zeta_1,\zeta_2)\otimes I\right)\left(I\otimes\check{R}(\zeta_1,\zeta_3)\right)\left(\check{R}(\zeta_2,\zeta_3)\otimes I\right).\end{aligned}$$

Let us compute them explicitly. Note that they cannot be obtained from the universal R matrix $\mathcal{R}\in U_q\mathfrak{b}^+\otimes U_q\mathfrak{b}^-$, because all modules concerned are $U_q\mathfrak{b}^+$ modules.

We want to have the intertwiner $R_{A,B}(\zeta_1/\zeta_2)\in \operatorname{End}\left(W_{\zeta_1}^+\otimes W_{\zeta_2}^+\right)$ such that

$$R_{A,B}(\zeta_1/\zeta_2)R_A^+(\zeta_1)R_B^+(\zeta_2) = R_B^+(\zeta_2)R_A^+(\zeta_1)R_{A,B}(\zeta_1/\zeta_2).$$

Due to (2.57), the same $R_{A,B}(\zeta)$ intertwines also the product $R_A^-(\zeta_1)R_B^-(\zeta_2)$ with the one in the opposite order.

Using the permutation $P_{A,B} \in \mathrm{End}\left(W^+_{\zeta_1} \otimes W^+_{\zeta_2}\right)$, $P_{A,B}(x \otimes y) = y \otimes x$, we normalise $R_{A,B}(\zeta_1/\zeta_2)$ so that $R_{A,B}(1) = P_{A,B}$. Note that

$$P_{A,B} R^+_B(\zeta_2) R^+_A(\zeta_1) P_{A,B} = R^+_A(\zeta_2) R^+_B(\zeta_1).$$

Introduce the notation

$$M(\zeta_1, \zeta_2) = R^+_A(\zeta_1) R^+_B(\zeta_2)\,.$$

Its matrix elements are given by

$$\begin{aligned}
M(\zeta_1, \zeta_2)_{1,1} &= q^{-D_A - D_B} + (\zeta_1\zeta_2)^2 q^{D_A + D_B + 4} \\
&\quad - \zeta_1\zeta_2\big(\zeta q^{D_A - D_B + 2} + \zeta^{-1} q^{D_B - D_A + 2} - \mathbf{a}_A \mathbf{a}^*_B q^{D_A - D_B}\big)\,, \\
M(\zeta_1, \zeta_2)_{1,2} &= \zeta_1 \Big(\zeta_1\zeta_2 \mathbf{a}_B q^{D_A + D_B + 2} - \big(\mathbf{a}_A q^{D_A + D_B} + \zeta^{-1} \mathbf{a}_B q^{D_B - D_A}\big)\Big)\,, \\
M(\zeta_1, \zeta_2)_{2,1} &= \zeta_2 \Big(\zeta_1\zeta_2 \mathbf{a}^*_A q^{D_B - D_A + 2} - \big(\zeta \mathbf{a}^*_A q^{-D_A - D_B} + \mathbf{a}^*_B q^{D_A - D_B}\big)\Big)\,, \\
M(\zeta_1, \zeta_2)_{2,2} &= q^{D_A + D_B} + \zeta_1\zeta_2 \mathbf{a}^*_A \mathbf{a}_B q^{D_B - D_A}\,.
\end{aligned}$$

Here and after, we set $\zeta = \zeta_1/\zeta_2$.

Let us extract the permutation:

$$R_{A,B}(\zeta) = P_{A,B} \check{R}_{A,B}(\zeta)\,.$$

For $\check{R}_{A,B}(\zeta)$ we have

$$\check{R}_{A,B}(\zeta) M(\zeta_1, \zeta_2)_{i,j} = M(\zeta_2, \zeta_1)_{i,j} \check{R}_{A,B}(\zeta)\,, \quad i, j = 1, 2. \tag{2.58}$$

The coefficients of these equations are functions of ζ and $\zeta_1\zeta_2$, so one can equate like powers of $\zeta_1\zeta_2$. This leads us to the ansatz

$$\check{R}_{A,B}(\zeta) = h(\zeta, u_{A,B}) \zeta^{D_A + D_B}\,, \qquad u_{A,B} = \mathbf{a}^*_A q^{-2D_A} \mathbf{a}_B\,. \tag{2.59}$$

Note that $u_{A,B}$ commutes with $\mathbf{a}_B$, $q^{D_A+D_B}$ and $\mathbf{a}^*_A q^{D_B - D_A}$, which implies (2.58) for $(i, j) = (2, 2)$. Next consider the case $(i, j) = (1, 2)$. The equation reduces to

$$h(\zeta, u_{A,B})(\mathbf{a}_A + \zeta^{-1} \mathbf{a}_B q^{-2D_A}) = (\mathbf{a}_A + \zeta \mathbf{a}_B q^{-2D_A}) h(\zeta, u_{A,B}).$$

This is equivalent to the intertwining relation

$$R_{A,B}(\zeta)(\pi^+_\zeta \otimes \pi^+_1)\Delta(e_0) = (\pi^+_\zeta \otimes \pi^+_1)\Delta'(e_0) R_{A,B}(\zeta).$$

Multiplying $\mathbf{a}^*_A$ from the left, and using the q^2-commutativity $\mathbf{a}^*_A u_{A,B} = q^2 u_{A,B} \mathbf{a}^*_A$, we have

$$h(\zeta, u_{A,B} q^2)(1 + \zeta^{-1} u_{A,B}) = (1 + \zeta u_{A,B}) h(\zeta, u_{A,B}).$$

Looking for $h(\zeta, u)$ as power series in u, we find that it must satisfy the equation

$$(1 + \zeta u) h(\zeta, u) = (1 + \zeta^{-1} u) h(\zeta, u q^2)\,, \tag{2.60}$$

which can be uniquely solved as

$$h(\zeta, u) = \sum_{n=0}^{\infty} (-u)^n \prod_{j=1}^{n} \frac{\zeta - \zeta^{-1} q^{2j-2}}{1 - q^{2j}}\,. \tag{2.61}$$

Note also that

$$\frac{h(\zeta q, u q)}{h(\zeta, u)} = 1 + \zeta u\,, \tag{2.62}$$

because $h(\zeta q, u q)/(1 + \zeta u)$ is a power series in u satisfying the same functional relation (2.60).

Before passing we do not forget other cases: the case $(i,j)=(2,1)$ is equivalent to the intertwining relation

$$R_{A,B}(\zeta)(\pi_\zeta^+\otimes\pi_1^+)\Delta(e_1)=(\pi_\zeta^+\otimes\pi_1^+)\Delta'(e_1)R_{A,B}(\zeta),$$

and is written as

$$h(\zeta,u_{A,B})(\zeta\mathbf{a}_A^*+\mathbf{a}_B^*q^{2D_A})=(\zeta^{-1}\mathbf{a}_A^*+\mathbf{a}_B^*q^{2D_A})h(\zeta,u_{A,B}).$$

Multiplying $\mathbf{a}_Bq^{-2D_A}$ from the right we get (2.60). Finally the case $(i,j)=(1,1)$ is written as

$$\begin{aligned}&h(\zeta,u_{A,B})\left(\zeta q^{-2D_B+2}+\zeta^{-1}q^{-2D_A+2}-\mathbf{a}_A\mathbf{a}_B^*q^{-2D_B}\right)\\&=\left(\zeta^{-1}q^{-2D_B+2}+\zeta q^{-2D_A+2}-\mathbf{a}_A\mathbf{a}_B^*q^{-2D_B}\right)h(\zeta,u_{A,B})\,.\end{aligned}$$

Multiplying $\mathbf{a}_A^*\mathbf{a}_B$ from the left and using (2.60) we get

$$h(\zeta,q^2u_{A,B})\left(\zeta q^2u_{A,B}+1\right)q^{2D_A-2D_B}=\left(\zeta^{-1}q^2u_{A,B}+1\right)q^{2D_A-2D_B}h(\zeta,u_{A,B}).$$

Due to $q^{2D_A-2D_B}u_{A,B}=q^4u_{A,B}q^{2D_A-2D_B}$ this reduces to (2.60).

The equation for the intertwiner (2.60) is similar to the one for the lattice sine-Gordon model [**FV**]. However there is an important difference. In the latter case, a power series solution diverges for $|q|=1$ and an analytic solution to (2.60) is a transcendental function. This is a serious complication when representations without highest or lowest weight are considered. However, in our case the power series (2.61) does make sense, because for both $W_{\zeta_1}^+\otimes W_{\zeta_2}^+$ and $W_{\zeta_1}^-\otimes W_{\zeta_2}^-$ the element $u_{A,B}$ is locally nilpotent.

2.3.3. Triangular Structures of $W^+\otimes W^-$. Unlike the case $W_{\zeta_1}^\pm\otimes W_{\zeta_2}^\pm$ discussed in the previous subsection, the tensor products $W_{\zeta_1}^\pm\otimes W_{\zeta_2}^\mp$ have a more complicated structure. We study them in this subsection.

Consider the product of R matrices $R_A^+(\zeta_1)R_B^-(\zeta_2)$, which intertwines the action of $x\in U_q\mathfrak{b}^+$ on $(W_{\zeta_1}^+\otimes W_{\zeta_2}^-)\otimes V_1^{(1)}$ with that on $V_1^{(1)}\otimes(W_{\zeta_1}^+\otimes W_{\zeta_2}^-)$. We will compare it with (2.46) after this. The product has the following structure:

$$\begin{aligned}&\zeta^{(D_A-D_B)/2}R_A^+(\zeta_1)R_B^-(\zeta_2)\zeta^{-(D_A-D_B)/2}\\&=\begin{pmatrix}1-\overline{\zeta}^{\,2}\,Y(\zeta)&-\overline{\zeta}\,W(\zeta)-c(\zeta_1,\zeta_2)V(\zeta)\\-\overline{\zeta}\,U(\zeta)&1-\overline{\zeta}^{\,2}Z(\zeta)\end{pmatrix}q^{-(D_A-D_B)\sigma^3}\,.\end{aligned}\tag{2.63}$$

Here and after

$$\zeta=\zeta_1/\zeta_2\,,\quad\overline{\zeta}=(\zeta_1\zeta_2)^{1/2}\,,\tag{2.64}$$

and $c(\zeta_1,\zeta_2)=\zeta_1^{-3/2}\zeta_2^{5/2}(1-\zeta_1^2q^2)$. The entries are given by

$$\begin{aligned}U(\zeta)&=\zeta(\mathbf{a}^*\otimes1)+q^{2D}\otimes\mathbf{a}\,,\\V(\zeta)&=\zeta(1\otimes\mathbf{a}^*)+\mathbf{a}\otimes q^{2D}\,,\\Y(\zeta)&=(\zeta q^2-\mathbf{a}\otimes\mathbf{a})(q^{2D}\otimes1)\,,\\Z(\zeta)&=\zeta^{-1}(1\otimes q^{2D+2})-\mathbf{a}^*q^{-2D}\otimes\mathbf{a}^*\,,\\W(\zeta)&=(1-\zeta^{-1}q^{-2}Z(\zeta))(\mathbf{a}\otimes1)\,.\end{aligned}\tag{2.65}$$

They satisfy the following relations.

$$
\begin{aligned}
(2.66)\qquad & Y(\zeta)Z(\zeta) = q^2 - \zeta^{-1}q^4U(\zeta)V(\zeta), \quad Z(\zeta)Y(\zeta) = q^2 - \zeta^{-1}q^2U(\zeta)V(\zeta)\,, \\
& Z(\zeta)U(\zeta) = q^{-2}U(\zeta)Z(\zeta), \quad Y(\zeta)U(\zeta) = q^2U(\zeta)Y(\zeta)\,, \\
& Z(\zeta)W(\zeta) = q^2W(\zeta)Z(\zeta) + (1-q^2)\zeta^{-2}(q^2\zeta - Z(\zeta))V(\zeta)\,, \\
& Y(\zeta)W(\zeta) = q^{-2}W(\zeta)Y(\zeta) + (1-q^2)(q\zeta)^{-2}(Y(\zeta) - q^2\zeta)V(\zeta)\,, \\
& U(\zeta)W(\zeta) = (\zeta - q^{-2}Y(\zeta))(1-\zeta^{-1}Z(\zeta)) - (1-q^2)\zeta^{-2}U(\zeta)V(\zeta)\,, \\
& W(\zeta)U(\zeta) = (1-\zeta^{-1}q^{-2}Z(\zeta))(\zeta - Y(\zeta))\,, \\
& [V(\zeta), U(\zeta)] = [V(\zeta), Y(\zeta)] = [V(\zeta), Z(\zeta)] = [V(\zeta), W(\zeta)] = 0\,.
\end{aligned}
$$

We introduce a subalgebra $\mathfrak{A}^{\mathrm{left}}(\zeta) \subset Osc \otimes Osc$ generated by the elements $U(\zeta)$, $V(\zeta)$, $Y(\zeta)$, $Z(\zeta)$, $W(\zeta)$. The element $V(\zeta)$ is central in $\mathfrak{A}^{\mathrm{left}}(\zeta)$. Let us explain that the quotient algebra $\mathfrak{A}^{\mathrm{left}}(\zeta)/\langle V(\zeta)\rangle$ obtained by setting $V(\zeta) = 0$ has a simple structure.

Consider the following quotient algebra

$$
(2.67)\qquad \mathfrak{A}(\zeta) = U_q(\mathfrak{sl}_2)/\langle C - \frac{\zeta+\zeta^{-1}}{(q-q^{-1})^2}\rangle\,.
$$

Algebra $\mathfrak{A}(\zeta)$ is defined by generators $E, F, K^{\pm1}$ and relations

$$
\begin{aligned}
(2.68)\qquad & KE = q^2EK\,, \quad KF = q^{-2}FK\,, \\
& (q-q^{-1})^2EF = \zeta + \zeta^{-1} - q^{-1}K - qK^{-1}\,, \\
& (q-q^{-1})^2FE = \zeta + \zeta^{-1} - qK - q^{-1}K^{-1}\,.
\end{aligned}
$$

Now we adjoin $V(\zeta) = 0$ to the relations (2.66) and compare them with (2.68). We conclude that there is a homomorphism $\boldsymbol{\pi}^{\mathrm{left}} : \mathfrak{A}(\zeta) \to \mathfrak{A}^{\mathrm{left}}(\zeta)/\langle V(\zeta)\rangle$

$$
\begin{aligned}
(2.69)\qquad & \boldsymbol{\pi}^{\mathrm{left}}(E) = \frac{1}{q-q^{-1}}U(\zeta)\,, \quad \boldsymbol{\pi}^{\mathrm{left}}(F) = \frac{1}{q-q^{-1}}W(\zeta)\,, \\
& \boldsymbol{\pi}^{\mathrm{left}}(K) = q^{-1}Y(\zeta)\,, \quad \boldsymbol{\pi}^{\mathrm{left}}(K^{-1}) = q^{-1}Z(\zeta)\,.
\end{aligned}
$$

Similarly, for the product of R matrices in the opposite order we have

$$
\begin{aligned}
(2.70)\qquad & \zeta^{(D_A-D_B)/2}R^-_{\bar B}(\zeta_2)R^+_{\bar A}(\zeta_1)\zeta^{-(D_A-D_B)/2} \\
& = \begin{pmatrix} 1-\overline{\zeta}^{\,2}\,\overline{Z}(\zeta) & -\overline{\zeta}\,\overline{U}(\zeta) \\ -\overline{\zeta}\,\overline{W}(\zeta) - c(\zeta_2,\zeta_1)\overline{V}(\zeta) & 1-\overline{\zeta}^{\,2}\,\overline{Y}(\zeta) \end{pmatrix} q^{-(D_A-D_B)\sigma^3}\,,
\end{aligned}
$$

where

$$
(2.71)\qquad \overline{X}(\zeta) = \sigma(X(\zeta^{-1})) \quad \text{for} \quad X = U, V, Y, Z, W\,,
$$

and $\sigma(a\otimes b) = b\otimes a$. Proceeding in the same way as above, we introduce another subalgebra $\mathfrak{A}^{\mathrm{right}}(\zeta)$ generated by the symbols $\overline{X}(\zeta)$ ($X = U, V, Y, Z, W$.) We have a homomorphism $\boldsymbol{\pi}^{\mathrm{right}} : \mathfrak{A}(\zeta) \to \mathfrak{A}^{\mathrm{right}}(\zeta)/\langle \overline{V}(\zeta)\rangle$ given by

$$
\begin{aligned}
& \boldsymbol{\pi}^{\mathrm{right}}(F) = \frac{1}{q-q^{-1}}\overline{U}(\zeta)\,, \quad \boldsymbol{\pi}^{\mathrm{right}}(E) = \frac{1}{q-q^{-1}}\overline{W}(\zeta)\,, \\
& \boldsymbol{\pi}^{\mathrm{right}}(K^{-1}) = q^{-1}\overline{Y}(\zeta)\,, \quad \boldsymbol{\pi}^{\mathrm{right}}(K) = q^{-1}\overline{Z}(\zeta)\,.
\end{aligned}
$$

So, 'left' and 'right' differ by the automorphism $K \to K^{-1}, E \to F, F \to E, \zeta \to \zeta^{-1}$ of $\mathfrak{A}(\zeta)$.

Now we come to a beautiful part of the BLZ construction: under the trace, one is led automatically to the quotient by $V(\zeta)$ or $\overline{V}(\zeta)$. Let us explain this point.

It is easier to start with the 'right' case. In the tensor product $W^+ \otimes W^-$, we introduce a basis

$$f_{j,m} = \overline{W}(\zeta)^j \overline{V}(\zeta)^m (|0\rangle \otimes |-1\rangle)\,.$$

To see that they form a basis, first we observe that they form a spanning set of $\mathfrak{A}^{\mathrm{right}}(\zeta)(|0\rangle \otimes |-1\rangle) \subset W^+ \otimes W^-$. We use the relations (2.66) and the following equalities:

$$\begin{aligned}
&\overline{U}(\zeta)(|0\rangle \otimes |-1\rangle) = 0,\\
&\overline{Y}(\zeta)(|0\rangle \otimes |-1\rangle) = \zeta^{-1}(|0\rangle \otimes |-1\rangle),\\
&\overline{Z}(\zeta)(|0\rangle \otimes |-1\rangle) = \zeta q^2(|0\rangle \otimes |-1\rangle).
\end{aligned}$$

We set $\mathcal{F}^m = \mathrm{Span}\{f_{j,p} \mid j \geq 0\,, p \geq m\}$, then we have a decreasing sequence of subspaces of $W^+ \otimes W^-$:

$$\mathfrak{A}^{\mathrm{right}}(\zeta)(|0\rangle \otimes |-1\rangle) = \mathcal{F}^0 \supset \cdots \supset \mathcal{F}^m \supset \mathcal{F}^{m+1} \supset \cdots\,.$$

The algebra $\mathfrak{A}^{\mathrm{right}}(\zeta)$ acts on each $\mathcal{F}^m$ and, in particular, we have $\overline{V}(\zeta)\mathcal{F}^m \subset \mathcal{F}^{m+1}$. The operator $D \otimes 1 - 1 \otimes D \in \mathrm{End}(W^+ \otimes W^-)$ satisfies

$$(D \otimes 1 - 1 \otimes D) f_{j,m} = (j + m + 1) f_{j,m}\,. \tag{2.72}$$

In conclusion, $\boldsymbol{\pi}^{\mathrm{right}}$ induces an $\mathfrak{A}(\zeta)$-action on the m-th block $\mathcal{F}^m/\mathcal{F}^{m+1}$. As $U_q(\mathfrak{sl}_2)$-module it is isomorphic to π^{Λ} with $q^{\Lambda} = (\zeta q)^{-1}$. In addition, up to a normalisation factor the action of the product of R-matrices (2.63) on each block agrees with the R-matrix $\left(ev_{\zeta_1} \otimes \pi^{(1)}_{\zeta_2}\right)\mathcal{R}$ with the identification $\zeta_1/\zeta_2 = \bar{\zeta}$. From these facts it follows that $f_{j,m}$'s are linearly independent and form a basis of $\mathfrak{A}^{\mathrm{right}}(\zeta)(|0\rangle \otimes |-1\rangle)$, and moreover that $W^+ \otimes W^- = \mathfrak{A}^{\mathrm{right}}(\zeta)(|0\rangle \otimes |-1\rangle)$.

In the 'left' case we choose the following basis of $W^+ \otimes W^-$

$$e_{j,m} = U(\zeta)^j (1 \otimes \mathbf{a})^m (|0\rangle \otimes |-1\rangle)\,.$$

To see that they form a basis we use, in addition to (2.66), the following equalities

$$\begin{aligned}
&V(\zeta)(|0\rangle \otimes |-1\rangle) = 0, \quad W(\zeta)(|0\rangle \otimes |-1\rangle) = 0,\\
&Y(\zeta)(|0\rangle \otimes |-1\rangle) = \zeta q^2(|0\rangle \otimes |-1\rangle),\\
&Z(\zeta)(|0\rangle \otimes |-1\rangle) = \zeta^{-1}(|0\rangle \otimes |-1\rangle),,
\end{aligned}$$

and the commutation relations

$$\begin{aligned}
&[U(\zeta), 1 \otimes \mathbf{a}] = [Y(\zeta), 1 \otimes \mathbf{a}] = 0\,,\\
&[Z(\zeta), 1 \otimes \mathbf{a}] = \zeta^{-1}(1 - q^2) U(\zeta) q^{-2(D \otimes 1 - 1 \otimes D)}\,,\\
&[W(\zeta), 1 \otimes \mathbf{a}] = -\zeta^{-2}(1 - q^2)(\zeta - q^{-2} Y(\zeta)) q^{-2(D \otimes 1 - 1 \otimes D)}\,,\\
&V(\zeta)(1 \otimes \mathbf{a}) = q^{-2}(1 \otimes \mathbf{a}) V(\zeta) + \zeta(1 - q^{-2}).
\end{aligned} \tag{2.73}$$

We set $\mathcal{F}_m = \mathrm{Span}\{e_{j,p} \mid j \geq 0\,, p \leq m\}$, then we have an increasing sequence of subspaces of $W^+ \otimes W^-$:

$$\{0\} = \mathcal{F}_{-1} \subset \mathcal{F}_0 \subset \cdots \subset \mathcal{F}_{m-1} \subset \mathcal{F}_m \subset \cdots \subset W^+ \otimes W^-.$$

The algebra $\mathfrak{A}^{\mathrm{left}}(\zeta)$ acts on each $\mathcal{F}_m$ and, in particular, we have $V(\zeta)\mathcal{F}_m\subset\mathcal{F}_{m-1}$. We have

$$(D\otimes 1-1\otimes D)e_{j,m}=(j+m+1)e_{j,m}\,. \tag{2.74}$$

In conclusion, $\boldsymbol{\pi}^{\mathrm{left}}$ induces an $\mathfrak{A}(\zeta)$-action on the m-th block $\mathcal{F}_m/\mathcal{F}_{m-1}$. It is isomorphic, as $U_q(\mathfrak{sl}_2)$-module, to π^Λ with $q^\Lambda=(\zeta q)^{-1}$. In addition, up to a normalisation factor the action of the product of R-matrices (2.63) on each block agrees with the R-matrix $\left(ev_{\zeta_1}\otimes\pi^{(1)}_{\zeta_2}\right)\mathcal{R}$ with the identification $\zeta_1/\zeta_2=\bar{\zeta}$. From these facts it follows that $e_{j,m}$'s are linearly independent and form a basis of $W^+\otimes W^-$.

From the above consideration we arrive at the following.

LEMMA 2.2. *The maps $\boldsymbol{\pi}^{\mathrm{left}}$ and $\boldsymbol{\pi}^{\mathrm{right}}$ are isomorphisms of algebras.*

The statement follows from the faithfulness of the representation π^Λ of $\mathfrak{A}(\zeta)$.

LEMMA 2.3. *Define $\boldsymbol{\rho}^{\mathrm{left}}:\mathfrak{A}^{\mathrm{left}}(\zeta)\to\mathfrak{A}^{\mathrm{left}}(\zeta)/\langle V(\zeta)\rangle\to\mathfrak{A}(\zeta)$ to be the composition of the projection and $\left(\boldsymbol{\pi}^{\mathrm{left}}\right)^{-1}$. For any $X\in\mathfrak{A}^{\mathrm{left}}(\zeta)$ we have*

$$\mathrm{Tr}\big(q^{2\kappa(D\otimes 1-1\otimes D)}X\big)=\frac{\zeta^{-\kappa}}{q^\kappa-q^{-\kappa}}\mathrm{Tr}^\Lambda\big(q^{\kappa H}\boldsymbol{\rho}^{\mathrm{left}}(X)\big)\,, \tag{2.75}$$

where the trace in the right hand side is taken over the Verma module V^Λ with $q^\Lambda=(\zeta q)^{-1}$.

Similarly, define $\boldsymbol{\rho}^{\mathrm{right}}:\mathfrak{A}^{\mathrm{right}}(\zeta)\to\mathfrak{A}^{\mathrm{right}}(\zeta)/\langle V(\zeta)\rangle\to\mathfrak{A}(\zeta)$ be the composition of the projection and $\left(\boldsymbol{\pi}^{\mathrm{right}}\right)^{-1}$. Then, for any $X\in\mathfrak{A}^{\mathrm{right}}(\zeta)$ we have

$$\mathrm{Tr}\big(q^{2\kappa(D\otimes 1-1\otimes D)}X\big)=\frac{\zeta^{-\kappa}}{q^\kappa-q^{-\kappa}}\mathrm{Tr}^\Lambda\big(q^{\kappa H}\boldsymbol{\rho}^{\mathrm{right}}(X)\big)\,, \tag{2.76}$$

with the same notation as in the previous case.

PROOF. We begin with the case $|q|=1$ and $|q^\kappa|<1$ when the traces are defined as usual ones and use the above constructions, then continue analytically with respect to q and q^κ. The only thing which probably needs explanation is the multiplier in the right hand side. Rewrite it as

$$-\frac{1}{1-q^{2\kappa}}(\zeta q^{-1})^{-\kappa}\,.$$

The minus comes from the difference between Tr and $\mathrm{Tr}_{W^+\otimes W^-}$ in the equality (2.51), the factor $\frac{1}{1-q^{2\kappa}}$ comes from summing up the geometric progression over the irreducible blocks parametrised by $m\geq 0$, and $(\zeta q^{-1})^{-\kappa}$ comes from (2.72), (2.74), which are rephrased as the statement that the operator $q^{2(D\otimes 1-1\otimes D)}$ induces the action of $q^{\Lambda+2+2m}K$ on the m-th block. □

Notice an amusing circumstance. There is a freedom to shift κ by an integer. We will make use of it in the next chapter:

LEMMA 2.4.

(2.77)

$$\mathrm{Tr}\left(Y(\zeta)q^{2\kappa(D\otimes 1-1\otimes D)}X^{\mathrm{left}}\right)=q\zeta\frac{q^{\kappa+1}-q^{-\kappa-1}}{q^{\kappa}-q^{-\kappa}}\mathrm{Tr}\left(q^{2(\kappa+1)(D\otimes 1-1\otimes D)}X^{\mathrm{left}}\right),$$
$$\mathrm{Tr}\left(Z(\zeta)q^{2\kappa(D\otimes 1-1\otimes D)}X^{\mathrm{left}}\right)=q\zeta^{-1}\frac{q^{\kappa-1}-q^{-\kappa+1}}{q^{\kappa}-q^{-\kappa}}\mathrm{Tr}\left(q^{2(\kappa-1)(D\otimes 1-1\otimes D)}X^{\mathrm{left}}\right),$$
$$\mathrm{Tr}\left(\overline{Y}(\zeta)q^{2\kappa(D\otimes 1-1\otimes D)}X^{\mathrm{right}}\right)=q\zeta^{-1}\frac{q^{\kappa-1}-q^{-\kappa+1}}{q^{\kappa}-q^{-\kappa}}\mathrm{Tr}\left(q^{2(\kappa-1)(D\otimes 1-1\otimes D)}X^{\mathrm{right}}\right),$$
$$\mathrm{Tr}\left(\overline{Z}(\zeta)q^{2\kappa(D\otimes 1-1\otimes D)}X^{\mathrm{right}}\right)=q\zeta\frac{q^{\kappa+1}-q^{-\kappa-1}}{q^{\kappa}-q^{-\kappa}}\mathrm{Tr}\left(q^{2(\kappa+1)(D\otimes 1-1\otimes D)}X^{\mathrm{right}}\right),$$

for any $X^{\mathrm{left}}\in\mathfrak{A}^{\mathrm{left}}(\zeta)$, $X^{\mathrm{right}}\in\mathfrak{A}^{\mathrm{right}}(\zeta)$.

PROOF. We prove the first identity. The rest are similar. Note that

$$\boldsymbol{\rho}^{\mathrm{left}}(Y(\zeta))=q^{H+1},$$

by (2.69). Then, by applying (2.75), we obtain

$$(\mathrm{LHS})=\frac{q\zeta^{-\kappa}}{q^{\kappa}-q^{-\kappa}}\mathrm{Tr}^{\mathrm{A}}\left(q^{(\kappa+1)H}\boldsymbol{\rho}^{\mathrm{left}}(X^{\mathrm{left}})\right)=(\mathrm{RHS}).$$

□

2.3.4. Triangular Structure of $\pi^{(1)}_\zeta\otimes\pi^{\pm}_\zeta$. Another useful property of the q-oscillator representations was discovered by Bazhanov-Lukyanov-Zamolodchikov: the tensor products $\pi^{(1)}_\zeta\otimes\pi^{\pm}_\zeta$ have a certain triangular structure. Consider first $\pi^{(1)}_\zeta\otimes\pi^{+}_\zeta$. The statement is that it has a submodule isomorphic to $\pi^{+}_{\zeta q^{-1}}$, and the corresponding quotient module is isomorphic to $\pi^{+}_{\zeta q}$. We shall need explicit formulas; so let us formulate this as an identity of R-matrices. A direct calculation shows that

(2.78)
$$\begin{aligned}&F^{-1}_{a,A}R_a(\zeta)R^{+}_A(\zeta)F_{a,A}\\&=\begin{pmatrix}1-\zeta^2 & 0\\ (q^{-1}-q)\zeta\sigma^{+} & 1-\zeta^2q^2\end{pmatrix}_a\begin{pmatrix}R^{+}_A(\zeta q)q^{-\sigma^3/2} & 0\\ 0 & R^{+}_A(\zeta q^{-1})q^{\sigma^3/2}\end{pmatrix}_a,\end{aligned}$$

where

$$F_{a,A}=1-\mathbf{a}_A\sigma^{+}_a\,. \tag{2.79}$$

Similarly the product $R_a(\zeta)R^{-}_A(\zeta)$ can be brought to an upper-triangular form by applying a similarity transformation with $\sigma^1_a\sigma^1$ to (2.78).

2.4. Higher Spins in Matsubara Space

Using the comultiplication it is straightforward to generalise the results of this section to the *Matsubara* space

$$\mathfrak{H}_{\mathbf{M}}=V^{(1)}_{\tau_1 q^{1/2}}\otimes\cdots\otimes V^{(1)}_{\tau_{\mathbf{n}} q^{1/2}}\,. \tag{2.80}$$

It will be useful in future to consider the *Matsubara* chains composed of representations of different spins,

$$\mathfrak{H}_{\mathbf{M}} = V^{(2s_1)}_{\tau_1 q^{1/2}} \otimes \cdots \otimes V^{(2s_n)}_{\tau_n q^{1/2}} \,. \tag{2.81}$$

Since these considerations will be of auxiliary character, the presentation in this subsection is brief and some proofs are omitted. They can be found in the literature [**KRS**], [**Ji2**].

Similarly to the definitions (2.48), (2.52) evaluating the universal R-matrix on representations we present the result as product of a transcendental scalar function and a polynomial R-matrix. For example,

(2.82)
$$(ev_{\zeta_1} \otimes \pi^{(2s)}_{\zeta_2})\mathcal{R} = \mathsf{r}_s(\zeta, C) R^{(ev,2s)}(\zeta) \,, \quad R^{(ev,2s)}(0) = (ev_{\zeta_1} \otimes \pi^{(2s)}_{\zeta_2})(q^{-\frac{h_1 \otimes h_1}{2}}) \,,$$

where $R^{(ev,2s)}(\zeta)$ is a polynomial of degree $4s$ in ζ,

$$\mathsf{r}_s(\zeta, C)\mathsf{r}_s(\zeta q^{-1}, C) = \prod_{j=0}^{2s-1} (q^{2j-2s+1} - (q - 1/q)^2 \zeta^2 \cdot C + \zeta^4 q^{-2j+2s-1}) \,. \tag{2.83}$$

We do not give an explicit formula for $R^{(ev,2s)}(\zeta)$ which can be obtained either from (2.39) or using the fusion procedure discussed below. Specialising further to the Verma module and recalling the corresponding value of the Casimir operator

$$C = \frac{q^{\Lambda+1} + q^{-\Lambda-1}}{(q - q^{-1})^2} \,,$$

one observes further factorisation of the right hand side of (2.83).

For the q-oscillator auxiliary representation π^+ one derives from (2.39) a rather simple formula

$$\begin{aligned} &(\pi^+_{\zeta_1} \otimes \pi^{(2s)}_{\zeta_2})\mathcal{R} = \mathsf{s}_s(\zeta) R^{(+,2s)}(\zeta) \,, \\ &R^{(+,2s)}(\zeta) = \exp_{q^2}(-\zeta \mathbf{a} E)\,\mathrm{diag}\,(\delta_0, \ldots, \delta_{2s}) \exp_{q^2}(-\zeta \mathbf{a}^* F) \cdot q^{-DH} \,, \end{aligned} \tag{2.84}$$

where $\zeta = \zeta_1/\zeta_2$,

$$\delta_j = \prod_{p=0}^{2s-j-1} (1 - \zeta^2 q^{2p-2s+1}) \,,$$

where the basis $(v_{2s}, \cdots, v_0)$ (see (2.28)) is used for $\pi^{(2s)}$, and $\mathsf{s}_s(\zeta)$ satisfies the equation

$$\mathsf{s}_s(\zeta)\mathsf{s}_s(\zeta q^{-1}) \prod_{j=0}^{2s-1} (1 - \zeta^2 q^{2s-2j-1}) = 1 \,. \tag{2.85}$$

Comparing (2.83) and (2.85) we find a relation which will be important later on

$$\mathsf{r}_s\left(\zeta, \frac{q^{\Lambda+1} + q^{-\Lambda-1}}{(q - q^{-1})^2}\right) = \mathsf{s}_s(\zeta q^{\frac{\Lambda+1}{2}})\mathsf{s}_s(\zeta q^{-\frac{\Lambda+1}{2}}) \,.$$

As has been mentioned, finite dimensional representations can be obtained as quotients of Verma modules. A subtlety occurs at this point.

REMARK 2.5. Define

$$(\pi^{(2s_1)}_{\zeta_1}\otimes\pi^{(2s_2)}_{\zeta_2})\mathcal{R}=\mathsf{r}_{s_1,s_2}(\zeta)R^{(2s_1,2s_2)}(\zeta)\,. \tag{2.86}$$

In (2.82) substitute $s=s_2$. The right hand side belongs to $U_q(\mathfrak{sl}_2)\otimes\mathrm{End}(V^{(2s_2)})$. taking the irreducible representation $\pi^{(2s_1)}$ of $U_q(\mathfrak{sl}_2)$ one obtains (2.86). The subtle point is that for $s_2>s_1$ the image of $R^{(ev,2s_2)}(\zeta)$ happens to be divisible by

$$P(\zeta,s_1,s_2)=\prod_{j=1}^{2(s_2-s_1)}\left(1-q^{2j-2(s_2-s_1)}\zeta^2\right).$$

We define $R^{(s_1,s_2)}(\zeta)$ dividing by this multiplier which we move to $\mathsf{r}_{s_1,s_2}(\zeta)$:

$$\mathsf{r}_{s_1,s_2}(\zeta)=P(\zeta,s_1,s_2)\cdot\mathsf{r}_{s_2}\big(\zeta,\frac{q^{2s_1+1}+q^{-2s_1-1}}{(q-q^{-1})^2}\big)\,. \tag{2.87}$$

$$(\pi^{(2s_1)}\otimes\mathrm{id})R^{(ev,2s_2)}(\zeta)=P(\zeta,s_1,s_2)R^{(2s_1,2s_2)}(\zeta)\,. \tag{2.88}$$

Let us discuss the fusion procedure which allows us to prove many statements for the generalised *Matsubara* space (2.81) by simple reference to similar statements for the usual one (2.80).

The fusion procedure is a recursive method to calculate R matrices for finite dimensional evaluation representations starting from that of spin 1/2 representation. It is based on the following statement: the tensor product $V^{(2s)}_{\tau q}\otimes V^{(1)}_{\tau q^{1/2-s}}$ contains a submodule isomorphic to $V^{(2s-1)}_{\tau q^{3/2}}$, and the corresponding quotient module is isomorphic to $V^{(2s+1)}_{\tau q^{1/2}}$. Therefore, if we denote by $\iota^{2s,1}_{2s-1}:V^{(2s-1)}_{\tau q^{3/2}}\to V^{(2s)}_{\tau q}\otimes V^{(1)}_{\tau q^{1/2-s}}$ the embedding and by $\mathcal{P}^{2s+1}_{2s,1}:V^{(2s)}_{\tau q}\otimes V^{(1)}_{\tau q^{1/2-s}}\to V^{(2s+1)}_{\tau q^{1/2}}$ the projection, then we have the following identities.

$$\begin{aligned}(\mathrm{id}\otimes\pi^{(2s)}_{\tau q}\otimes\pi^{(1)}_{\tau q^{1/2-s}})\big((\mathrm{id}\otimes\Delta)\mathcal{R}\big)\cdot(\mathrm{id}\otimes\iota^{2s,1}_{2s-1})\\=(\mathrm{id}\otimes\iota^{2s,1}_{2s-1})(\mathrm{id}\otimes\pi^{(2s-1)}_{\tau q^{3/2}})(\mathcal{R})\,,\end{aligned} \tag{2.89}$$

$$\begin{aligned}(\mathrm{id}\otimes\mathcal{P}^{2s+1}_{2s,1})\cdot\big((\mathrm{id}\otimes\pi^{(2s)}_{\tau q}\otimes\pi^{(1)}_{\tau q^{1/2-s}})(\mathrm{id}\otimes\Delta)\mathcal{R}\big)\\=(\mathrm{id}\otimes\pi^{(2s+1)}_{\tau q^{1/2}})(\mathcal{R})\cdot(\mathrm{id}\otimes\mathcal{P}^{2s+1}_{2s,1})\,.\end{aligned} \tag{2.90}$$

In most cases these identities will be sufficient for our goals. However, for readers interested in practical applications, let us briefly explain how to handle (2.90).

We shall abuse the symbols $|\uparrow\rangle$ (resp. $|\downarrow\rangle$) to denote the highest (resp. lowest) vector in the tensor product of an arbitrary number of representations. Similar notation is used for covectors: $\langle\uparrow|$, $\langle\downarrow|$.

Recalling that

$$ev_\zeta(e_1)=\zeta E,\qquad ev_\zeta(f_1)=\zeta^{-1}F\,,$$

we define

$$\mathcal{E}=(\tau q^{1/2})^{-1}\big(\pi^{(2s)}_{\tau q}\otimes\pi^{(1)}_{\tau q^{1/2-s}}\big)\Delta e_1\,,\quad \mathcal{F}=(\tau q^{1/2})\big(\pi^{(2s)}_{\tau q}\otimes\pi^{(1)}_{\tau q^{1/2-s}}\big)\Delta f_1\,.$$

Choosing the normalisation $\mathcal{P}^{2s+1}_{2s,1}|\downarrow\rangle=v^{(2s+1)}_0$, we have for $v\in V^{(2s)}_{\tau q}\otimes V^{(1)}_{\tau q^{1/2-s}}$

$$\mathcal{P}^{2s+1}_{2s,1}v=\sum_{j=0}^{2s+1}\frac{\langle\uparrow|\mathcal{E}^jv\rangle}{\langle\uparrow|\mathcal{E}^{2s+1}|\downarrow\rangle}v^{(2s+1)}_{2s+1-j}\quad\in V^{(2s+1)}_{\tau q^{1/2}}\,. \tag{2.91}$$

Obviously,

$$\mathcal{P}^{2s+1}_{2s,1}\mathcal{E}=\pi^{(2s+1)}(E)\mathcal{P}^{2s+1}_{2s,1}\,,\quad \mathcal{P}^{2s+1}_{2s,1}\mathcal{F}=\pi^{(2s+1)}(F)\mathcal{P}^{2s+1}_{2s,1}\,.$$

Applying ev_ζ to (2.90), using (2.22) and $\mathsf{r}_{1/2}(q^{s-1/2}\zeta)\mathsf{r}_s(q^{-1}\zeta)=\mathsf{r}_{s+1/2}(q^{-1/2}\zeta)$, we obtain

$$\begin{aligned}\mathcal{P}^{2s+1}_{2s,1}(I_{2s+1}\otimes R^{(ev,1)}(q^{s-1/2}\zeta/\tau))(R^{(ev,2s)}(q^{-1}\zeta/\tau)\otimes I_2)\\ =R^{(ev,2s+1)}(q^{-1/2}\zeta/\tau)\mathcal{P}^{2s+1}_{2s,1}\,.\end{aligned}\tag{2.92}$$

In this formula we regard $R^{(ev,2s)}(\zeta)$ as a $(2s+1)\times(2s+1)$ matrix with values in $U_q(\mathfrak{sl}_2)$. We regard also $\mathcal{P}^{2s+1}_{2s,1}$ as a $(2s+2)\times 2(2s+1)$ matrix, which can be found from (2.91).

Take some basis $\{x_i\}$ of $U_q(\mathfrak{sl}_2)$ and expand both sides of (2.92) in this basis :

$$\sum x_i\mathcal{P}^{2s+1}_{2s,1}Y_i=\sum x_iZ_i\mathcal{P}^{2s+1}_{2s,1}\,,$$

where $Y_i\in\mathrm{End}(\mathbb{C}^{4s+2})$ are known and $Z_i\in\mathrm{End}(\mathbb{C}^{2s+2})$ are considered as unknowns. This produces an overdetermined system of equations for every Z_i, which is guaranteed to have a solution by virtue of the fusion procedure. One easily convinces oneself that this is a very easy and fast way of computation. It is instructive to check that

$$R^{(ev,2)}(\zeta)\cdot\mathrm{diag}\left(q^{-H},1\,,\,q^{H}\right)$$

$$=\begin{pmatrix}(\zeta^2K-1)(q^2\zeta^2K-1) & (q-q^{-1})\zeta F(\zeta^2K-1) & (1-q^2)^2q^{-2}(1+q^2)^{-1}\zeta^2F^2\\ (q^2-q^{-2})\zeta(\zeta^2K-1)E & \begin{matrix}1+\zeta^4q^2+(1-q^2)^2q^{-1}\zeta^2C\\ -\zeta^2(1+q^2)(K+K^{-1})\end{matrix} & (q-q^{-1})\zeta(\zeta^2K^{-1}-1)F\\ q^{-4}(1-q^2)^2(1+q^2)\zeta^2E^2 & (q^2-q^{-2})\zeta E(\zeta^2K^{-1}-1) & (\zeta^2K^{-1}-1)(q^2\zeta^2K^{-1}-1)\end{pmatrix}.$$

Consider the transfer-matrix on *Matsubara* space (2.81) and any auxiliary representation:

$$T_{\mathrm{aux},\mathbf{M}}(\zeta_a)=\prod_{\mathbf{m}=1}^{\overset{\curvearrowleft}{\mathbf{n}}}L_{\mathrm{aux},\mathbf{m}}(\zeta/\tau_{\mathbf{m}})\,,$$

where

$$L_{\mathrm{aux},\mathbf{m}}(\zeta/\tau_{\mathbf{m}})=R_{\mathrm{aux},\mathbf{m}}(q^{-1/2}\zeta/\tau_{\mathbf{m}})\,.$$

In the last two formulas we do not indicate explicitly the representations in which the R-matrix and the L-operator act since they are already indirectly encoded. Using (2.92), and (2.88) for finite-dimensional auxiliary representations, one obtains $T_{\mathrm{aux},\mathbf{M}}(\zeta_a)$ by fusion starting from the transfer-matrix whose quantum space is the tensor product of

$$\mathbf{N}=2\sum_{\mathbf{m}=1}^{\mathbf{n}}s_{\mathbf{m}}$$

two-dimensional evaluation representations with the parameters

$$\{\theta_1,\cdots\theta_{\mathbf{N}}\}=\bigcup_{\mathbf{m}=1}^{\mathbf{n}}\{\tau_{\mathbf{m}}q^{-s_{\mathbf{m}}-\frac{1}{2}+p}\}_{p=1}^{2s_{\mathbf{m}}}\,.$$

The Algebraic Bethe Ansatz goes smoothly, the functions $a(\xi), d(\xi)$ change to

$$a(\xi) = \prod_{\mathbf{m}=1}^{\mathbf{n}} (1 - q^{-2s_{\mathbf{m}}}(\xi/\tau_{\mathbf{m}})^2)\,, \quad d(\xi) = \prod_{\mathbf{m}=1}^{\mathbf{n}} (1 - q^{2s_{\mathbf{m}}}(\xi/\tau_{\mathbf{m}})^2)\,. \tag{2.93}$$

At several places $\mathbf{n}$ from Section 2.1 changes to $\mathbf{N}$ which is the number of spin-1/2 representations before fusion. For given spin s we introduce the union of Bethe roots

$$\{\lambda_j\}_{j=1}^{\mathbf{N}} = \{\lambda_j^-(\kappa+\alpha)\}_{j=1}^{\frac{\mathbf{N}}{2}-s} \cup \{\lambda_j^+(\kappa)\}_{j=1}^{\frac{\mathbf{N}}{2}+s}\,.$$

Using

$$\mathcal{P}_{2s_2,1}^{2s_2+1}|\downarrow\rangle = |\downarrow\rangle\,, \quad \langle\uparrow|\mathcal{P}_{2s_2,1}^{2s_2+1} = \langle\uparrow|$$

one derives for the scalar product

(2.94)

$$\langle\uparrow|\prod_{j=1}^{\mathbf{N}} C(\lambda_j)|\downarrow\rangle = \prod_{j=1}^{\mathbf{N}}\prod_{\mathbf{m}=1}^{\mathbf{n}} P(\lambda_j/\tau_{\mathbf{m}}, 1/2, s_{\mathbf{m}})\; M_{\mathbf{N}}\big(\lambda_1,\cdots,\lambda_{\mathbf{N}}|\theta_{\mathbf{1}},\cdots\theta_{\mathbf{N}}\big)\,.$$

Graphically the left hand side is represented as follows

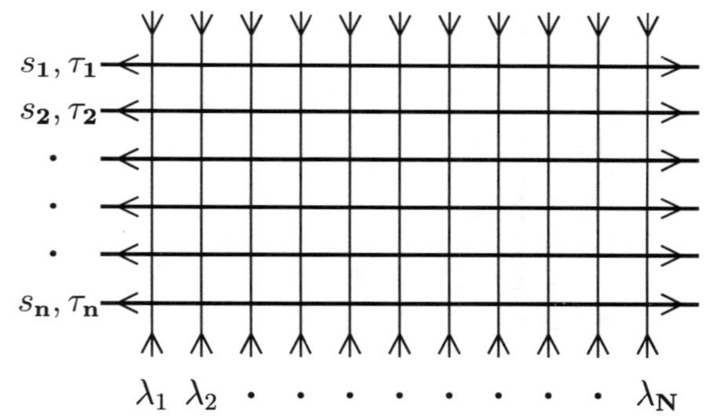

FIG. 7.

This does not correspond to the six-vertex model anymore: there are $2s_{\mathbf{m}}+1$ states on horizontal lines now. However, the boundary conditions still correspond to lowest and highest vectors, hence they are drawn as left and right arrows.

For higher spins the formula (2.94) gives an expression for the scalar product $\langle\kappa+\alpha|\kappa\rangle$ in terms of a huge determinant of the size $\mathbf{N}\times\mathbf{N}$. Later we shall see that a more economic formula exists, the determinant being of the size $\mathbf{n}\times\mathbf{n}$.

2.4.1. Summary. We consider the transfer-matrices with different auxiliary spaces introduced above. For quantum space we can take generalised *Matsubara* space (2.81). Another example, used in the next chapter, is the space

$$\mathrm{End}\big(V_{\xi_1}^{(1)}\big)\otimes\cdots\otimes\mathrm{End}\big(V_{\xi_n}^{(1)}\big) \tag{2.95}$$

on which the adjoint R-matrices act, see (3.11). The following theorem summarises the results of the present section. It holds irrespective of the nature of the quantum space (for example, (2.81) or (2.95)) which will be denoted by $\mathfrak{H}_{\mathrm{qtm}}$.

We shall denote by S_{qtm} the total spin in the quantum space. In (2.97) and (2.98) below, we use S_{qtm} in this sense. When we deal with R-matrices in adjoint action, S_{qtm} denotes its adjoint action.

THEOREM 2.6. *The following hold for the R-matrices with q-oscillator auxiliary space.*

1. *For* $W^{\pm}_{\zeta_1}\otimes W^{\pm}_{\zeta_2}$ *we have the intertwiner*

$$R_{A,B}(\zeta)=P_{A,B}h(\zeta,u_{A,B})\zeta^{D_A+D_B}\,,\qquad \zeta=\zeta_1/\zeta_2\,,$$

where $u_{A,B}$ *and* $h_{A,B}(\zeta,u)$ *are given by* (2.59) *and* (2.61), *respectively. It satisfies*

$$\begin{aligned}R_{A,B}(\zeta)R^{\pm}_{A,\mathrm{qtm}}(\zeta_1)R^{\pm}_{B,\mathrm{qtm}}(\zeta_2)&=R^{\pm}_{B,\mathrm{qtm}}(\zeta_2)R^{\pm}_{A,\mathrm{qtm}}(\zeta_1)R_{A,B}(\zeta)\,,\\ R_{A,B}(\zeta)\Delta(x)&=\Delta'(x)R_{A,B}(\zeta)\ \text{for } x\in U_q\mathfrak{b}^+.\end{aligned}\tag{2.96}$$

2. *We have the identities between traces*

For $X(\zeta)\in\mathfrak{A}^{\mathrm{left}}(\zeta)$

$$\begin{aligned}&\mathrm{Tr}_{A,B}\Bigl(q^{2\kappa(D_A-D_B)}X(\zeta)R^{+}_{A,\mathrm{qtm}}(\zeta_1)R^{-}_{B,\mathrm{qtm}}(\zeta_2)\Bigr)\\ &\quad=\mathrm{Tr}^{\Lambda}\Bigl(q^{\kappa H}(\zeta^{H/4}\boldsymbol{\rho}^{\mathrm{left}}(X(\zeta))\zeta^{-H/4})R_{\star,\mathrm{qtm}}(\overline{\zeta})\Bigr)\frac{\zeta^{-(\kappa-S_{\mathrm{qtm}})}}{q^{(\kappa-S_{\mathrm{qtm}})}-q^{-(\kappa-S_{\mathrm{qtm}})}}\,.\end{aligned}\tag{2.97}$$

For $X(\zeta)\in\mathfrak{A}^{\mathrm{right}}(\zeta)$

$$\begin{aligned}&\mathrm{Tr}_{A,B}\Bigl(q^{2\kappa(D_A-D_B)}X(\zeta)R^{-}_{B,\mathrm{qtm}}(\zeta_2)R^{+}_{A,\mathrm{qtm}}(\zeta_1)\Bigr)\\ &\quad=\mathrm{Tr}^{\Lambda}\Bigl(q^{\kappa H}(\zeta^{H/4}\boldsymbol{\rho}^{\mathrm{right}}(X(\zeta))\zeta^{-H/4})R_{\star,\mathrm{qtm}}(\overline{\zeta})\Bigr)\frac{\zeta^{-(\kappa-S_{\mathrm{qtm}})}}{q^{(\kappa-S_{\mathrm{qtm}})}-q^{-(\kappa-S_{\mathrm{qtm}})}}\,.\end{aligned}\tag{2.98}$$

In both cases $\zeta=\zeta_1/\zeta_2$, $\overline{\zeta}=(\zeta_1\zeta_2)^{1/2}$, *and* $q^{\Lambda}=(\zeta q)^{-1}$.

3. *For* $V^{(1)}_{\zeta}\otimes W^{+}_{\zeta}$ *we have*

$$\begin{aligned}&F^{-1}_{a,A}R_{a,\mathrm{qtm}}(\zeta)R^{+}_{A,\mathrm{qtm}}(\zeta)F_{a,A}\\ &\qquad=\begin{pmatrix}\mathsf{a}(\zeta)R^{+}_{A,\mathrm{qtm}}(\zeta q)q^{-S_{\mathrm{qtm}}} & 0\\ C_{A,\mathrm{qtm}}(\zeta) & \mathsf{d}(\zeta)R^{+}_{A,\mathrm{qtm}}(\zeta q^{-1})q^{S_{\mathrm{qtm}}}\end{pmatrix}_a\,,\end{aligned}$$

where $\mathsf{a}(\zeta)$, $\mathsf{d}(\zeta)$ *are functions depending on the choice of* $\mathfrak{H}_{\mathrm{qtm}}$, *and* $C_{A,\mathrm{qtm}}(\zeta)$ *is an operator, all of which can be computed using* (2.78), *and the fusion procedure in the quantum space if needed. A similar, upper-triangular formula for* R^{-}_{A} *is obtained by a similarity transformation by* $\sigma^1_a J_{\mathrm{qtm}}$, *where* J_{qtm} *is the spin reversal.*

PROOF. There is nothing to prove for items 1 and 3: they follow directly from the formulas in Subsec.2.3.2 and (2.78).

To prove the item 2 it suffices to look attentively at the formulas (2.63), (2.70), (2.46), and (2.75), (2.76). We use also $\mathsf{r}_{\mathsf{s}}\bigl(\overline{\zeta},\frac{\zeta+\zeta^{-1}}{(q-q^{-1})^2}\bigr)=\mathsf{s}_s(\zeta_1)\mathsf{s}_s(\zeta_2)$. Notice that the multipliers $q^{-(D_A-D_B)\sigma^3}$ in (2.63), (2.70) change κ to $\kappa-S_{\mathrm{qtm}}$ in (2.75). □

Although there is no intertwiner for $W^{+}_{\zeta_1}$ and $W^{-}_{\zeta_2}$, comparing (2.97) and (2.98) we see that we can interchange $R^{+}_{A,\mathrm{qtm}}(\zeta_1)$ and $R^{-}_{B,\mathrm{qtm}}(\zeta_2)$ under the trace with suitable insertions of elements belonging to $\mathfrak{A}^{\mathrm{left}}(\zeta)$ or $\mathfrak{A}^{\mathrm{right}}(\zeta)$, respectively. Since this fact will be used in the future, let us describe it.

Before formulating this it is convenient to enlarge the algebra $\mathfrak{A}^{\mathrm{left}}(\zeta)$ slightly. For generic Λ, the operator $(1-q^{\Lambda}K^{-1})^{-1}F$ is well defined on the Verma module. The above results concerning $\mathfrak{A}^{\mathrm{left}}(\zeta)$ remain valid if we replace it by a larger algebra $\mathfrak{A}^{\mathrm{left}}_{\mathrm{loc}}(\zeta)$ generated by $U(\zeta),V(\zeta),Y(\zeta),Z(\zeta)$ and

$$\mathbf{a}\otimes 1=(1-\zeta^{-1}q^{-2}Z(\zeta))^{-1}W(\zeta)\,.\tag{2.99}$$

Similarly we introduce $\mathfrak{A}^{\rm right}_{\rm loc}(\zeta)$ generated by $\overline{U}(\zeta),\overline{V}(\zeta),\overline{Y}(\zeta),\overline{Z}(\zeta)$ and $1\otimes\mathbf{a}$.

DEFINITION 2.7. Elements of $\mathfrak{A}^{\rm left}_{\rm loc}(\zeta)$ (resp. $\mathfrak{A}^{\rm right}_{\rm loc}(\zeta)$) are called left (resp. right) admissible operators.

Let $\boldsymbol{\rho}$ denote the composition $\mathfrak{A}^{\rm left}_{\rm loc}(\zeta)\to\mathfrak{A}^{\rm left}_{\rm loc}(\zeta)/\langle V(\zeta)\rangle\stackrel{\sim}{\to}\mathfrak{A}^{\rm right}_{\rm loc}(\zeta)/\langle\overline{V}(\zeta)\rangle$. Explicitly,

$$\begin{aligned}&\boldsymbol{\rho}(U(\zeta))=\overline{W}(\zeta)\,,\quad \boldsymbol{\rho}(W(\zeta))=\overline{U}(\zeta)\,,\quad \boldsymbol{\rho}(V(\zeta))=0\,,\\&\boldsymbol{\rho}(Y(\zeta))=\overline{Z}(\zeta)\,,\quad \boldsymbol{\rho}(Z(\zeta))=\overline{Y}(\zeta)\,.\end{aligned}$$

Then we have

COROLLARY 2.8. *For a left admissible $X(\zeta)$ we have*

$$\begin{aligned}&\mathrm{Tr}_{A,B}\Bigl(q^{2\kappa(D_A-D_B)}X(\zeta)R^{+}_{A,\mathrm{qtm}}(\zeta_1)R^{-}_{B,\mathrm{qtm}}(\zeta_2)\Bigr)\\&\quad=\mathrm{Tr}_{A,B}\Bigl(q^{2\kappa(D_A-D_B)}\boldsymbol{\rho}(X(\zeta))\;R^{-}_{B,\mathrm{qtm}}(\zeta_2)R^{+}_{A,\mathrm{qtm}}(\zeta_1)\Bigr),\end{aligned}\tag{2.100}$$

where $\zeta=\zeta_1/\zeta_2$ as usual.

For $X(\zeta)$ containing $\mathbf{a}\otimes 1$ we have to use (2.99) obtaining $(1-\zeta^{-1}q^{-2}\overline{Y}(\zeta))^{-1}$ in the right hand side, but corresponding traces are well-defined automatically.

In the usual situation when there is an intertwiner R, $\boldsymbol{\rho}(\bullet)$ would be just $R{\bullet}R^{-1}$. So, we have a replacement of the adjoint R-matrix whose use is restricted: it is defined only on $\mathfrak{A}^{\rm left}_{\rm loc}(\zeta)$.

2.5. Q-Operators

We start a new section, but essentially we continue the Bazhanov-Lukyanov-Zamolodchikov construction.

Let us return to the generalised *Matsubara* chain. We had the definitions

$$T_{a,\mathbf{M}}(\zeta)=\prod_{\mathbf{m}=1}^{\overset{\curvearrowleft}{\mathbf{n}}}L_{a,\mathbf{m}}(\zeta/\tau_{\mathbf{m}})\,,\quad T_{a,\mathbf{M}}(\zeta,\kappa)=T_{a,\mathbf{M}}(\zeta)q^{\kappa\sigma^3_a}\,.\tag{2.101}$$

Comparing with the previous section we see that $T_{a,\mathbf{M}}(\zeta)$ is the image of the universal R-matrix (up to a scalar factor $\prod_{\mathbf{m}=1}^{\mathbf{n}}\mathsf{r}(\zeta q^{-1/2}/\tau_{\mathbf{m}},\pi^{(s_{\mathbf{m}})}(C))$) where the auxiliary space is $V^{(1)}_\zeta$ and the quantum space is (2.81). Choosing the q-oscillator representations as the auxiliary space we define

$$T^{\pm}_{A,\mathbf{M}}(\zeta,\kappa)=\prod_{\mathbf{m}=1}^{\overset{\curvearrowleft}{\mathbf{n}}}L^{\pm}_{A,\mathbf{m}}(\zeta/\tau_{\mathbf{m}})\cdot q^{\pm2\kappa D_A}\,,\quad L^{\pm}_{A,\mathbf{m}}(\zeta)=R^{\pm}_{A,\mathbf{m}}(\zeta q^{-1/2})\,.$$

Motivated by the relation

$$T_{\mathbf{M}}(\zeta,\kappa)=\mathrm{Tr}_a\left(T_{a,\mathbf{M}}(\zeta,\kappa)\right)\,,$$

we define Q-operators

$$Q^{\pm}_{\mathbf{M}}(\zeta,\kappa)=\zeta^{\pm(\kappa-S_{\mathbf{M}})}(1-q^{\pm2(\kappa-S_{\mathbf{M}})})\mathrm{Tr}_A\left(T^{\pm}_{A,\mathbf{M}}(\zeta,\kappa)\right)\,.\tag{2.102}$$

The reason for introducing the multiplier $\zeta^{\pm(\kappa-S_{\mathbf{M}})}$ will be clear soon. The multiplier $(1-q^{\pm2(\kappa-S_{\mathbf{M}})})$ is introduced for normalisation:

$$Q^{\pm}_{\mathbf{M}}(\zeta,\kappa)=\zeta^{\pm(\kappa-S_{\mathbf{M}})}\left(1+O(\zeta^2)\right)\,,\quad \zeta\to0\,;$$

see (2.53).

Let us discuss the commutation relations. From item 1 of Theorem 2.6 we conclude that

$$[Q^{\pm}_{\mathbf{M}}(\zeta_1,\kappa),Q^{\pm}_{\mathbf{M}}(\zeta_2,\kappa)]=0\,. \tag{2.103}$$

From item 2 with the choice $X(\zeta)=1$, we conclude that

$$[Q^{+}_{\mathbf{M}}(\zeta_1,\kappa),Q^{-}_{\mathbf{M}}(\zeta_2,\kappa)]=0\,. \tag{2.104}$$

Finally, since $R^{\pm}_{a,A}(\zeta_1/\zeta_2)$ are intertwiners for $V^{(1)}_{\zeta_1}\otimes W^{\pm}_{\zeta_2}$, we conclude that

$$[T_{\mathbf{M}}(\zeta_1,\kappa),Q^{\pm}_{\mathbf{M}}(\zeta_2,\kappa)]=0\,.$$

Consider item 3 of Theorem 2.6. Multiplying over the Mastubara chain we obtain

$$\begin{aligned}F^{-1}_{a,A}&T_{a,\mathbf{M}}(\zeta,\kappa)T^{+}_{A,\mathbf{M}}(\zeta,\kappa)F_{a,A}\\&=\begin{pmatrix}a(\zeta)T^{+}_{A,\mathbf{M}}(\zeta q,\kappa)q^{\kappa-S_{\mathbf{M}}} & 0\\ \star & d(\zeta)T^{+}_{A,\mathbf{M}}(\zeta q^{-1},\kappa)q^{-\kappa+S_{\mathbf{M}}}\end{pmatrix}_a,\end{aligned}$$

where $a(\zeta)$, $d(\zeta)$ coincide with (2.93), and the lower triangular element is not important here. Taking trace with respect to both auxiliary spaces we obtain the Baxter equation:

$$T_{\mathbf{M}}(\zeta,\kappa)Q_{\mathbf{M}}(\zeta,\kappa)=a(\zeta)Q_{\mathbf{M}}(\zeta q,\kappa)+d(\zeta)Q_{\mathbf{M}}(\zeta q^{-1},\kappa)\,. \tag{2.105}$$

We omit the upper index $+$ in Q because Q^{-} satisfies exactly the same equation. To see that it suffices to make a similarity transformation with the spin reversal $J_{\mathbf{M}}$, noticing that

$$J_{\mathbf{M}}T_{\mathbf{M}}(\zeta,\kappa)J_{\mathbf{M}}=T_{\mathbf{M}}(\zeta,-\kappa)\,,\quad J_{\mathbf{M}}Q^{+}_{\mathbf{M}}(\zeta,\kappa)J_{\mathbf{M}}=Q^{-}_{\mathbf{M}}(\zeta,-\kappa)\,. \tag{2.106}$$

Deriving (2.105) we see the role of $\zeta^{\pm(\kappa-S_{\mathbf{M}})}$ in the definition of $Q^{\pm}_{\mathbf{M}}(\zeta,\kappa)$. Equation (2.105) is an operator version of (2.12) which was obtained formally for the eigenvalues. This promotion to the operator level is a very strong result.

Up to now we considered the transfer matrices associated with either the two-dimensional or the q-oscillator representation. Let us introduce also transfer matrices $T^{(s)}(\zeta,\kappa)$ with $(2s+1)$-dimensional auxiliary representations defined in the same way as (2.101), replacing $L_{a,\mathbf{m}}(\zeta/\tau_{\mathbf{m}})$ by

$$L^{(s)}_{a,\mathbf{m}}(\zeta/\tau_{\mathbf{m}})=R^{(s)}_{a,\mathbf{m}}(q^{-1/2}\zeta/\tau_{\mathbf{m}})$$

and $q^{\kappa\sigma_a}$ by $q^{\kappa H_a}$.

For $\Lambda=2s$ the Verma module V^{Λ} contains a submodule $V^{-\Lambda-2}$, the quotient space being $V^{(2s)}$. We use (2.97) with $X_{A,B}(\zeta)=1$ for two different values of parameters: $\zeta_1=\xi$, $\zeta_2=\xi q^{2s+1}$ and $\zeta_1=\xi q^{2s+1}$, $\zeta_2=\xi$. In the former case, the identity reads as

$$Q^{+}_{\mathbf{M}}(\xi,\kappa)Q^{-}_{\mathbf{M}}(\xi q^{2s+1},\kappa)=\left(q^{-\kappa+S_M}-q^{\kappa-S_M}\right)\mathrm{Tr}^{\Lambda}\left(q^{\kappa H}R_{\star,\mathrm{qtm}}(\xi q^{1/2})\right),$$

and in the latter case it reads as

$$Q^{+}_{\mathbf{M}}(\xi q^{2s+1},\kappa)Q^{-}_{\mathbf{M}}(\xi,\kappa)=\left(q^{-\kappa+S_M}-q^{\kappa-S_M}\right)\mathrm{Tr}^{-\Lambda-2}\left(q^{\kappa H}R_{\star,\mathrm{qtm}}(\xi q^{1/2})\right).$$

Having in mind (2.31) and (2.87) one derives

$$(2.107)\qquad Q^+_{\mathbf{M}}(\xi,\kappa)Q^-_{\mathbf{M}}(\xi q^{2s+1},\kappa)-Q^+_{\mathbf{M}}(\xi q^{2s+1},\kappa)Q^-_{\mathbf{M}}(\xi,\kappa)$$
$$=(q^{-\kappa+S_{\mathbf{M}}}-q^{\kappa-S_{\mathbf{M}}})\prod_{\mathbf{m}=\mathbf{1}}^{\mathbf{n}}P(s,s_{\mathbf{m}},\zeta/\tau_{\mathbf{m}}q^s)\cdot T^{(s)}_{\mathbf{M}}(\xi q^{s+1/2},\kappa)\,.$$

The most important case for us is $s=0$. In that case we rewrite (2.107) as

$$(2.108)\qquad Q^+_{\mathbf{M}}(\xi,\kappa)Q^-_{\mathbf{M}}(\xi q,\kappa)-Q^+_{\mathbf{M}}(\xi q,\kappa)Q^-_{\mathbf{M}}(\xi,\kappa)=(q^{-\kappa+S_{\mathbf{M}}}-q^{\kappa-S_{\mathbf{M}}})W(\xi)\,,$$

where

$$(2.109)\qquad W(\xi)=\prod_{\mathbf{m}=\mathbf{1}}^{\mathbf{n}}P(0,s_{\mathbf{m}},\zeta/\tau_{\mathbf{m}})\,.$$

Instead of using this explicit formula, it is often more convenient to characterise $W(\xi)$ by two conditions

$$(2.110)\qquad \frac{W(\xi q)}{W(\xi)}=\frac{d(\xi q)}{a(\xi q)}\,,\quad W(0)=1\,,$$

(if the *Matsubara* spaces all have spin 1/2 we have $W(\zeta)=d(\zeta)$, but it is not the case in general.)

The relation (2.108) is called the quantum Wronskian relation because it plays the same role for two solutions to the second order difference equation (2.105). This equation is sufficient for deriving all other functional equations (for example the Baxter equation).

REMARK 2.9. There is a direct way to get (2.108). The operators $Q^{\pm}_{\mathbf{M}}(\xi,\kappa)$ are two solutions to (2.105), and their commutativity with $T_{\mathbf{M}}(\zeta)$ follows from the existence of an intertwiner. From this we immediately conclude that the left hand side of (2.108) equals $W(\xi)F(\xi)$ with $F(\xi)=F(\xi q)$. The latter equation does not allow rational solutions other than a constant. The constant is easily computed at $\xi=0$.

At this point one might doubt the necessity of the complicated part of the Bazhanov-Lukyanov-Zamolodchikov construction, i.e., the one concerning the tensor product $W^+_{\zeta_1}\otimes W^-_{\zeta_2}$. Indeed, according to Remark 2.9 we can obtain the q-Wronskian relation directly. On the other hand this relation is a first order difference equation for $Q^-_{\mathbf{M}}(\zeta,\kappa)$ which allows in principle to express it in terms of $Q^+_{\mathbf{M}}(\zeta,\kappa)$. Hence the commutativity (2.104) follows from the commutativity (2.103).

However, we shall see that the study of $W^+_{\zeta_1}\otimes W^-_{\zeta_2}$ cannot be avoided in the more complicated situation, where we construct creation-annihilation operators in the *Space* direction.

2.6. Destri-deVega Equation

2.6.1. General Procedure. For practical calculations in the limit $\mathbf{n}\to\infty$, it is very convenient to transform the Baxter equation into a non-linear integral equation.

Consider the eigenvector of $T_{\mathbf{M}}(\zeta,\kappa)$ corresponding to the maximal eigenvalue of $T_{\mathbf{M}}(\zeta)$. For this vector we have $S_{\mathbf{M}}=0$.

The corresponding eigenvalues of $T_{\mathbf{M}}(\zeta,\kappa)$ and $Q^-_{\mathbf{M}}(\zeta,\kappa)$ will be denoted by $T(\zeta,\kappa)$ and $Q^-(\zeta,\kappa)$, respectively. They are polynomials of ζ^2 (apart from an overall power of ζ for the latter). Consider the function

$$\mathfrak{a}(\zeta|\kappa)=\frac{a(\zeta)Q^-(\zeta q,\kappa)}{d(\zeta)Q^-(\zeta q^{-1},\kappa)}\,. \tag{2.111}$$

This is a rational function of ζ^2. Obviously the solutions to the equation

$$1+\mathfrak{a}(\zeta|\kappa)=0$$

are the zeroes of either $T(\zeta,\kappa)$ or $Q^-(\zeta,\kappa)$. Let γ be a contour in the η^2-plane going counterclockwise around the zeroes $\lambda_j^-(\kappa)^2$ of $Q^-(\zeta,\kappa)$, see (2.15). We assume that $\zeta^2q^{\pm 2},0$ and the zeroes of $T(\zeta,\kappa)$ are outside γ.

Then we have

$$\log\mathfrak{a}(\zeta|\kappa)=-2\pi i\nu\kappa+\log\frac{a(\zeta)}{d(\zeta)}+\frac{1}{2\pi i}\int_\gamma\log\left(\frac{\zeta^2q^2-\eta^2}{\zeta^2q^{-2}-\eta^2}\right)d\log(1+\mathfrak{a}(\eta|\kappa))\,.$$

Integrating by parts we arrive at the equation

$$\log\mathfrak{a}(\zeta|\kappa)=-2\pi i\nu\kappa+\log\frac{a(\zeta)}{d(\zeta)}+\int_\gamma K_0(\zeta/\eta)\log(1+\mathfrak{a}(\eta|\kappa))\frac{d\eta^2}{\eta^2}\,, \tag{2.112}$$

where

$$K_0(\zeta)=-\frac{1}{2\pi i}\frac{\zeta^2(q^2-q^{-2})}{(1-\zeta^2q^2)(1-\zeta^2q^{-2})}\,. \tag{2.113}$$

In order to proceed further we need some knowledge about the zeroes of $T(\zeta,\kappa)$ and $Q^-(\zeta,\kappa)$. We shall consider the simplest case $s_{\mathbf{m}}=1/2$ for all $\mathbf{m}$, and all $\tau_{\mathbf{m}}$ and κ are real. Then the typical picture of zeroes is given in the following figure.

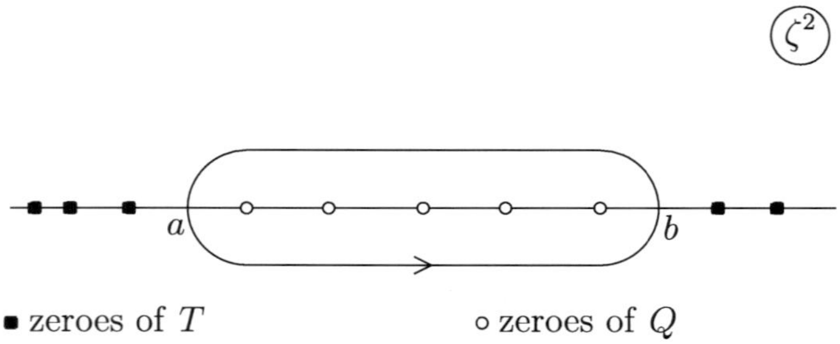

FIG. 8.

The Bethe roots are real on Fig. 8. This is the case for the ground state if κ is real. If κ is imaginary (real magnetic field) the locus of the Bethe roots will deform, but then the contours of integration should be deformed correspondingly. The exact position of a and b is not important as long as they separate the zeroes of $T(\zeta,\kappa)$ from those of $Q^-(\zeta,\kappa)$. Certainly in this case $\mathfrak{a}(\zeta|\kappa)$ obeys the Schwarz principle,

$$\mathfrak{a}(\bar\zeta|\kappa)=\frac{1}{\overline{\mathfrak{a}(\zeta|\kappa)}}\,. \tag{2.114}$$

For large $\mathbf{n}$ we have $|\mathfrak{a}(\zeta|\kappa)| \ll 1$ for $\zeta^2 \in \mathbb{C}^+$, and correspondingly $|\mathfrak{a}(\zeta|\kappa)| \gg 1$ for $\zeta^2 \in \mathbb{C}^-$, where $\mathbb{C}^\pm = \{z \in \mathbb{C} \mid \pm \mathrm{Im}\, z > 0\}$. The mnemonic rule here is that $\mathfrak{a}(\zeta|\kappa)$ is small (resp. large) in the half-plane where $a(\zeta)$ (resp. $d(\zeta)$) has zeroes. Let us rewrite the integrand on the lower part of the contour, using $\overline{K_0(\zeta)} = K_0(\bar{\zeta})$ and

$$\log(1+\mathfrak{a}(\zeta|\kappa)) = \log\mathfrak{a}(\zeta|\kappa) + \log\Bigl(1+\frac{1}{\mathfrak{a}(\zeta|\kappa)}\Bigr) = \log\mathfrak{a}(\zeta|\kappa) + \log(1+\overline{\mathfrak{a}(\bar{\zeta}|\kappa)})\,.$$

Then (2.112) takes the form

$$\begin{aligned}\log\mathfrak{a}(\zeta|\kappa) &- \int_a^b K_0(\zeta/\eta)\,\log\mathfrak{a}(\eta|\kappa)\frac{d\eta^2}{\eta^2}\\ &= -2\pi i\nu\kappa + \log\frac{a(\zeta)}{d(\zeta)} - 2i\int_a^b K_0(\zeta/\eta)\,\mathrm{Im}\log\bigl(1+\mathfrak{a}(\eta+i0|\kappa)\bigr)\frac{d\eta^2}{\eta^2}\,.\end{aligned}$$

Invert the operator in the left hand side by introducing the resolvent $R_0(\zeta)$:

$$R_0(\zeta/\xi) - \int_a^b K_0(\zeta/\eta)\,R_0(\eta/\xi)\frac{d\eta^2}{\eta^2} = K_0(\zeta/\xi)\,. \tag{2.115}$$

We arrive at

$$\log\mathfrak{a}(\zeta|\kappa) = f(\zeta,\kappa) - 2i\int_a^b R_0(\zeta/\eta)\,\mathrm{Im}\log\bigl(1+\mathfrak{a}(\eta+i0|\kappa)\bigr)\frac{d\eta^2}{\eta^2}\,, \tag{2.116}$$

where

$$f(\zeta,\kappa) = (I+R_0)(-2\pi i\nu\kappa + \log(a/d))(\zeta)\,.$$

The equation (2.116) is the sought-for Destri-de Vega (DdV) equation [**KBP**, **DDV**]. In spite of its almost tautological derivation, it happens to be a very powerful tool for studying the $\mathbf{n}\to\infty$ limit.

2.6.2. Finite temperature case. Let us make a comment about the case of finite temperature. According to the discussion at the end of Section 1.6, we are to take

$$a(\zeta) = \bigl((1-\tau^2\zeta^2)(1-\zeta^2q^{-2}/\tau^2)\bigr)^{\mathbf{n}/2}, \quad d(\zeta) = \bigl((1-\tau^2q^2\zeta^2)(1-\zeta^2/\tau^2)\bigr)^{\mathbf{n}/2}.$$

We are interested in the limit $\tau^2-1 = O(1/\mathbf{n})$, $\mathbf{n}\to\infty$. The function $1+\mathfrak{a}(\eta|\kappa)$ has a singularity which approaches to $\eta^2=1$:

$$d\log\bigl(1+\mathfrak{a}(\eta|\kappa)\bigr) = -\frac{\mathbf{n}}{2}\frac{1}{(\eta^2-\tau^2)}\frac{d\eta^2}{\eta^2} + O(1)\,.$$

In order to take it into account, we change the contour γ to γ_1 which includes $\zeta^2=\tau^2$ inside. This amounts to changing $\log a(\zeta)/d(\zeta)$ to

$$\log\frac{a(\zeta)}{d(\zeta)} + \frac{\mathbf{n}}{2}\log\Bigl(\frac{1-\zeta^2q^2/\tau^2}{1-\zeta^2q^{-2}/\tau^2}\Bigr)\,.$$

Setting $u=-\beta/\mathbf{n}$ and letting $\mathbf{n}\to\infty$, we obtain

(2.117)

$$\log\mathfrak{a}(\zeta|\kappa) = -2\pi i\nu\kappa - \frac{(1-q^2)^2\zeta^2q^{-1}}{(1-\zeta^2)(1-\zeta^2q^2)}\beta - \int_{\gamma_1} K_0(\zeta/\eta)\log(1+\mathfrak{a}(\eta|\kappa))\frac{d\eta^2}{\eta^2}\,.$$

The procedure of inverting $I-K_0$ can be applied without change, which yields the DdV equation at finite temperature.

CHAPTER 3

Fermions

Recall that we are interested in operators of the form $q^{2\alpha S(0)}\mathcal{O}$, where $\mathcal{O}$ is a local operator of spin zero,

$$\mathbb{S}(\mathcal{O}) = 0, \qquad \mathbb{S}(X) = [S, X] .$$

Let us denote the space of all such operators by $\mathcal{W}_{\alpha,0}$. At first glance, the space $\mathcal{W}_{\alpha,0}$ seems quite a structureless object. The goal of this chapter is to introduce some order in it. Namely we shall describe $\mathcal{W}_{\alpha,0}$ as a space created from the primary field $q^{2\alpha S(0)}$ by certain operators. The analogy with CFT is clear.

To this aim, it will be convenient to introduce also intermediate spaces $\mathcal{W}_{\alpha,s}$ of operators $q^{2\alpha S(0)}\mathcal{O}$, with $\mathcal{O}$ being of spin s:

$$\mathbb{S}(\mathcal{O}) = s\mathcal{O} .$$

Elements of $\mathcal{W}_{\alpha,s}$ will be said to have spin s and tail α. Altogether we shall consider a huge space consisting of all quasi-local operators

$$\mathcal{W} = \bigoplus_{\alpha\in\mathbb{C}, s\in\mathbb{Z}} \mathcal{W}_{\alpha,s} . \tag{3.1}$$

Any operator $q^{\alpha S(0)}\mathcal{O}$ stabilises as I_j for $j \gg 0$ and as $q^{\alpha\sigma_j^3}$ for $j \ll 0$. In other words, it can be presented as $q^{2\alpha S(k-1)}X_{[k,l]}$ where $X_{[k,l]}$ is an operator which acts as identity outside the interval $[k, l]$. By definition, the length and the support of an operator are

$$\text{length}(q^{2\alpha S(0)}\mathcal{O}) = \min(l - k + 1) , \tag{3.2}$$

$$\text{support}(q^{2\alpha S(0)}\mathcal{O}) = \min\,[k, l] , \tag{3.3}$$

where min is taken among all such representations.

We shall introduce the following families of operators on this space: the creation operators $\mathbf{t}^*(\zeta)$, $\mathbf{b}^*(\zeta)$, $\mathbf{c}^*(\zeta)$, and the annihilation operators $\mathbf{b}(\zeta)$, $\mathbf{c}(\zeta)$. The operator $\mathbf{t}^*(\zeta)$ is bosonic, and it does not change spin and tail. The rest of operators are fermionic. Operators $\mathbf{b}^*(\zeta), \mathbf{c}(\zeta)$ raise the spin by 1 and lower the tail by 1, while $\mathbf{b}(\zeta), \mathbf{c}^*(\zeta)$ lower the spin by 1 and raise the tail by 1. So, denoting

$$\mathcal{W}_{\alpha} = \bigoplus_{s\in\mathbb{Z}} \mathcal{W}_{\alpha,s} \tag{3.4}$$

and considering α as a block diagonal operator with eigenvalue α on $\mathcal{W}_{\alpha}$, we see that $\alpha + \mathbb{S}$ is a quantity conserved by the action of all our operators.

Our operators naturally split into blocks. It will be convenient for us to begin with the definition of blocks which have $\mathcal{W}_\alpha$ as the *target*. Namely, we shall define

$$
\begin{aligned}
&\mathbf{t}^*(\zeta,\alpha): \ \mathcal{W}_\alpha \to \mathcal{W}_\alpha\,,\\
&\mathbf{b}^*(\zeta,\alpha),\ \mathbf{c}(\zeta,\alpha): \ \mathcal{W}_{\alpha+1} \to \mathcal{W}_\alpha\,,\\
&\mathbf{b}(\zeta,\alpha),\ \mathbf{c}^*(\zeta,\alpha): \ \mathcal{W}_{\alpha-1} \to \mathcal{W}_\alpha\,.
\end{aligned}
$$

Certainly, operators on the entire space $\mathcal{W}$ are defined by collecting these blocks.

We begin this chapter with some additional information on the fusion of two-dimensional and q-oscillator representations. We introduce a basic object $\mathbf{k}(\zeta,\alpha)$, and describe its properties. Then we define annihilation and creation operators, and prove the reduction and commutation relations for them. This chapter is necessarily rather technical. For readers not to be lost in technical details, we formulate all important results as lemmas and theorems.

3.1. Intertwiner and Quasi-intertwiner for Fused Modules

3.1.1. Intertwiner for Representations of the Same Kind. Recalling the formula (2.78) we define

$$
R^+_{\{a,A\}}(\zeta) = F^{-1}_{a,A}R_a(\zeta)R^+_A(\zeta)F_{a,A}\,, \tag{3.5}
$$

by using (2.55), (2.53) and (2.79). Similarly we define $R^-_{\{a,A\}}(\zeta)$ using $R^-_A(\zeta)$ in place of $R^+_A(\zeta)$ and conjugating by $\sigma^1_a\sigma^1$. We set also $R^\pm_{\mathrm{aux}}(\zeta,\alpha) = R^\pm_{\mathrm{aux}}(\zeta)q^{\alpha H_{\mathrm{aux}}}$.

In what follows V_a and V_b denote $V^{(1)}_{\zeta_1}$ and $V^{(1)}_{\zeta_2}$, respectively.

LEMMA 3.1. *We have the intertwining relation*

$$
\begin{aligned}
&R_{\{a,A\},\{b,B\}}(\zeta_1/\zeta_2)R^+_{\{a,A\}}(\zeta_1,\alpha)R^+_{\{b,B\}}(\zeta_2,\alpha)\\
&= R^+_{\{b,B\}}(\zeta_2,\alpha)R^+_{\{a,A\}}(\zeta_1,\alpha)R_{\{a,A\},\{b,B\}}(\zeta_1/\zeta_2)\,,
\end{aligned}
$$

where the matrix $R_{\{a,A\},\{b,B\}}(\zeta)$ is block triangular as a matrix on the tensor product $V_a\otimes V_b$:

$$
R_{\{a,A\},\{b,B\}}(\zeta) = \begin{pmatrix} R_{11} & 0 & 0 & 0\\ R_{21} & R_{22} & 0 & 0\\ R_{31} & 0 & R_{33} & 0\\ R_{41} & R_{42} & R_{43} & R_{44}\end{pmatrix}. \tag{3.6}
$$

Its entries are given as follows.

$$R_{11} = -\zeta^2 q^{D_A} R_{A,B}(\zeta) q^{-D_B}, \qquad R_{21} = \frac{q\zeta^3}{1-q^2\zeta^2} R_{A,B}(q\zeta)\mathbf{a}^*_A q^{-D_A-D_B},$$

$$R_{22} = -\frac{q\zeta^2}{1-q^2\zeta^2} q^{-D_A} R_{A,B}(q^2\zeta) q^{-D_B},$$

$$R_{33} = -(\zeta^2-q^2) q^{D_A} R_{A,B}(\zeta/q^2) q^{D_B-1}, \quad R_{43} = -\zeta R_{A,B}(\zeta/q)\mathbf{a}^*_A q^{-D_A+D_B},$$

$$R_{44} = q^{-D_A} R_{A,B}(\zeta) q^{D_B},$$

$$R_{31} = -q\zeta\, q^{-2D_B}\mathbf{a}^*_B R_{A,B}(\zeta/q) q^{D_A+D_B}, \quad R_{41} = \frac{q\zeta^2}{1-q^2\zeta^2} q^{-D_B}\mathbf{a}^*_B R_{A,B}(\zeta)\mathbf{a}^*_A q^{-D_A},$$

$$R_{42} = -\frac{\zeta}{1-q^2\zeta^2}\mathbf{a}^*_B R_{A,B}(q\zeta) q^{-D_A-D_B}.$$

PROOF. Using

$$R_{\{a,A\},\{b,B\}}(\zeta) = (F_{a,A}F_{b,B})^{-1} R^+_{B,a}(\zeta^{-1})^{-1} R_{a,b}(\zeta) R_{A,B}(\zeta) R^+_{A,b}(\zeta) F_{a,A}F_{b,B}\,,$$

we arrive at the announced result after some computation. In the process we use the identity (2.62) for $h(\zeta,u)$. □

The most important feature of the formulas above is the relative minus sign between R_{11} and R_{44}. This will eventually give rise to anti-commutators.

By a similarity transformation with $\sigma^1_a\sigma^1_b$, we obtain the intertwiner between $R^-_{\{a,A\}}(\zeta_1,\alpha)R^-_{\{b,B\}}(\zeta_2,\alpha)$ and $R^-_{\{b,B\}}(\zeta_2,\alpha)R^-_{\{a,A\}}(\zeta_1,\alpha)$.

3.1.2. Quasi-intertwiner for Operators of Different Kind.

Consider now the product $R^-_{\{a,A\},\mathrm{qtm}}(\zeta_1,\alpha)R^+_{\{b,B\},\mathrm{qtm}}(\zeta_2,\alpha)$. In the usual sense, there is no intertwiner which reverses the order of this product, but there is a substitute which can be used under the trace. We retain the notation in the Summary of Section 2.3.

LEMMA 3.2. *Let $X_{a,b,A,B}(\zeta)$ be a matrix in $V_a\otimes V_b$ with left admissible matrix elements. (For the definition of admissible operators, see Section 2.3.3.) Then we have*

$$\begin{aligned}(3.7)\quad &\mathrm{Tr}_{a,b,A,B}\Bigl(q^{\sigma^3_b D_A} X_{a,b,A,B}(\zeta) q^{-\sigma^3_b D_A} R^+_{\{a,A\},\mathrm{qtm}}(\zeta_1,\alpha) R^-_{\{b,B\},\mathrm{qtm}}(\zeta_2,\alpha)\Bigr)\\ &= \mathrm{Tr}_{a,b,A,B}\Bigl(q^{-\sigma^3_a D_B} Y_{a,b,A,B}(\zeta) q^{\sigma^3_a D_B} R^-_{\{b,B\},\mathrm{qtm}}(\zeta_2,\alpha) R^+_{\{a,A\},\mathrm{qtm}}(\zeta_1,\alpha)\Bigr)\,,\\ &Y_{a,b,A,B}(\zeta) = \boldsymbol{\rho}\Bigl(\mathcal{F}_{\{a,A\},\{b,B\}}(\zeta)(X_{a,b,A,B}(\zeta))\mathcal{F}_{\{a,A\},\{b,B\}}(\zeta)^{-1}\Bigr)\,.\end{aligned}$$

The matrix $\mathcal{F}_{\{a,A\},\{b,B\}}(\zeta)$ is given by

$$\mathcal{F}_{\{a,A\},\{b,B\}}(\zeta) = \begin{pmatrix}\mathcal{F}_{11} & \mathcal{F}_{12} & 0 & 0\\ 0 & \mathcal{F}_{22} & 0 & 0\\ \mathcal{F}_{3,1} & \mathcal{F}_{3,2} & \mathcal{F}_{3,3} & \mathcal{F}_{3,4}\\ 0 & \mathcal{F}_{4,2} & 0 & \mathcal{F}_{4,4}\end{pmatrix}_{a,b}, \tag{3.8}$$

where, setting $Y = Y_{A,B}(\zeta)$ *and* $U = U_{A,B}(\zeta)$, *we have*

$$\mathcal{F}_{11} = Y^{-1}(1-\zeta Y)(1-\zeta q^2 Y)\,, \quad \mathcal{F}_{12} = -\zeta q^2 Y^{-1}(1-\zeta Y)\mathbf{a}_A\,,$$
$$\mathcal{F}_{22} = q(1-\zeta^2 q^2)Y^{-1}\,,$$

$$\mathcal{F}_{33} = -(1-\zeta^2 q^2)q^{-1}Y\,, \qquad \mathcal{F}_{34} = (1-\zeta^2 q^2)q^{-3}\zeta\frac{Y}{1-\zeta q^{-2}Y}\mathbf{a}_A\,,$$
$$\mathcal{F}_{44} = -q^{-2}(1-\zeta^2 q^2)(1-\zeta^2 q^{-2})\frac{Y}{(1-\zeta q^{-2}Y)(1-\zeta q^{-4}Y)},$$

$$\mathcal{F}_{31} = -U, \qquad \mathcal{F}_{32} = -\zeta q\frac{q^{-3}Y-\zeta q}{1-\zeta q^{-2}Y}\,,$$
$$\mathcal{F}_{42} = -q\frac{1-\zeta^2 q^2}{(1-\zeta q^{-2}Y)(1-\zeta q^{-4}Y)}U\,.$$

PROOF. Let us omit the index qtm. In the left hand side undo the fusion:

$$\begin{aligned}&R^+_{\{a,A\}}(\zeta_1,\alpha)R^-_{\{b,B\}}(\zeta_2,\alpha)\\ &= (F^+_{a,A}F^-_{b,B})^{-1}R_a(\zeta_1,\alpha)R^+_A(\zeta_1,\alpha)R_b(\zeta_2,\alpha)R^-_B(\zeta_2,\alpha)F^+_{a,A}F^-_{b,B}\,.\end{aligned}$$

Now move $R^+_A(\zeta_1,\alpha)$ and $R^-_B(\zeta_2,\alpha)$ next to each other using the Yang-Baxter equation:

$$R^+_A(\zeta_1,\alpha)R_b(\zeta_2,\alpha) = R^+_{A,b}(\zeta)^{-1}R_b(\zeta_2,\alpha)R^+_A(\zeta_1,\alpha)R^+_{A,b}(\zeta)\,, \quad \zeta=\zeta_1/\zeta_2\,.$$

Using the cyclicity of the trace, we see that the LHS of (3.7) becomes

$$\mathrm{Tr}_{a,b,A,B}\Bigl(V_{a,b,A,B}(\zeta)X_{a,b,A,B}(\zeta)V^{-1}_{a,b,A,B}(\zeta)R_a(\zeta_1,\alpha)R_b(\zeta_2,\alpha)R^+_A(\zeta_1,\alpha)R^-_B(\zeta_2,\alpha)\Bigr)$$

where we have set

$$\begin{aligned}V_{a,b,A,B}(\zeta) &= R^+_{A,b}(\zeta)F^+_{a,A}F^-_{b,B}q^{D_A\sigma^3_b}\\ &= \begin{pmatrix}1-\zeta^2 Y_{A,B}(\zeta) & -\zeta\mathbf{a}_A\\ -U_{A,B}(\zeta) & 1\end{pmatrix}_b\left(1-q^{\sigma^3_b}\mathbf{a}_A\sigma^+_a\right)\,.\end{aligned}$$

Notice that $V_{a,b,A,B}(\zeta)$ has left admissible matrix elements. Then after some calculations we arrive at $\mathcal{F}_{\{a,A\},\{b,B\}}(\zeta)$. □

We call the matrix $\mathcal{F}_{\{a,A\},\{b,B\}}(\zeta)$ *quasi-intertwiner.* Since it is used only under the trace, the operators $U(\zeta), W(\zeta), Y(\zeta), Z(\zeta)$ can be treated as elements of $U_q(\mathfrak{sl}_2)$ in accordance with the rule (2.69), disregarding all terms containing $V(\zeta)$.

3.2. Operators $\mathbf{k}(\zeta,\alpha)$ and $\phi(\mathbf{k})(\zeta,\alpha)$

In this section we introduce an operator $\mathbf{k}(\zeta,\alpha)$, an auxiliary object which is a cornerstone of our construction.

3.2.1. Adjoint Action of R Matrices. We start considering quasi-local operators in the *Space* direction. We shall mainly deal with the homogeneous case. First, let us consider a finite interval $J=[k,l]$, and operators X_J acting on $(\mathbb{C}^2)^{\otimes(l-k+1)}$. We shall write

$$|J|=l-k+1\,. \tag{3.9}$$

Our basic tool is the adjoint monodromy matrix. Let 'aux' stand for one of the auxiliary spaces $a, A, \{a,A\},\cdots$. We define

$$\begin{aligned}\mathbb{T}_{\mathrm{aux},J}(\zeta)(X_J)&=T_{\mathrm{aux},J}(\zeta)\cdot X_J\cdot T_{\mathrm{aux},J}(\zeta)^{-1}\,,\\ T_{\mathrm{aux},J}(\zeta)&=R_{\mathrm{aux},l}(\zeta)R_{\mathrm{aux},l-1}(\zeta)\cdots R_{\mathrm{aux},k}(\zeta)\,.\end{aligned} \tag{3.10}$$

In other words,

$$\begin{aligned}\mathbb{T}_{\mathrm{aux},J}(\zeta)&=\mathbb{R}_{\mathrm{aux},l}(\zeta)\mathbb{R}_{\mathrm{aux},l-1}(\zeta)\cdots\mathbb{R}_{\mathrm{aux},k}(\zeta)\,,\\ \mathbb{R}_{\mathrm{aux},j}(\zeta)(X_J)&=R_{\mathrm{aux},j}(\zeta)\cdot X_J\cdot R_{\mathrm{aux},j}(\zeta)^{-1}\,.\end{aligned} \tag{3.11}$$

So these are operators acting on operators.

We shall also use the adjoint monodromy matrix

$$\mathbb{T}_{\mathrm{aux},J}(\zeta,\alpha)(X_J)=\mathbb{T}_{\mathrm{aux},J}(\zeta)(q^{\alpha H_{\mathrm{aux}}}X_J)\,,$$

where $H_{\mathrm{aux}}=\pi_\zeta^{aux}(h)$, $t_1=q^h$ (see (2.36)). As for the q-oscillator representations, we deal mostly with W^+ and skip the superscript $+$. When W^- is also involved we shall write the superscripts explicitly.

Later we shall also consider the inhomogeneous case, changing the definition (3.10) to

$$T_{\mathrm{aux},J}(\zeta)=R_{\mathrm{aux},l}(\zeta/\xi_l)R_{\mathrm{aux},l-1}(\zeta/\xi_{l-1})\cdots R_{\mathrm{aux},k}(\zeta/\xi_k)\,. \tag{3.12}$$

The rest of the definition remains the same.

From (2.78) we find

$$\begin{aligned}T_{\{a,A\},J}(\zeta)&=\begin{pmatrix}\mathsf{f}_J(\zeta)T_{A,J}(\zeta q)q^{-S_J} & 0\\ C_{A,J}(\zeta) & \mathsf{f}_J(\zeta q)T_{A,J}(\zeta q^{-1})q^{S_J}\end{pmatrix}_a,\\ T_{\{a,A\},J}(\zeta)^{-1}&=\begin{pmatrix}\mathsf{f}_J(\zeta)^{-1}q^{S_J}T_{A,J}(\zeta q)^{-1} & 0\\ -\overline{C}_{A,J}(\zeta) & \mathsf{f}_J(\zeta q)^{-1}q^{-S_J}T_{A,J}(\zeta q^{-1})^{-1}\end{pmatrix}_a,\end{aligned} \tag{3.13}$$

where $\mathsf{f}_J(\zeta)$ is a scalar function and $C_{A,J}(\zeta)$, $\overline{C}_{A,J}(\zeta)$ are certain operators. We do not need the explicit formula for the latter. We note only the obvious relation

$$C_{A,J}(\zeta)q^{S_J}T_{A,J}(\zeta q)^{-1}=\mathsf{f}_J(\zeta)\mathsf{f}_J(\zeta q)T_{A,J}(\zeta q^{-1})q^{S_J}\overline{C}_{A,J}(\zeta)\,. \tag{3.14}$$

In our homogeneous case we have

$$\mathsf{f}_J(\zeta)=(1-\zeta^2)^{|J|}\,,$$

while in the inhomogeneous case we have

$$\mathsf{f}_J(\zeta)=\prod_{j\in J}(1-(\zeta/\xi_j)^2)\,.$$

In order to keep the exposition applicable to both homogeneous and inhomogeneous cases, it will be convenient to borrow a terminology from algebraic geometry. Consider a function of the form $g(\zeta)=\zeta^\lambda\tilde{g}(\zeta^2)$ where $\lambda\in\mathbb{C}$ and $\tilde{g}(\zeta^2)$ is a rational function of ζ^2. We use the term 'divisor (g)' to mean the divisor of $\tilde{g}(\zeta^2)$ in $\mathbb{C}^\times$ as a function of ζ^2. Thus the divisor (f_J) is $(l-k+1)\cdot(1)$ in the homogeneous case, and $(\xi_k^2)+\cdots+(\xi_l^2)$ in the inhomogeneous case. We multiply divisors from the right; for example, the divisor $(\mathsf{f}_J)q^{-2}$ is the divisor of $\mathsf{f}_J(\zeta q)$.

We have

$$\mathbb{T}_{\{a,A\},J}(\zeta,\alpha)=\begin{pmatrix}\mathbb{A}_{A,J}(\zeta,\alpha) & 0\\ \mathbb{C}_{A,J}(\zeta,\alpha) & \mathbb{D}_{A,J}(\zeta,\alpha)\end{pmatrix}_a, \tag{3.15}$$
$$\mathbb{A}_{A,J}(\zeta,\alpha)=\mathbb{T}_{A,J}(\zeta q,\alpha)q^{\alpha-\mathbb{S}},\quad \mathbb{D}_{A,J}(\zeta,\alpha)=\mathbb{T}_{A,J}(\zeta q^{-1},\alpha)q^{-\alpha+\mathbb{S}}.$$

We never put the index J to $\mathbb{S}$: it just measures the spin.

Clearly $\mathbb{T}_{a,J}(\zeta,\alpha)$ is singular on $(\mathsf{f}_J)q^2+(\mathsf{f}_J)q^{-2}$ while $\mathbb{T}_{A,J}(\zeta,\alpha)$ is singular on (f_J). However, among the three blocks of $\mathbb{T}_{\{a,A\},J}(\zeta,\alpha)$, only $\mathbb{C}_{A,J}(\zeta,\alpha)$ is singular on the sum of divisors $(\mathsf{f}_J)q^2+(\mathsf{f}_J)+(\mathsf{f}_J)q^{-2}$; $\mathbb{A}_{A,J}(\zeta,\alpha)$ and $\mathbb{D}_{A,J}(\zeta,\alpha)$ are singular only on $(\mathsf{f}_J)q^{-2}$ and $(\mathsf{f}_J)q^2$, respectively. We shall see that this simple fact has far-reaching consequences. Notice also that there are no additional multipliers in the second line of (3.15) because we consider the adjoint action.

3.2.2. Definition and Reduction Properties. We introduce an operator $\mathbf{k}_J(\zeta,\alpha)$ acting on X_J by

$$\mathbf{k}_J(\zeta,\alpha)(X_J):=\mathrm{Tr}_A\Big\{\mathbb{C}_{A,J}(\zeta,\alpha)\zeta^{\alpha-\mathbb{S}}\big(q^{-2S_J}X_J\big)\Big\}. \tag{3.16}$$

Since $[F_{a,A},\sigma_a^+]=0$, it can also be written as

$$\mathbf{k}_J(\zeta,\alpha)(X_J)=\mathrm{Tr}_{A,a}\left\{\sigma_a^+\mathbb{T}_{a,J}(\zeta,\alpha)\mathbb{T}_{A,J}(\zeta,\alpha)\zeta^{\alpha-\mathbb{S}}(q^{-2S_J}X_J)\right\}. \tag{3.17}$$

The operator $\mathbf{k}_J(\zeta,\alpha)$ possesses the following simple property which we call the left reduction. Suppose the interval $J=[k,l]$ is divided into the left and the right pieces as

$$J=L\sqcup R,\quad L=[k,m],\ R=[m+1,l]. \tag{3.18}$$

LEMMA 3.3. *Suppose $X_J=q^{2(\alpha+1)S_L}Y_R$. Then we have*

$$\mathbf{k}_J(\zeta,\alpha)(q^{2(\alpha+1)S_L}Y_R)=q^{2\alpha S_L}\mathbf{k}_R(\zeta,\alpha)(Y_R). \tag{3.19}$$

PROOF. The proof of this and similar statements follows from

$$\mathbb{T}_{\mathrm{aux},J}(\zeta,\alpha)=\mathbb{T}_{\mathrm{aux},R}(\zeta)\mathbb{T}_{\mathrm{aux},L}(\zeta,\alpha),$$
$$\mathbb{T}_{\mathrm{aux},L}(\zeta,\alpha)(q^{2\alpha S_L})=\mathbb{T}_{\mathrm{aux},L}(\zeta)(q^{\alpha(H_{\mathrm{aux}}+2S_L)})=q^{\alpha(H_{\mathrm{aux}}+2S_L)}.$$

In the present case $\mathrm{aux}=\{a,A\}$. □

The left reduction property tells that, in the limit $k\to-\infty$, we obtain an operator acting as $\mathcal{W}_{\alpha+1,s-1}\to\mathcal{W}_{\alpha,s}$. A much more complicated property is the right reduction. The operator $\mathbf{k}_J(\zeta,\alpha)$ does not literally satisfy the right reduction, but it does up to a 'q-exact form' defined as follows.

DEFINITION 3.4. Consider the finite-difference operator

$$\Delta_\zeta(g(\zeta))=g(\zeta q)-g(\zeta q^{-1}). \tag{3.20}$$

A q-exact form on an interval J is an expression $\Delta_\zeta(g(\zeta))$, where $g(\zeta)=\zeta^{\pm\alpha}\tilde{g}(\zeta^2)$ and $\tilde{g}(\zeta^2)$ is a rational function whose divisor (g) is contained in (f_J).

We have the following statement.

LEMMA 3.5. *Suppose $J = L \sqcup R$ as in (3.18), and assume now that $X_J = Y_L I_R$, i.e., it acts as identity on the interval R. Then*

$$\mathbf{k}_J(\zeta,\alpha)(X_J) = \mathbf{k}_L(\zeta,\alpha)(Y_L)I_R + \Delta_\zeta \mathbf{v}_J(\zeta,\alpha)(Y_L), \tag{3.21}$$

where

$$\begin{aligned} \mathbf{v}_J(\zeta,\alpha)(Y_L) &= \mathrm{Tr}_A\left(V_{A,R}(\zeta)\mathbb{T}_{A,L}(\zeta,\alpha)\zeta^{\alpha-\mathbb{S}}(q^{-2S_L}Y_L)\right), \\ V_{A,R}(\zeta) &= -\frac{1}{\mathsf{f}_R(\zeta)}\theta_R\left(C_{A,R}(\zeta)q^{S_R}T_{A,R}(q\zeta)^{-1}\right). \end{aligned} \tag{3.22}$$

Moreover $\mathbf{v}_J(\zeta,\alpha)(Y_L)$ is singular only on the divisor (f_J), so that $\Delta_\zeta \mathbf{v}_J(\zeta,\alpha)(Y_L)$ is a q-exact form.

PROOF. We have

$$\begin{aligned} &\mathbf{k}_J(\zeta,\alpha)(Y_L) \\ &= \mathrm{Tr}_{a,A}\Big(\sigma_a^+ T_{\{a,A\},R}(\zeta)q^{-2S_R}\mathbb{T}_{\{a,A\},L}(\zeta,\alpha)\zeta^{\alpha-\mathbb{S}}(q^{-2S_L}Y_L)T_{\{a,A\},R}(\zeta)^{-1}\Big) \\ &= \mathrm{Tr}_{a,A}\Big(W_{a,A,R}(\zeta)\mathbb{T}_{\{a,A\},L}(\zeta,\alpha)\zeta^{\alpha-\mathbb{S}}(q^{-2S_L}Y_L)\Big), \end{aligned} \tag{3.23}$$

where

$$W_{a,A,R}(\zeta) = \theta_R\big(\theta_R(T_{\{a,A\},R}(\zeta)^{-1})\sigma_a^+\theta_R(T_{\{a,A\},R}(\zeta)q^{-2S_R})\big).$$

We use here a trick which will be repeated over and over again. In (3.23) we wish to bring $T_{\{a,A\},R}(\zeta)^{-1}$ to the front using cyclicity of the trace in the auxiliary space. In order not to spoil the order of multiplication in the R-part of the quantum space, we first apply the anti-involution θ_R, use cyclicity and apply θ_R once again.

Now we use the crossing symmetry relations (2.56) and (2.54) and obtain

$$\begin{aligned} \theta_R\big(T_{\{a,A\},R}(\zeta)\big) &= \mathsf{f}_R(\zeta)\mathsf{f}_R(\zeta q)\mathsf{f}_R(\zeta q^2)T_{\{a,A\},R}(\zeta q)^{-1}, \\ \theta_R\big(T_{\{a,A\},R}(\zeta)^{-1}\big) &= \frac{1}{\mathsf{f}_R(\zeta q^{-1})\mathsf{f}_R(\zeta)\mathsf{f}_R(\zeta q)}T_{\{a,A\},R}(\zeta q^{-1}), \end{aligned}$$

and therefore

$$\begin{aligned} &\theta_R\big(W_{a,A,R}(\zeta)\big) = \frac{\mathsf{f}_R(\zeta q^2)}{\mathsf{f}_R(\zeta q^{-1})}T_{\{a,A\},R}(\zeta q^{-1})\sigma_a^+ q^{2S_R}T_{\{a,A\},R}(\zeta q)^{-1} \\ &= \begin{pmatrix} -\mathsf{f}_R(\zeta q)^{-1}C_{A,R}(\zeta q)q^{S_R}T_{A,R}(\zeta q^2)^{-1} & 1 \\ \star & \mathsf{f}_R(\zeta q^{-1})^{-1}C_{A,R}(\zeta q^{-1})q^{S_R}T_{A,R}(\zeta)^{-1} \end{pmatrix}_a. \end{aligned}$$

Here we use (3.14) in order to eliminate $\overline{C}_{A,R}(\zeta)$. Substituting $W_{a,A,R}(\zeta)$ into (3.23) we obtain the required identity (3.21). Note that the $(2,1)$ block does not contribute to the trace because the matrix $\mathbb{T}_{\{a,A\},L}(\zeta,\alpha)$ is lower triangular with respect to a.

It remains to show that $\mathbf{v}_J(\zeta,\alpha)$ is singular on the divisor (f_J) alone. The function

$$V_{A,R}(\zeta) = -\frac{1}{\mathsf{f}_R(\zeta)}\theta_R\left(C_{A,R}(\zeta)q^{S_R}T_{A,R}(q\zeta)^{-1}\right)$$

may have singularities on $(\mathsf{f}_R) + (\mathsf{f}_R)q^{-2}$. However, the singularity on $(\mathsf{f}_R)q^{-2}$ is fictitious in view of (3.14). On the other hand $\mathbb{T}_{A,L}(\zeta,\alpha)$ is singular on (f_L). Putting these together we see that $\mathbf{v}_J(\zeta,\alpha)$ is singular only on (f_J). □

From now on we omit writing I_R in (3.21).

The conclusion of this Lemma is ambivalent. On the one hand, there is no real right reduction. The operator $\mathbf{k}_J(\zeta,\alpha)$ acts on Y_L to produce, in addition to the nice piece $\mathbf{k}_L(\zeta,\alpha)(Y_L)$, a hardly controllable expression $\Delta_\zeta \mathbf{v}_J(\zeta,\alpha)(Y_L)$. On the other hand, the latter expression is a q-exact form, which leaves a hope that our construction is still useful.

The operator $\mathbf{k}_J(\zeta,\alpha)$ was constructed using W_ζ^+. Another operator can be constructed using W_ζ^-; in fact it can be obtained from $\mathbf{k}_J(\zeta,\alpha)$ by applying the spin reversal combined with the reflection $\alpha\to-\alpha$. We give a general definition which will be often used later for different operators.

DEFINITION 3.6. For an operator $\mathbf{x}(\zeta,\alpha)$ we define an operator $\phi(\mathbf{x})(\zeta,\alpha)$ by

$$\phi(\mathbf{x})(\zeta,\alpha)=q^{-1}N(\alpha-\mathbb{S}-1)\circ\mathbb{J}\circ\mathbf{x}(\zeta,-\alpha)\circ\mathbb{J}\,,$$

where $\mathbb{J}$ denotes the adjoint of the spin reversal

$$\mathbb{J}(X_{[k,l]})=J_{[k,l]}X_{[k,l]}J_{[k,l]}\,,\quad J_{[k,l]}=\prod_{j=k}^{l}\sigma_j^1\,,$$

and the multiplier $q^{-1}N(\alpha-\mathbb{S}-1)$ with $N(x)=q^{-x}-q^{x}$ is introduced for future convenience.

The operator $\mathbf{k}_J(\zeta,\alpha)$ raises the spin while the operator $\phi(\mathbf{k})_J(\zeta,\alpha)$ lowers the spin. Explicitly we have

(3.24)

$$\phi(\mathbf{k})_J(\zeta,\alpha)(X_J):=q^{-1}N(\alpha-\mathbb{S}-1)\mathrm{Tr}_{a,A}\Bigl\{\sigma_a^-\mathbb{T}^-_{\{a,A\},J}(\zeta,\alpha)\zeta^{-\alpha+\mathbb{S}}\bigl(q^{2S_J}X_J\bigr)\Bigr\}.$$

3.2.3. Commutation Relations. Before proceeding to the commutation relations of operators $\mathbf{k}_J(\zeta,\alpha)$ and $\phi(\mathbf{k})_J(\zeta,\alpha)$, let us give one more definition.

DEFINITION 3.7. A q-exact 2-form on an interval J is an expression of the form

$$\Delta_{\zeta_1}Q(\zeta_1,\zeta_2)+\Delta_{\zeta_2}P(\zeta_1,\zeta_2)\,,$$

where $Q(\zeta_1,\zeta_2)$ (resp. $P(\zeta_1,\zeta_2)$) has singularities in the variable ζ_1^2 (resp. on ζ_2^2) only on (f_J).

Let us study the commutation relations of operators of the same kind. In the following we fix J and skip it.

LEMMA 3.8. *The commutation relation for $\mathbf{k}(\zeta,\alpha)$ is given by a q-exact 2-form:*

$$\begin{aligned}(3.25)\qquad &\mathbf{k}(\zeta_1,\alpha)\mathbf{k}(\zeta_2,\alpha+1)+\mathbf{k}(\zeta_2,\alpha)\mathbf{k}(\zeta_1,\alpha+1)\\ &=\Delta_{\zeta_1}\mathbf{m}^{(++)}(\zeta_1,\zeta_2,\alpha)+\Delta_{\zeta_2}\mathbf{m}^{(++)}(\zeta_2,\zeta_1,\alpha),\end{aligned}$$

where

(3.26)

$$\begin{aligned}&\mathbf{m}^{(++)}(\zeta_1,\zeta_2,\alpha)(X)\\ &=\mathrm{Tr}_{b,A,B}\Bigl(M^{(++)}_{b,A,B}(\zeta_1/\zeta_2)\mathbb{T}_A(\zeta_1,\alpha)\mathbb{T}_{\{b,B\}}(\zeta_2,\alpha)(\zeta_1\zeta_2)^{\alpha-\mathbb{S}}(q^{-4S}X)\Bigr),\\ &M^{(++)}_{b,A,B}(\zeta)=\frac{\zeta^{-1}q^{-1}}{\zeta-\zeta^{-1}}\begin{pmatrix}q^{2D_B+1}\mathbf{a}_A^*q^{-2D_A}\mathbf{a}_A^* & -\zeta^{-1}q^{D_B}(1+\zeta u_{A,B})\mathbf{a}_A^*q^{D_B}\\ 0 & -q^{2D_B-1}\mathbf{a}_A^*q^{-2D_A}\mathbf{a}_A^*\end{pmatrix}_b,\end{aligned}$$

with $u_{A,B}=\mathbf{a}_A^*q^{-2D_A}\mathbf{a}_B$. The pole of $\mathbf{m}^{(++)}(\zeta_1,\zeta_2,\alpha)$ at $\zeta_1^2=\zeta_2^2$ is fictitious.

PROOF. We use the letters A_1, C_1, D_1, A_2, C_2, D_2 to denote $\mathbb{A}_A(\zeta_1,\alpha)$, $\mathbb{C}_A(\zeta_1,\alpha)$, $\mathbb{D}_A(\zeta_1,\alpha)$, $\mathbb{A}_B(\zeta_2,\alpha)$, $\mathbb{C}_B(\zeta_2,\alpha)$, $\mathbb{D}_B(\zeta_2,\alpha)$. We use also the formulas for $R_{i,j}$ in (3.6) where $\zeta=\zeta_1/\zeta_2$.

First consider (RHS) of (3.25). There are some cancellations. Namely, the term in $\mathbf{m}^{(++)}(q^{-1}\zeta_1,\zeta_2)$ which comes from the $(1,1)$ element of $M^{(++)}_{b,A,B}(\zeta)$ cancels with the one in $\mathbf{m}^{(++)}(q\zeta_2,\zeta_1)$ from the $(2,2)$ element. This is a consequence of the Yang-Baxter relation $R_{33}D_1A_2=A_2D_1R_{33}$ and

$$R_{3,3}q^{2D_B}\mathbf{a}^*_Aq^{-2D_A}\mathbf{a}^*_A=q^{-4}\zeta^2\cdot q^{2D_A}\mathbf{a}^*_Bq^{-2D_B}\mathbf{a}^*_BR_{3,3}\,.$$

Similarly, cancellations take place between the term in $\mathbf{m}^{(++)}(q\zeta_1,\zeta_2)$, which comes from the $(2,2)$ element of $M^{(++)}_{b,A,B}(\zeta)$, and the one in $\mathbf{m}^{(++)}(q^{-1}\zeta_2,\zeta_1)$ from the $(1,1)$ element. So, we will prove the equality (3.25) for the rest.

Rewrite (LHS) of (3.25), e.g.,

(3.27)
$$\begin{aligned}
&\mathbf{k}(\zeta_1,\alpha)\mathbf{k}(\zeta_2,\alpha+1)(X_J)\\
&\quad=\mathrm{Tr}_{a,b,A,B}\Bigl\{\sigma^+_a\sigma^+_b\mathbb{T}_{\{a,A\}}(\zeta_1,\alpha)\zeta_1^{\alpha-\mathbb{S}}q^{-2S}\mathbb{T}_{\{b,B\}}(\zeta_2,\alpha+1)\zeta_2^{\alpha-\mathbb{S}+1}(q^{-2S}X)\Bigr\}\\
&\quad=\frac{\zeta_2}{q\zeta_1}\mathrm{Tr}_{a,b,A,B}\Bigl\{q^{2D_B}\sigma^+_a\sigma^+_b\mathbb{T}_{\{a,A\}}(\zeta_1,\alpha)\mathbb{T}_{\{b,B\}}(\zeta_2,\alpha)(\zeta_1\zeta_2)^{\alpha-\mathbb{S}}(q^{-4S}X)\Bigr\}.
\end{aligned}$$

Interchanging the roles of ζ_1,ζ_2, and using $R_{11}=-\zeta^2q^{2D_A}R_{44}q^{-2D_B}$, we obtain

$$(LHS)=(\zeta q)^{-1}\mathrm{Tr}_{A,B}\Bigl\{q^{2D_B}(C_1C_2-R_{44}^{-1}C_2C_1R_{11})(\zeta_1\zeta_2)^{\alpha-\mathbb{S}}(q^{-4S}X)\Bigr\}.$$

We then substitute

(3.28)
$$\begin{aligned}
&C_1C_2-R_{44}^{-1}C_2C_1R_{11}\\
&\quad=-R_{44}^{-1}R_{42}A_1C_2+D_1C_2R_{33}^{-1}R_{31}-R_{44}^{-1}R_{43}R_{33}^{-1}A_2C_1R_{11}+R_{44}^{-1}D_2C_1R_{21}\\
&\qquad+R_{44}^{-1}(R_{43}R_{33}^{-1}R_{31}-R_{41})A_1A_2-D_1D_2R_{44}^{-1}(R_{43}R_{33}^{-1}R_{31}-R_{41}),
\end{aligned}$$

which follows from the Yang-Baxter relation (Lemma 3.1). As an example, consider the term $R_{44}^{-1}D_2C_1R_{21}$. The corresponding contribution is

$$\begin{aligned}
&(\zeta q)^{-1}\mathrm{Tr}_{A,B}\Bigl\{q^{2D_B}(R_{44}^{-1}D_2C_1R_{21})(\zeta_1\zeta_2)^{\alpha-\mathbb{S}}(q^{-4S}X)\Bigr\}\\
&=(\zeta q)^{-1}\mathrm{Tr}_{A,B}\Bigl\{(R_{21}q^{2D_B}R_{44}^{-1})D_2C_1(\zeta_1\zeta_2)^{\alpha-\mathbb{S}}(q^{-4S}X)\Bigr\}\\
&=\zeta^{-2}\mathrm{Tr}_{A,B}\Bigl\{(q^{D_A}R_{21}q^{2D_B}R_{44}^{-1}q^{-D_A})\mathbb{T}_B(\zeta_2q^{-1},\alpha)\mathbb{C}_A(\zeta_1,\alpha)(\zeta_1\zeta_2)^{\alpha-\mathbb{S}+1}(q^{-4S}X)\Bigr\},
\end{aligned}$$

where we use

$$q^{\mathbb{S}}\,\mathbb{T}_{\{a,A\}}(\zeta_1,\alpha)=q^{-D_A-\frac12\sigma^3_a}\mathbb{T}_{\{a,A\}}(\zeta_1,\alpha)q^{\mathbb{S}+D_A+\frac12\sigma^3_a}\,. \tag{3.29}$$

Now we compute to obtain

$$q^{D_A}R_{21}q^{2D_B}R_{44}^{-1}q^{-D_A}=\frac{q\zeta^3}{1-\zeta^2}q^{D_A}R_{A,B}(\zeta q)\mathbf{a}^*_Aq^{-D_A}R_{A,B}(\zeta)^{-1}$$
$$=\frac{q\zeta^3}{1-\zeta^2q^2}q^{D_A}P_{A,B}h(\zeta q,u_{A,B})(\zeta q)^{D_A+D_B}\mathbf{a}^*_Aq^{-D_A}R_{A,B}(\zeta)^{-1}$$
$$=\frac{q^2\zeta^4}{1-\zeta^2q^2}q^{D_A}\mathbf{a}^*_Bq^{-D_B}P_{A,B}h(\zeta q,u_{A,B}q^{-1})\zeta^{D_A+D_B}R_{A,B}(\zeta)^{-1}q^{D_A+D_B}$$
$$=\frac{q^2\zeta^4}{1-\zeta^2}q^{D_A}\mathbf{a}^*_Bq^{-D_B}(1+q^{-2}\zeta^{-1}u_{B,A})q^{D_A+D_B}$$
$$=\frac{q^2\zeta^4}{1-\zeta^2q^2}q^{D_A}(1+q^{-1}\zeta^{-1}u_{B,A})\mathbf{a}^*_Bq^{D_A}\,,$$

which is exactly as needed. We use here

$$\frac{h(\zeta q,uq^{-1})}{h(\zeta,u)}=1+q^{-2}\zeta^{-1}u\,,$$

which follows from (2.60) and (2.62). The other terms in (3.28) are treated similarly.

It remains to prove that the singularity of $\mathbf{m}^{(++)}(\zeta_1,\zeta_2,\alpha)$ at $\zeta_1^2=\zeta_2^2$ is fictitious. Having had enough of algebra we give a simple analytic proof in the spirit of Remark 2.9. We have

$$\mathbf{m}^{(++)}(\zeta_1,\zeta_2,\alpha)=\frac{1}{1-\zeta^2}F(\zeta_1,\zeta_2,\alpha)(\zeta_1\zeta_2)^{\alpha-\mathbb{S}}\,,$$

where $F(\zeta_1,\zeta_2,\alpha)$ is a rational function of ζ_1^2,ζ_2^2. The left hand side of (3.25) is regular at $\zeta_1^2=\zeta_2^2q^2$. In the right hand side we have near this point

$$\frac{1}{\zeta^2q^{-2}-1}\left(F(\zeta_2q,\zeta_2q,\alpha)q^{\alpha-\mathbb{S}}+F(\zeta_2,\zeta_2,\alpha)q^{-\alpha+\mathbb{S}}\right)\zeta_2^{2(\alpha-\mathbb{S})}+O(1)\,.$$

Requiring that the singularity vanishes, we obtain an equation for $F(\zeta_2,\zeta_2,\alpha)$. For generic α, this equation does not have solutions rational in ζ_2^2. Since everything is analytic in α, the function $F(\zeta_2,\zeta_2,\alpha)$ vanishes identically. □

Now we proceed to the more complicated commutation relations between operators of different kinds. Here we deal with $W^+\otimes W^-$, so, we shall put superscripts $\pm$ to $\mathbb{T}$'s. First let us introduce a function which will be extensively used later,

$$\psi(\zeta,\alpha)=\zeta^\alpha\frac{\zeta^2+1}{2(\zeta^2-1)}\,. \tag{3.30}$$

Often we shall use $\psi(\zeta,\alpha+\mathbb{S})$; it has been said that the combination $\alpha+\mathbb{S}$ is constant under the action of all our operators.

LEMMA 3.9. *The anti-commutator of* $\mathbf{k}$ *and* $\phi(\mathbf{k})$ *is given by a* q*-exact two form*

$$\mathbf{k}(\zeta_1,\alpha)\phi(\mathbf{k})(\zeta_2,\alpha+1)+\phi(\mathbf{k})(\zeta_2,\alpha)\mathbf{k}(\zeta_1,\alpha-1)=\Delta_{\zeta_1}\mathbf{m}^{(+-)}(\zeta_1,\zeta_2,\alpha)+\Delta_{\zeta_2}\mathbf{m}^{(-+)}(\zeta_2,\zeta_1,\alpha)\,. \tag{3.31}$$

The operators in the right hand side are given by

$$\begin{aligned}\mathbf{m}^{(+-)}(\zeta_1,\zeta_2,\alpha) &= N(\alpha-\mathbb{S})\mathrm{Tr}_{b,A,B}\left(M^{(+-)}_{b,A,B}(\zeta)\mathbb{T}^+_A(\zeta_1,\alpha)\mathbb{T}^-_{\{b,B\}}(\zeta_2,\alpha)\right)\zeta^{\alpha-\mathbb{S}}\\ &\quad -\psi(\zeta_1/\zeta_2,\alpha+\mathbb{S})\,,\\ \mathbf{m}^{(-+)}(\zeta_2,\zeta_1,\alpha) &= -\mathbb{J}\,\mathbf{m}^{(+-)}(\zeta_2,\zeta_1,-\alpha)\,\mathbb{J}\,,\end{aligned}$$

where

$$M^{(+-)}_{b,A,B}(\zeta)=\frac{q}{\zeta^2-1}q^{\sigma^3_bD_A}\left(\frac{1}{2}(\zeta^2+1)\sigma^3_b+U_{AB}(\zeta)\sigma^-_b\right)q^{-\sigma^3_bD_A}\,.$$

The functions $\mathbf{m}^{(\pm\mp)}(\zeta_1,\zeta_2,\alpha)$ *are regular at* $\zeta_1=\zeta_2$.

PROOF. Denote

$$\widetilde{\mathbf{m}}^{(+-)}(\zeta_1,\zeta_2,\alpha)=\mathbf{m}^{(+-)}(\zeta_1,\zeta_2,\alpha)+\psi(\zeta_1/\zeta_2,\alpha+\mathbb{S})\,.$$

Since

$$-\mathbb{J}\,\psi(\zeta_2/\zeta_1,-\alpha+\mathbb{S})\,\mathbb{J}=\psi(\zeta_1/\zeta_2,\alpha+\mathbb{S})\,,$$

and

$$\Delta_{\zeta_1}\psi(\zeta_1/\zeta_2,\alpha+\mathbb{S})+\Delta_{\zeta_2}\psi(\zeta_1/\zeta_2,\alpha+\mathbb{S})=0\,,$$

the formula (3.31) holds if we replace $\mathbf{m}^{(+-)}$ by $\widetilde{\mathbf{m}}^{(+-)}$.

First we show that

$$\begin{aligned}&\mathbf{k}(\zeta_1,\alpha)\phi(\mathbf{k})(\zeta_2,\alpha+1)-\Delta_{\zeta_1}\widetilde{\mathbf{m}}^{(+-)}(\zeta_1,\zeta_2,\alpha)\\ &\quad=-\big(\phi(\mathbf{k})(\zeta_2,\alpha)\mathbf{k}(\zeta_1,\alpha-1)-\Delta_{\zeta_2}\widetilde{\mathbf{m}}^{(-+)}(\zeta_2,\zeta_1,\alpha)\big)\,.\end{aligned}$$

In the left hand side the monodromy matrices under the trace are ordered as $\mathbb{T}^+_{\{a,A\}}(\zeta_1,\alpha)\mathbb{T}^-_{\{b,B\}}(\zeta_2,\alpha)$, while in the right hand side the order is opposite. Applying (2.77) to $-\phi(\mathbf{k})(\zeta_2,\alpha)\mathbf{k}(\zeta_1,\alpha-1)$ and using (3.29) and suitable analogs of (3.27), we arrive at the following identity to be proved:

$$\begin{aligned}&\mathrm{Tr}_{a,b,A,B}\left(q^{\sigma^3_bD_A}X_{a,b,A,B}(\zeta)q^{-\sigma^3_bD_A}\mathbb{T}^+_{\{a,A\}}(\zeta_1,\alpha)\mathbb{T}^-_{\{b,B\}}(\zeta_2,\alpha)\right)\zeta^{\alpha-\mathbb{S}}\,,\\ &=-\mathrm{Tr}_{a,b,A,B}\left(q^{-\sigma^3_aD_B}\overline{X}_{a,b,A,B}(\zeta)q^{\sigma^3_aD_B}\mathbb{T}^-_{\{b,B\}}(\zeta_2,\alpha)\mathbb{T}^+_{\{a,A\}}(\zeta_1,\alpha)\right)\zeta^{\alpha-\mathbb{S}}\,,\end{aligned}$$

where

(3.32)

$$\begin{aligned}X_{a,b,A,B}(\zeta)&=Y_{A,B}(\zeta)\sigma^+_a\sigma^-_b-q\sigma^3_a\Big(\frac{\zeta^2q^{2\sigma^3_a}+1}{2(\zeta^2q^{2\sigma^3_a}-1)}\sigma^3_b+\frac{1}{\zeta^2q^{2\sigma^3_a}-1}U_{A,B}(\zeta)\sigma^-_b\Big)\,,\\ \overline{X}_{a,b,A,B}(\zeta)&=\overline{Y}_{A,B}(\zeta)\sigma^+_a\sigma^-_b-q\sigma^3_b\Big(\frac{\zeta^{-2}q^{-2\sigma^3_b}+1}{2(\zeta^{-2}q^{-2\sigma^3_b}-1)}\sigma^3_a+\frac{1}{\zeta^{-2}q^{-2\sigma^3_b}-1}\overline{U}_{A,B}(\zeta)\sigma^+_a\Big).\end{aligned}$$

Deriving (3.32) we used the relations

$$\begin{aligned}&\mathrm{Tr}_{b,B,A}\Big\{Y_{b,B,A}\mathbb{T}^+_A(q^{\pm1}\zeta_1,\alpha)\mathbb{T}^-_{\{b,B\}}(\zeta_2,\alpha)q^{\pm(\alpha-\mathbb{S})}\Big\}\\ &\quad=\mathrm{Tr}_{b,B,a,A}\Big\{q^{\pm(\frac{1}{2}\sigma^3_b-D_B)}Y_{b,B,A}q^{\mp(\frac{1}{2}\sigma^3_b-D_B)}\tau^\pm_a\mathbb{T}^+_{\{a,A\}}(\zeta_1,\alpha)\mathbb{T}^-_{\{b,B\}}(\zeta_2,\alpha)\Big\}\,,\end{aligned}\tag{3.33}$$

together with

$$q^{\mp D_B}U_{AB}(\zeta q^{\pm1})q^{\pm D_B}=q^{\pm1}U_{AB}(\zeta)\,.$$

Now apply (3.7), and verify that

$$\boldsymbol{\rho}\Big(\mathcal{F}X_{a,b,A,B}(\zeta)\mathcal{F}^{-1}\Big)=-\overline{X}_{a,b,A,B}(\zeta)+(\star)\sigma_a^-+(\star)\sigma_b^+\,,$$

the last two terms are irrelevant under the trace.

Let us investigate the singularity. We have

$$\widetilde{\mathbf{m}}^{(+-)}(\zeta_1,\zeta_2,\alpha)=\psi(\zeta,\alpha+\mathbb{S})F(\zeta_1,\zeta_2),$$
$$\widetilde{\mathbf{m}}^{(+-)}(\zeta_2,\zeta_1,\alpha)=\psi(\zeta,\alpha+\mathbb{S})G(\zeta_1,\zeta_2)\,,$$

where $F(\zeta_1,\zeta_2)$ and $G(\zeta_1,\zeta_2)$ are rational functions of ζ_1^2, ζ_2^2 which are regular on the diagonal. Since the left hand side of (3.31) is regular, we obtain two equations

$$F(\zeta q,\zeta q)=G(\zeta,\zeta)\,,\quad F(\zeta,\zeta)=G(\zeta q,\zeta q)\,,$$

which means for $F(\zeta,\zeta)$

$$F(\zeta q^2,\zeta q^2)=F(\zeta,\zeta)\,.$$

Since $F(\zeta,\zeta)$ is rational in ζ^2, the only solution is $F(\zeta,\zeta)=Const.$ The constant is easy to compute at $\zeta_1=\zeta_2=0$,

$$\begin{aligned}&q^{-1}N(\alpha-\mathbb{S})\mathrm{Tr}_{b,a,B}\Big\{q\Big(\sigma_b^3+(\star)\sigma_b^-\Big)q^{(\alpha-\mathbb{S})(2D_A-2D_B+\sigma_b^3)}\Big\}\\&=N(\alpha-\mathbb{S})\frac{q^{\alpha-\mathbb{S}}-q^{-\alpha+\mathbb{S}}}{(1-q^{2(\alpha-\mathbb{S})})(1-q^{-2(\alpha-\mathbb{S})})}\\&=1\,.\end{aligned}$$

(The multiplier $q^{-1}N(\alpha-\mathbb{S})$ in Definition 3.6 of ϕ is so chosen as to satisfy the above normalisation.) This implies that $m^{(\pm\mp)}(\zeta_1,\zeta_2,\alpha)$ are regular on the diagonal. The proof is over. □

3.2.4. Analytic Properties. Let us discuss the analytic properties of $\mathbf{k}_J(\zeta,\alpha)$ with respect to ζ in some detail. To this end it is convenient to consider the inhomogeneous case where all poles are simple.

Let $J=[k,l]$. We have an obvious symmetry property

$$\begin{aligned}&\mathbb{R}_{i,i+1}(\xi_i/\xi_{i+1})\mathbf{k}_J(\zeta,\alpha|\xi_k,\dots,\xi_i,\xi_{i+1},\dots,\xi_l)\mathbb{R}_{i,i+1}(\xi_i/\xi_{i+1})^{-1}\\&\qquad=\mathbb{P}_{i,i+1}\mathbf{k}_J(\zeta,\alpha|\xi_k,\dots,\xi_{i+1},\xi_i,\dots,\xi_l)\mathbb{P}_{i,i+1}\,,\end{aligned}\tag{3.34}$$

where we indicate the dependence on the inhomogeneities explicitly. In view of (3.34), in order to study the singularities of $\mathbf{k}_J(\zeta,\alpha)$ at $\zeta=\xi_jq^{\pm1}$, it is sufficient to consider $\zeta=\xi_lq^{\pm1}$.

LEMMA 3.10. *For $J=[k,l]$, we have*

$$\mathrm{res}_{\zeta=q^{-1}\xi_l}\mathbf{k}_J(\zeta,\alpha)(X_J)\frac{d\zeta^2}{\zeta^2}=-\mathrm{res}_{\zeta=\xi_l}\Big(\sigma_l^+\,\mathbf{q}_J(\zeta,\alpha)(q^{-2S_J}X_J)\Big)\frac{d\zeta^2}{\zeta^2}\,,$$
$$\mathrm{res}_{\zeta=q\xi_l}\mathbf{k}_J(\zeta,\alpha)(X_J)\frac{d\zeta^2}{\zeta^2}=-\mathrm{res}_{\zeta=\xi_l}\Big(\mathbf{q}_J(\zeta,\alpha)(q^{-2S_J}X_J)\,\sigma_l^+\Big)\frac{d\zeta^2}{\zeta^2}\,,$$

where the adjoint Q-operator is defined by [1]

$$\mathbf{q}_J(\zeta,\alpha)(X_J)=\mathrm{Tr}_A\left[\mathbb{T}_A(\zeta,\alpha)\zeta^{\alpha-\mathbb{S}}(X_J)\right]\,.\tag{3.35}$$

[1] The definition of $\mathbf{q}_J(\zeta,\alpha)$ is slightly changed from [**BJMST2**].

PROOF. We prove the first formula. The other one is similar. We start from

$$\mathbf{k}_J(\zeta,\alpha)(X_J) = \mathrm{Tr}_{A,a}\Big\{\sigma_a^+ R_{\{a,A\},l}(\zeta/\xi_l) \times \mathbb{T}_{\{a,A\},J'}(\zeta,\alpha)\zeta^{\alpha-\mathbb{S}}(q^{-2S_J}X_J)R_{\{a,A\},l}(\zeta/\xi_l)^{-1}\Big\},$$

where $J' = [k, l-1]$. We use (2.78), (3.5) and (3.15) for $\mathbb{T}_{\{a,A\},J'}(\zeta,\alpha)$. Taking the residue at $\zeta = q^{-1}\xi_l$ we obtain

$$\operatorname*{res}_{\zeta=q^{-1}\xi_l} \mathbf{k}_J(\zeta,\alpha)(X_J) = \sigma_l^+ \mathrm{Tr}_A\Big\{R_{A,l}(1)\mathbb{T}_{A,J'}(\xi_l,\alpha)\xi_l^{\alpha-\mathbb{S}}(q^{-2S_J}X_J) \operatorname*{res}_{\zeta=\xi_l}(R_{A,l}(\zeta/\xi_l)^{-1})\Big\}.$$

Calculating similarly with $\mathbf{q}_J(\zeta,\alpha)(X_J)$ and comparing the results, we obtain the first formula. □

Let us discuss the behaviour of $\mathbf{k}_J(\zeta,\alpha)$ as $\zeta \to 0$ or $\zeta \to \infty$. Here it is convenient to use the formula (3.17). Both $\mathbb{T}_{a,J}(\zeta,\alpha)$ and $\mathbb{T}_{A,J}(\zeta,\alpha)$ are regular at $\zeta = 0$, and $\mathbb{T}_{a,J}(0,\alpha)$ is diagonal in a. Hence we have

$$\mathbf{k}_J(\zeta,\alpha) = \zeta^{\alpha-\mathbb{S}+2}\big(\mathbf{k}_{J,0} + O(\zeta^2)\big)\,, \quad \zeta \to 0\,, \tag{3.36}$$

with some $\mathbf{k}_{J,0}$ independent of ζ^2.

In order to describe the behaviour for $\zeta \to \infty$ we use the gauge transformation (1.32)

$$\widetilde{R}_a(\zeta) = \zeta^{-\sigma^3/2}R_a(\zeta)\zeta^{\sigma^3/2}\,,$$

together with

$$\widetilde{R}_A(\zeta) = \zeta^{-\sigma^3/2-1}R_A(\zeta)\zeta^{-\sigma^3/2}\,.$$

Notice that the second one is not a similarity transformation. Then $\widetilde{R}_a(\zeta)^{\pm1}$, $\widetilde{R}_A(\zeta)^{\pm1}$ depend only on ζ^2 and are regular at $\zeta^2 = \infty$. Let $\widetilde{\mathbb{T}}_{a,J}(\zeta)$, $\widetilde{\mathbb{T}}_{A,J}(\zeta)$ be the corresponding adjoint monodromy matrices. Denoting further the adjoint action of $G = \prod_{j\in J}\xi_j^{\sigma_j^3/2}$ by $\mathbb{G}$, we can rewrite

$$\mathbf{k}_J(\zeta,\alpha)(X_J) = \zeta^{\alpha+\mathbb{S}}\mathbb{G}^{-1}\mathrm{Tr}_{a,A}\Big\{\sigma_a^+\widetilde{\mathbb{T}}_{a,J}(\zeta,\alpha)\widetilde{\mathbb{T}}_{A,J}(\zeta,\alpha)\mathbb{G}^{-1}(q^{-2S_J}X_J)\Big\}\,.$$

This implies

$$\mathbf{k}_J(\zeta,\alpha) = \zeta^{\alpha+\mathbb{S}}\big(\mathbf{k}_{J,\infty} + O(\zeta^{-2})\big)\,, \quad \zeta \to \infty\,, \tag{3.37}$$

with some $\mathbf{k}_{J,\infty}$ independent of ζ^2.

3.3. Annihilation Operators

Introduce the operators

$$\overline{\mathbf{c}}_J(\zeta,\alpha) = \frac{1}{2\pi i}\oint_\Gamma \psi(\zeta/\xi,\alpha+\mathbb{S})\mathbf{k}_J(\xi,\alpha)\frac{d\xi^2}{\xi^2}\,, \tag{3.38}$$

$$\mathbf{c}_J(\zeta,\alpha) = \frac{1}{4\pi i}\oint_\Gamma \psi(\zeta/\xi,\alpha+\mathbb{S})\left\{\mathbf{k}_J(q\xi,\alpha) + \mathbf{k}_J(q^{-1}\xi,\alpha)\right\}\frac{d\xi^2}{\xi^2}\,, \tag{3.39}$$

where the contour Γ goes around the divisor (f_J) and ζ^2 is taken outside the contour. Essentially $\overline{\mathbf{c}}_J(\zeta,\alpha)$ is the singular part of $\mathbf{k}_J(\zeta,\alpha)$ on (f_J) while $\mathbf{c}_J(\zeta,\alpha)$ is the arithmetic mean of the singular parts on $(\mathsf{f}_J)q^2$ and $(\mathsf{f}_J)q^{-2}$. We put $\alpha+\mathbb{S}$ in the kernel in order that it commutes with our operators, as explained before.

Obviously, the left reduction for $\mathbf{k}$ implies that for $\mathbf{c}$, $\bar{\mathbf{c}}$. What is non-trivial is that the right reduction takes place for the latter.

THEOREM 3.11. *Suppose $J=L\sqcup R$ as in* (3.18). *Then we have the reduction relations*

$$\mathbf{c}_J(\zeta,\alpha)(Y_L)=\mathbf{c}_L(\zeta,\alpha)(Y_L),\quad \bar{\mathbf{c}}_J(\zeta,\alpha)(Y_L)=\bar{\mathbf{c}}_L(\zeta,\alpha)(Y_L)\,. \tag{3.40}$$

PROOF. We have shown in (3.21) that

$$\mathbf{k}_J(\zeta,\alpha)(Y_L)=\mathbf{k}_J(\zeta,\alpha)(Y_L)+\Delta_\zeta\mathbf{v}_J(\zeta,\alpha)(Y_L), \tag{3.41}$$

where $\mathbf{v}_J(\zeta,\alpha)(Y_L)$ has singularities only on (f_J). The second equality in (3.40) follows immediately because $\Delta_\zeta\mathbf{v}_J(\zeta,\alpha)(Y_L)$ is simply regular on (f_J). The first equality follows because the singularity on (f_J) cancels in the sum $\Delta_\zeta\mathbf{v}_J(\zeta q,\alpha)(Y_L)+\Delta_\zeta\mathbf{v}_J(\zeta q^{-1},\alpha)(Y_L)$. □

The meaning of our construction is clear; the reduction relation for $\mathbf{k}$ holds up to a q-exact form, so we integrate over 'q-closed cycles' in order to kill the q-exact forms.

Notice that the right hand sides in (3.40) are singular only on (f_L), so the singularity is defined by the length of a quasi-local operator on which $\mathbf{c}_J(\zeta,\alpha)$ or $\bar{\mathbf{c}}_J(\zeta,\alpha)$ act, and not by the length of the original interval J.

Due to the left and right reduction relations, the operators $\mathbf{c}_J(\zeta,\alpha)$, $\bar{\mathbf{c}}_J(\zeta,\alpha)$ have stable meaning as $J\to(-\infty,\infty)$. This allows us to define them acting on quasi-local operators on the whole line as we wanted,

$$\mathbf{c}(\zeta,\alpha):\ \mathcal{W}_{s-1,\alpha+1}\to\mathcal{W}_{s,\alpha}\,,\quad \bar{\mathbf{c}}(\zeta,\alpha):\ \mathcal{W}_{s-1,\alpha+1}\to\mathcal{W}_{s,\alpha}\,. \tag{3.42}$$

Using (3.6), we define two more operators

$$\mathbf{b}(\zeta,\alpha)=\phi(\mathbf{c})(\zeta,\alpha)\,,\quad \bar{\mathbf{b}}(\zeta,\alpha)=\phi(\bar{\mathbf{c}})(\zeta,\alpha)\,, \tag{3.43}$$

which act as

$$\mathbf{b}(\zeta,\alpha):\ \mathcal{W}_{s+1,\alpha-1}\to\mathcal{W}_{s,\alpha}\,,\quad \bar{\mathbf{b}}(\zeta,\alpha):\ \mathcal{W}_{s+1,\alpha-1}\to\mathcal{W}_{s,\alpha}\,. \tag{3.44}$$

We shall see later that $\bar{\mathbf{b}}(\zeta,\alpha)$ and $\bar{\mathbf{c}}(\zeta,\alpha)$ are not quite independent from $\mathbf{b}(\zeta,\alpha)$, $\mathbf{c}(\zeta,\alpha)$; see Theorem 3.23 below.

THEOREM 3.12. *We have the commutation relations*

$$\begin{aligned}\mathbf{c}(\zeta_1,\alpha)\mathbf{c}(\zeta_2,\alpha+1)+\mathbf{c}(\zeta_2,\alpha)\mathbf{c}(\zeta_1,\alpha+1)&=0\,,\\ \mathbf{b}(\zeta_1,\alpha)\mathbf{c}(\zeta_2,\alpha-1)+\mathbf{c}(\zeta_2,\alpha)\mathbf{b}(\zeta_1,\alpha+1)&=0\,,\\ \mathbf{b}(\zeta_1,\alpha)\mathbf{b}(\zeta_2,\alpha-1)+\mathbf{b}(\zeta_2,\alpha)\mathbf{b}(\zeta_1,\alpha-1)&=0\,.\end{aligned} \tag{3.45}$$

The same relations hold if some of $\mathbf{b}(\zeta,\alpha)$, $\mathbf{c}(\zeta,\alpha)$ are replaced by $\bar{\mathbf{b}}(\zeta,\alpha)$, $\bar{\mathbf{c}}(\zeta,\alpha)$.

PROOF. The first two relations follow immediately by integrating (3.25), (3.31) according to the definition (3.39). The third relation follows from the first by applying ϕ. □

Annihilation operators $\mathbf{c}(\zeta)$, $\bar{\mathbf{c}}(\zeta)$, $\mathbf{b}(\zeta)$, $\bar{\mathbf{b}}(\zeta)$ on the total space of quasi-local operators $\mathcal{W}$ (3.1) are defined by collecting their action on each block (3.42), (3.44). These operators mutually anticommute.

Recall the definition of the length of operators (3.2). The reduction relations (3.40) imply

$$\text{length}(\mathbf{c}(\zeta)(q^{2\alpha S(0)}\mathcal{O})) \leq \text{length}(q^{2\alpha S(0)}\mathcal{O}), \tag{3.46}$$

and similarly for $\bar{\mathbf{c}}(\zeta)$, $\mathbf{b}(\zeta)$, $\bar{\mathbf{b}}(\zeta)$. For the 'primary fields' $q^{2\alpha S(k)}$ we have

$$\mathbf{c}(\zeta)(q^{2\alpha S(k)}) = 0, \tag{3.47}$$

and similarly for others. More generally, the definition of the annihilation operators implies their local nilpotency; if $l = \text{length}(q^{2\alpha S(0)}\mathcal{O})$, then

$$\mathbf{c}(\zeta_{l+1}) \cdots \mathbf{c}(\zeta_2)\mathbf{c}(\zeta_1)(q^{2\alpha S(0)}\mathcal{O}) = 0. \tag{3.48}$$

Indeed, on a quasi-local operator of support J, the block $\mathbf{c}_J(\zeta, \alpha)$ is a linear combination of $l = |J|$ anti-commuting operators which are the coefficients of the singular part on the divisor (f_J). Therefore (3.48) holds.

REMARK 3.13. In the inhomogeneous case, the residues

$$\mathbf{x}^{(j)}(\alpha) = \underset{\zeta^2=\xi_j^2}{\text{res}}\, \mathbf{x}(\zeta, \alpha)\frac{d\zeta^2}{\zeta^2} \quad (\mathbf{x} = \mathbf{c}, \mathbf{b}) \tag{3.49}$$

are mutually anticommuting operators. On a site of length one, $\mathbf{c}^{(1)}(\alpha)$ (resp. $\mathbf{b}^{(1)}(\alpha)$) is non-trivial only on operators of spin -1 (resp. $+1$). Combining this with the R matrix symmetry, we obtain the relation

$$\mathbf{b}^{(j)}(\alpha + 1)\mathbf{c}^{(j)}(\alpha) = 0. \tag{3.50}$$

This means that, on finite chains, the annihilation operators do not generate a Grassmann algebra but rather a quotient of it.

3.4. Creation Operators

The annihilation operators introduced in the previous section send a quasi-local operator to one of the same or shorter length. We want now to define creation operators which can increase the length but preserve locality; in other words, they must keep the length finite. We begin with a simple but instructive example of the operator $\mathbf{t}^*(\zeta)$. Then we proceed to the much more complicated case of the fermionic creation operators $\mathbf{b}^*(\zeta)$, $\mathbf{c}^*(\zeta)$.

3.4.1. Operator $\mathbf{t}^*(\zeta)$. Recall the construction of the local integrals of motion for the XXZ spin chain given in Section 1.6. There we constructed the local integrals

$$\mathbf{I}_p = \sum_{j=-\infty}^{\infty} h_{[j,j+p]},$$

where the density $h_{[j,j+p]}$ acts nontrivially only on the interval $[j, j+p]$. As we shall see, the infinite sum is meaningful in the following discussion.

Consider a quasi-local operator $q^{2\alpha S(0)}\mathcal{O}$, and let $\mathbf{I}_p$ act on it by the adjoint. The result

$$\left[\mathbf{I}_p, q^{2\alpha S(0)}\mathcal{O}\right] \tag{3.51}$$

is quasi-local. Indeed, for sufficiently large $|j|$ the quasi-local operator stabilises, and

$$\left[h_{[j,j+p]}\,,\, q^{\alpha\sum_{k=j}^{j+p}\sigma_k^3}\right] = 0.$$

So the operator (3.51) is quasi-local, but its length is normally larger than that of $q^{2\alpha S(0)}\mathcal{O}$. Let us formalise this construction.

Define

$$\mathbf{t}^*_J(\zeta,\alpha)(X_J)=\mathrm{Tr}_a\Big(\mathbb{T}_{a,J}(\zeta,\alpha)(X_J)\Big)\,. \tag{3.52}$$

For this operator we have the simple left reduction: for $J=L\sqcup R$,

$$\mathbf{t}^*_J(\zeta,\alpha)(q^{2\alpha S_L}Y_R)=q^{2\alpha S_L}\mathbf{t}^*_R(\zeta,\alpha)(Y_R)\,.$$

Let us think about the right reduction for $X_J=Y_L$. Suppose $J=[k,l]$, $L=[k,m]$, $m<l$. Using (1.32), (1.33) in Section 1.6 and similar definitions for the adjoint R-matrices, we evaluate

$$\begin{aligned}\mathbf{t}^*_J(\zeta,\alpha)(Y_L)&=\zeta^{-\mathbb{S}}\mathrm{Tr}_a\Big(\widetilde{\mathbb{T}}_{a,J}(\zeta,\alpha)\zeta^{\mathbb{S}}(Y_L)\Big)=\mathrm{Tr}_a\Big(\widetilde{\mathbb{T}}_{a,J}(\zeta,\alpha)(Y_L)\Big)\\&=\mathrm{Tr}_a\Big(\check{\mathbb{R}}_{a,l}(\zeta)\check{\mathbb{R}}_{l,l-1}(\zeta)\cdots\check{\mathbb{R}}_{m+2,m+1}(\zeta)\mathbb{P}_{a,l}\cdots\mathbb{P}_{a,m+1}\widetilde{\mathbb{T}}_{a,L}(\zeta,\alpha)(Y_L)\Big)\\&=\mathrm{Tr}_a\Big(\check{\mathbb{R}}_{a,l}(\zeta)\check{\mathbb{R}}_{l,l-1}(\zeta)\cdots\check{\mathbb{R}}_{m+2,m+1}(\zeta)\widetilde{\mathbb{T}}_{m+1,L}(\zeta,\alpha)(Y_L)\Big)\,.\end{aligned} \tag{3.53}$$

Notice now that

$$\check{\mathbb{R}}_{i,j}(\zeta)=I+(\zeta^2-1)\mathbf{r}_{i,j}(\zeta^2)\,, \tag{3.54}$$

where $\mathbf{r}_{i,j}(\zeta^2)$ is regular at $\zeta^2=1$ and

$$\mathbf{r}_{i,j}(\zeta^2)(I_i\otimes I_j)=0\,. \tag{3.55}$$

Substitute (3.54) into the last line of (3.53). This gives under the trace a sum of 2^{l-m} terms of the form

$$\mathbf{r}_{j_p+1,j_p}(\zeta)\mathbf{r}_{j_{p-1}+1,j_{p-1}}(\zeta)\cdots\mathbf{r}_{j_1+1,j_1}(\zeta)\widetilde{\mathbb{T}}_{m+1,L}(\zeta,\alpha)(Y_L)\,, \tag{3.56}$$

where $j_p>j_{p-1}>\cdots>j_1\geq m+1$, and we set $a=l+1$. Here we realise that, due to (3.55), the only non-vanishing terms in (3.56) are those without gaps, i.e., $j_p=j_{p-1}+1,\dots,j_1=m+1$. Hence

(3.57)

$$\begin{aligned}\mathbf{t}^*_J(\zeta,\alpha)(Y_L)&=2\sum_{j=1}^{l-m-1}(\zeta^2-1)^j\mathbf{r}_{m+j+1,m+j}(\zeta)\cdots\mathbf{r}_{m+2,m+1}(\zeta)\widetilde{\mathbb{T}}_{m+1,L}(\zeta,\alpha)(Y_L)\\&+(\zeta^2-1)^{l-m}\mathrm{Tr}_a\Big\{\mathbf{r}_{a,l}(\zeta)\mathbf{r}_{l,l-i}(\zeta)\cdots\mathbf{r}_{m+2,m+1}(\zeta)\widetilde{\mathbb{T}}_{m+1,L}(\zeta,\alpha)(Y_L)\Big\}\\&\equiv 2\sum_{j=1}^{l-m-1}(\zeta^2-1)^j\mathbf{r}_{m+j+1,m+j}(\zeta)\cdots\mathbf{r}_{m+2,m+1}(\zeta)\widetilde{\mathbb{T}}_{m+1,L}(\zeta,\alpha)(Y_L)\\&\qquad\qquad\mod\ (\zeta^2-1)^{l-m}\,.\end{aligned}$$

Here, for power series $F(\zeta),G(\zeta)$ in ζ^2-1, $F(\zeta)\equiv G(\zeta)\mod (\zeta^2-1)^{l-m}$ means that $F(\zeta)-G(\zeta)$ is divisible by $(\zeta^2-1)^{l-m}$.

Eq.(3.57) says that the p-th Taylor coefficient of $\mathbf{t}^*_J(\zeta,\alpha)(Y_L)$ stabilises for $l>m+p$, giving rise to a formal power series in ζ^2-1:

(3.58)

$$\mathbf{t}^*_{[k,\infty)}(\zeta,\alpha)(Y_L)=2\sum_{j=1}^{\infty}(\zeta^2-1)^j\mathbf{r}_{m+j+1,m+j}(\zeta)\cdots\mathbf{r}_{m+2,m+1}(\zeta)\widetilde{\mathbb{T}}_{m+1,L}(\zeta,\alpha)(Y_L)\,.$$

This defines a family of operators $\mathbf{t}^*(\zeta)$ on $\mathcal{W}$ which sends $\mathcal{W}_{\alpha,s}$ to itself. Let us give a formal definition.

DEFINITION 3.14. We define the action of $\mathbf{t}^*(\zeta,\alpha)$ on an element $q^{2\alpha S(k-1)}X_{[k,l]}$ of $\mathcal{W}_{\alpha,s}$ by

$$\mathbf{t}^*(\zeta,\alpha)(q^{2\alpha S(k-1)}X_{[k,l]}) = q^{2\alpha S(k-1)}\mathbf{t}^*_{[k,\infty)}(\zeta,\alpha)(X_{[k,l]}) \,. \tag{3.59}$$

This definition is consistent due to the left reduction relation. The operator $\mathbf{t}^*(\zeta,\alpha)$ is given by a power series

$$\mathbf{t}^*(\zeta,\alpha) = \sum_{p=1}^{\infty}(\zeta^2-1)^{p-1}\mathbf{t}^*_p(\alpha)\,,$$

and we have obviously

$$\mathrm{lenght}\big(\mathbf{t}^*_p(\alpha)(q^{2\alpha S(0)}\mathcal{O})\big) \le \mathrm{lenght}(q^{2\alpha S(0)}\mathcal{O}) + p\,. \tag{3.60}$$

Using

$$\mathbb{R}_{i,j}(1) = \mathbb{P}_{i,j}\,,$$

it is easy to see that $\mathbf{t}^*_1$ is the right shift by one site of the lattice.

REMARK 3.15. Take a 2 by 2 matrix $\mathcal{K}$ and consider the following object:

$$\mathbf{t}^*_J(\zeta,\alpha,\mathcal{K})(X_J) = \mathrm{Tr}_a\left(\mathcal{K}_a\mathbb{T}_{a,J}(\zeta,\alpha)(X_J)\right)\,.$$

Then it is easy to conclude from the above definition that for $J = L \sqcup R$

$$\mathbf{t}^*_J(\zeta,\alpha,\mathcal{K})(Y_L) \equiv \frac{1}{2}\mathrm{Tr}(\mathcal{K})\ \mathbf{t}^*_J(\zeta,\alpha)(Y_L) \mod (\zeta^2-1)^{|R|-1}\,. \tag{3.61}$$

So we can freely put some matrix $\mathcal{K}$ under the trace when computing on the finite interval; the resulting inductive limit is essentially independent of $\mathcal{K}$. Note that $\mathcal{K}$ is allowed to depend on ζ^2 as long as the dependence is holomorphic at $\zeta^2 = 1$.

3.4.2. Commutation Relations with c, c̄. We begin with a lemma which is more general than we need here. This lemma will be extensively used later.

LEMMA 3.16. *Let $J = L \sqcup R$, and let $Y_{L,c} \in \mathrm{End}(V_L \otimes V_c)$. Then*

$$\begin{aligned}\mathbf{c}_J(\zeta_1,\alpha)\mathrm{Tr}_c&\Big\{\mathbb{T}_{c,R}(\zeta_2)(Y_{L,c})\Big\}\\ &\equiv \mathrm{Tr}_c\Big\{\mathbb{T}_{c,R}(\zeta_2)\mathbf{c}_{L\sqcup c}(\zeta_1,\alpha)(Y_{L,c})\Big\} \mod (\zeta_2^2-1)^{|R|-1}\,,\end{aligned} \tag{3.62}$$

where $\mathbf{c}_{L\sqcup c}(\zeta_1,\alpha)$ means the operator $\mathbf{c}$ constructed for the inhomogeneous chain $V_L \otimes V_{c,\zeta_2}$, in other words, starting in the definition (3.39) *with the monodromy matrix*

$$\mathbb{T}_{\{a,A\},L\sqcup c}(\xi)(X_J) = \mathbb{R}_{\{a,A\},c}(\xi/\zeta_2)\mathbb{T}_{\{a,A\},L}(\xi)\,.$$

The same relation holds for $\bar{\mathbf{c}}$.

PROOF. Here it is easier to consider $\bar{\mathbf{c}}$. Recalling the definition (3.38) we start with evaluation of the following object:

$$\mathcal{Z} = \mathrm{Tr}_c\big\{\mathbf{k}_{J\sqcup c}(\xi,\alpha)\mathbb{T}_{c,R}(\zeta_2)(Y_{L,c})\big\}\,.$$

We consider ζ_2^2 to be close to 1. The very meaning of the definition (3.38) implies that the contour of integration over ξ^2 contains 1 and ζ_2^2. So, we can take

$$|\xi^2-1| > |\zeta_2^2-1|\,. \tag{3.63}$$

There are two ways to compute $\mathcal{Z}$.

First way. Write explicitly

$$\mathcal{Z}=\mathrm{Tr}_{c,a,A}\Big\{Z_{c,a,A}\mathbb{T}_{\{a,A\},J}(\xi,\alpha)\xi^{\alpha-\mathbb{S}}\big(q^{-2S_J}\mathbb{T}_{c,R}(\zeta_2)(Y_{L,c})\big)\Big\}\,,$$

where

$$Z_{c,a,A}=\xi^{\mathbb{S}_c}\mathbb{R}_{\{a,A\},c}(\xi/\zeta_2)^{-1}(\sigma_a^+)q^{-\sigma_c^3}\,.$$

This object depends on ζ_2^2 and has a singularity at $\zeta_2^2=\xi^2$. However, due to (3.63) it can be understood as Taylor series in ζ_2^2-1. So, we proceed applying Remark 3.15

$$\mathcal{Z}\equiv\mathrm{Tr}_{a,A}\Big\{\mathrm{Tr}_c(Z_{c,a,A})\mathbb{T}_{\{a,A\},J}(\xi)\xi^{\alpha-\mathbb{S}}\big(q^{-2S_J}\frac{1}{2}\mathrm{Tr}_c\big(\mathbb{T}_{c,R}(\zeta_2)(Y_{L,c})\big)\big)\Big\}\\ \mod(\zeta_2^2-1)^{|R|-1}\,.$$

It is easy to compute

$$\mathrm{Tr}_c(Z_{c,a,A})=2\sigma_a^+\,,$$

hence

$$\mathcal{Z}\equiv\mathbf{k}_J(\xi,\alpha)\mathrm{Tr}_c\big(\mathbb{T}_{c,R}(\zeta_2)(Y_{L,c})\big)\mod(\zeta_2^2-1)^{|R|-1}\,. \tag{3.64}$$

Second way. Let us use another formula and apply the Yang-Baxter equation.

$$\begin{aligned}\mathcal{Z}&=\mathrm{Tr}_{c,a,A}\big\{\sigma_a^+\mathbb{R}_{\{a,A\},c}(\xi/\zeta_2)\mathbb{T}_{\{a,A\},R}(\xi)\mathbb{T}_{\{a,A\},L}(\xi,\alpha)\\&\qquad\times\mathbb{T}_{c,R}(\zeta_2)\xi^{\alpha-\mathbb{S}-\mathbb{S}_c}(q^{-2S_J-\sigma_c^3}Y_{L,c})\big\}\\&=\mathrm{Tr}_c\big\{\mathbb{T}_{c,R}(\zeta_2)\mathbf{k}_{L\sqcup c\sqcup R}(\xi,\alpha)(Y_{L,c})\big\}\\&\equiv\mathrm{Tr}_c\big\{\mathbb{T}_{c,R}(\zeta_2)\mathbf{k}_{L\sqcup c}(\xi,\alpha)(Y_{L,c})\big\}\mod q\text{-exact in }\xi\,,\end{aligned} \tag{3.65}$$

where we used the reduction (Lemma 3.5).

Comparing (3.64) and (3.65) we arrive at the statement of the Lemma for $\bar{\mathbf{c}}$. The case of $\mathbf{c}$ is treated similarly, but one has to be more careful with the contours of integration over ξ. □

REMARK 3.17. We can allow $Y_{L,c}$ to depend on ζ_2 holomorphically at $\zeta_2^2=1$.

THEOREM 3.18. *We have the commutation relations*

$$\begin{aligned}&[\mathbf{t}^*(\zeta_1),\mathbf{t}^*(\zeta_2)]=0\,,\\&[\mathbf{t}^*(\zeta_1),\mathbf{c}(\zeta_2)]=[\mathbf{t}^*(\zeta_1),\bar{\mathbf{c}}(\zeta_2)]=0\,,\\&[\mathbf{t}^*(\zeta_1),\mathbf{b}(\zeta_2)]=[\mathbf{t}^*(\zeta_1),\bar{\mathbf{b}}(\zeta_2)]=0\,.\end{aligned} \tag{3.66}$$

PROOF. The first relation is a trivial consequence of the Yang-Baxter equation.

To prove the second, take a semi-infinite interval $R=[m+1,\infty)$ in the relation (3.62) and set

$$Y_{L,c}=\mathbb{T}_{c,L}(\zeta_2,\alpha)(q^{\sigma_c^3}X_L)\,,$$

to obtain

$$\mathbf{c}(\zeta_1,\alpha)\mathbf{t}^*(\zeta_2,\alpha+1)(X_L)=\mathrm{Tr}_c\Big\{\mathbb{T}_{c,R}(\zeta_2)\mathbf{c}_{L\sqcup c}(\zeta_1,\alpha)\mathbb{T}_{c,L}(\zeta_2,\alpha)(q^{\sigma_c^3}X_L)\Big\}\,.$$

The right hand side is computed using

$$\begin{aligned}
&\mathbf{k}_{L\sqcup c}(\xi,\alpha)\mathbb{T}_{c,L}(\zeta_2,\alpha)(q^{\sigma^3_c}X_L)\\
&=\mathrm{Tr}_{a,A}\Big\{\sigma^+_a\mathbb{R}_{\{a,A\},c}(\xi/\zeta_2)\mathbb{T}_{\{a,A\},L}(\xi,\alpha)\zeta^{\alpha-\mathbb{S}-\mathbb{S}_c}q^{-2S_L-\sigma^3_c}\mathbb{T}_{c,L}(\zeta_2,\alpha)(q^{\sigma^3_c}X_L)\Big\}\\
&=\mathbb{T}_{c,L}(\zeta_2,\alpha)\mathrm{Tr}_{a,A}\Big\{\sigma^+_a\mathbb{T}_{\{a,A\},L}(\xi,\alpha)\mathbb{R}_{\{a,A\},c}(\xi/\zeta_2)\zeta^{\alpha-\mathbb{S}}(q^{-2S_L}X_L)\Big\}\\
&=\mathbb{T}_{c,L}(\zeta_2,\alpha)\mathbf{k}_L(\xi,\alpha)(X_L)\,,
\end{aligned}$$

where we use the Yang-Baxter equation, and the fact that $\mathbb{R}_{\{a,A\},c}(\xi/\zeta_2)$ collapses on an operator which acts as unity in $V_{\{a,A\}}\otimes V_c$. So we obtain

$$\mathbf{c}(\zeta_1,\alpha)\mathbf{t}^*(\zeta_2,\alpha+1)=\mathbf{t}^*(\zeta_2,\alpha)\mathbf{c}(\zeta_1,\alpha)\,,$$

which is the correct block version of the second relation in (3.66). The rest is similar. □

It has been said that $\mathbf{t}^*(\zeta)$ consists essentially of adjoint local integrals of motion. So, the commutation relations (3.66) should be considered as an evidence of consistency of our annihilation operators with the integrable structure.

3.5. Fermionic Creation Operators

Consider the formula (3.21)

$$\mathbf{k}_J(\zeta,\alpha)(Y_L)=\mathbf{k}_L(\zeta,\alpha)(Y_L)+\Delta_\zeta\mathbf{v}_J(\zeta,\alpha)(Y_L),$$

where $J=L\sqcup R$. It is tempting to introduce the q-primitive function

$$\mathbf{f}_J(\zeta,\alpha)=\Delta_\zeta^{-1}\mathbf{k}_J(\zeta,\alpha)\,, \tag{3.67}$$

for which

$$\mathbf{f}_J(\zeta,\alpha)(Y_L)=\mathbf{f}_L(\zeta,\alpha)(Y_L)+\mathbf{v}_J(\zeta,\alpha)(Y_L)\,. \tag{3.68}$$

There are several problems here. First, the primitive function is defined only up to quasi-constants (functions of $\zeta^{1/\nu}$). Second, as $\mathbf{k}_J(\zeta,\alpha)$ has singularities at $\zeta^2=1,q^{\pm2}$, we have to explain how we understand the primitive of a singular function. Actually these two problems are related. How to understand them?

Introduce

$$\mathbf{f}_{\mathrm{rat},J}(\zeta,\alpha)=\Delta_\zeta^{-1}\big(\mathbf{k}_J(\zeta,\alpha)-\bar{\mathbf{c}}_J(\zeta,\alpha)-\mathbf{c}_J(\zeta q,\alpha)-\mathbf{c}_J(\zeta q^{-1},\alpha)\big)\,. \tag{3.69}$$

The operator $\bar{\mathbf{c}}_J(\zeta,\alpha)$ is the singular part of $\mathbf{k}_J(\zeta,\alpha)$ on the divisor (f_J), the operator $\mathbf{c}_J(\zeta q,\alpha)+\mathbf{c}_J(\zeta q^{-1},\alpha)$ is the arithmetic mean of the singular parts of $\mathbf{k}_J(\zeta,\alpha)$ on the divisors $(\mathsf{f}_J)q^2$, $(\mathsf{f}_J)q^{-2}$. Hence, in the right hand side of (3.69), the q-primitive function can be chosen in the form $\zeta^\alpha F(\zeta^2)$, with $F(\zeta^2)$ having singularities only at 0, ∞ and (f_J). This requirement excludes the ambiguity of quasi-constants. Now that $\mathbf{f}_{\mathrm{rat},J}(\zeta,\alpha)$ is defined, we can fix $\mathbf{f}_J(\zeta,\alpha)$:

$$\begin{aligned}
\mathbf{f}_J(\zeta,\alpha)=\mathbf{f}_{\mathrm{rat},J}(\zeta,\alpha)+\frac{1}{2\pi i}\oint_\Gamma\Big(&\Delta_\zeta^{-1}\psi(\zeta/\xi,\alpha+\mathbb{S})\bar{\mathbf{c}}_J(\xi,\alpha)\\
&+\big(\Delta_\zeta^{-1}\psi(q\zeta/\xi,\alpha+\mathbb{S})+\Delta_\zeta^{-1}\psi(q^{-1}\zeta/\xi,\alpha+\mathbb{S})\big)\mathbf{c}_J(\xi,\alpha)\Big)\frac{d\xi^2}{\xi^2}\,,
\end{aligned} \tag{3.70}$$

where Γ goes around (f_J), and the singularities of $\Delta_\zeta^{-1}\psi(\zeta/\xi,\alpha+\mathbb{S})$ must be outside the contour Γ. Both $\mathbf{f}_J(\zeta,\alpha)$ and $\mathbf{f}_{\mathrm{rat},J}(\zeta,\alpha)$ will be useful for us. The reduction relation (3.68) is valid for both of them because $\mathbf{c}_J(\xi,\alpha), \overline{\mathbf{c}}_J(\xi,\alpha)$ satisfy the right reduction.

The q-primitive function $\Delta_\zeta^{-1}\psi(\zeta,\alpha)$ will be important, so let us make the definition precise in the physically relevant two cases: $0<q<1$ and $|q|=1$.

For $0<q<1$ we define for $|\operatorname{Re}\alpha|<2$

$$\Delta_\zeta^{-1}\psi(\zeta,\alpha) = \frac{1}{2}\,\zeta^\alpha \sum_{n=1}^{\infty}\Bigl(\frac{q^{(2+\alpha)(2n-1)}}{\zeta^{-2}-q^{4n-2}} + \frac{q^{(2-\alpha)(2n-1)}}{\zeta^{2}-q^{4n-2}}\Bigr). \tag{3.71}$$

In this case $\zeta^{-\alpha}\Delta_\zeta^{-1}\psi(\zeta,\alpha)$ is a meromorphic function of ζ^2 with essential singularities at $\zeta^2=0,\infty$.

For $|q|=1$ define $\Delta_\zeta^{-1}\psi(\zeta,\alpha)$ by the Mellin transform

$$\Delta_\zeta^{-1}\psi(\zeta,\alpha) = VP\int_C \zeta^{2ik}\coth\pi(k+\frac{i\alpha}{2})\frac{i}{4\sinh 2\pi\nu k}dk\,, \tag{3.72}$$

the contour C goes from $-\infty$ to ∞ through the points $k=0,-i\alpha/2$, and the principal value is taken with respect to both poles, and then continue analytically with respect to α. Contrary to the previous case $\zeta^{-\alpha}\Delta_\zeta^{-1}\psi(\zeta,\alpha)$ is not a single-valued function of ζ^2, but rather a meromorphic function of $\log\zeta$. It is bounded at $\log\zeta\to\pm\infty$, and its singularities closest to the real axis are the simple poles at $\log\zeta=\pm\pi i\nu$ with residues of opposite signs. The function (3.72) is symmetric with respect to the reflection

$$\alpha\to-\alpha,\quad \zeta\to\zeta^{-1}\,. \tag{3.73}$$

Our considerations up to Chapter 5 are algebraic, and they apply equally well to both cases $0<q<1$ and $|q|=1$. For the q-primitive function $\Delta_\zeta^{-1}\psi(\zeta,\alpha)$ we shall need only the symmetry (3.73), and the fact that the poles closest to the origin are situated at $\zeta=q^{\pm1}$.

Our goal is to construct a creation operator, i.e., an operator respecting the locality the same way as does the operator $\mathbf{t}^*(\zeta)$. Obviously $\mathbf{f}_J(\zeta,\alpha)$ and $\mathbf{f}_{\mathrm{rat},J}(\zeta,\alpha)$ are not what we need. The correct operator is defined by

$$\mathbf{b}_J^*(\zeta,\alpha) = \mathbf{f}_J(\zeta q,\alpha) + \mathbf{f}_J(\zeta q^{-1},\alpha) - \mathbf{t}_J^*(\zeta,\alpha)\mathbf{f}_J(\zeta,\alpha)\,. \tag{3.74}$$

We define also the operator $\mathbf{b}_{\mathrm{rat},J}^*(\zeta,\alpha)$ by a formula identical to (3.74), in which $\mathbf{f}_J(\zeta,\alpha)$ is replaced by $\mathbf{f}_{\mathrm{rat},J}^*(\zeta,\alpha)$. The formula (3.74) is a result of long experimental work; in spite of many attempts we have not been able to provide a scientific explanation for it. One reason is beauty: it looks a good idea to apply the Baxter operator to $\mathbf{f}_J(\zeta,\alpha)$. But the real reason is given by the following two lemmas which establish the right reduction relation for $\mathbf{b}_J^*(\zeta,\alpha)$. Since these lemmas are very important for our construction, we give rather detailed proofs.

LEMMA 3.19. *Consider as usual $J=L\sqcup R$. Then*

$$\mathbf{b}_J^*(\zeta,\alpha)(X_L) = \mathrm{Tr}_c\left\{\mathbb{T}_{c,R}(\zeta)\mathbf{g}_{c,L}(\zeta,\alpha)(X_L)\right\}, \tag{3.75}$$

where

$$\begin{aligned}\mathbf{g}_{c,L}(\zeta,\alpha) &= \frac{1}{2}\mathbf{f}_L(q\zeta,\alpha)+\frac{1}{2}\mathbf{f}_L(q^{-1}\zeta,\alpha)-\mathbb{T}_{c,L}(\zeta,\alpha)\mathbf{f}_L(\zeta,\alpha)+\mathbf{u}_{c,L}(\zeta,\alpha),\\ \mathbf{u}_{c,L}(\zeta,\alpha)(X_L) &= \mathrm{Tr}_{A,a}\left\{Y_{a,c,A}\mathbb{T}_{\{a,A\},L}(\zeta,\alpha)\zeta^{\alpha-\mathbb{S}}\left(q^{-2S_L}X_L\right)\right\},\end{aligned}$$

$$Y_{a,c,A} = -\frac{1}{2}\sigma_c^3\sigma_a^+ + \sigma_c^+\sigma_a^3 - \mathbf{a}_A\sigma_c^+\sigma_a^+\,. \tag{3.76}$$

PROOF. We decompose the left hand side of (3.75) as

$$\mathbf{b}^*_J(\zeta,\alpha)(X_L)=\mathbf{b}^*_L(\zeta,\alpha)(X_L)+I+II+III+IV,$$

where

$$\begin{aligned}
I&=\left(\mathbf{f}_J(q\zeta,\alpha)-\mathbf{f}_L(q\zeta,\alpha)\right)(X_L)\,,\\
II&=\left(\mathbf{f}_J(q^{-1}\zeta,\alpha)-\mathbf{f}_L(q^{-1}\zeta,\alpha)\right)(X_L)\,,\\
III&=-\mathbf{t}^*_J(\zeta,\alpha)\left(\mathbf{f}_J(\zeta,\alpha)-\mathbf{f}_L(\zeta,\alpha)\right)(X_L)\,,\\
IV&=-\left(\mathbf{t}^*_J(\zeta,\alpha)-\mathbf{t}^*_L(\zeta,\alpha)\right)\mathbf{f}_L(\zeta,\alpha)(X_L).
\end{aligned}$$

Note that

$$\mathrm{Tr}_c\left\{\mathbb{T}_{c,R}(\zeta)\left(\mathbf{g}_{c,L}(\zeta,\alpha)-\mathbf{u}_{c,L}(\zeta,\alpha)\right)(X_L)\right\}=\mathbf{b}^*_L(\zeta,\alpha)(X_L)+IV.$$

We want to show that

$$\mathrm{Tr}_c\left\{\mathbb{T}_{c,R}(\zeta)\mathbf{u}_c(\zeta,\alpha)(X_L)\right\}=I+II+III. \tag{3.77}$$

We use below the crossing symmetry relations intensively:

$$\begin{aligned}
\theta_R\big(T_{a,R}(\zeta)\big)&=\mathsf{f}_R(\zeta)\mathsf{f}_R(\zeta q^2)T_{a,R}(\zeta q)^{-1}\,,\\
\theta_R\big(T_{a,R}(\zeta)^{-1}\big)&=\frac{1}{\mathsf{f}_R(\zeta q)\mathsf{f}_R(\zeta q^{-1})}T_{a,R}(\zeta q^{-1})\,.
\end{aligned}$$

The main part of the proof consists in computing III. We use the reduction relation (3.68), the explicit formula (3.22) for $\mathbf{v}_J$, the trick of changing the order of multiplication by applying θ_R already familiar from the proof of Lemma 3.5, along with the fusion. We obtain

(3.78)

$$\begin{aligned}
III&=-\mathrm{Tr}_{a,A}\Big\{\mathbb{T}_{a,R}(\zeta)V_{A,R}(\zeta)\mathbb{T}_{a,L}(\zeta,\alpha)\mathbb{T}_{A,L}(\zeta,\alpha)\zeta^{\alpha-\mathbb{S}}(q^{-2S_L}X_L)\Big\}\\
&=\frac{\mathsf{f}_R(\zeta q^2)}{\mathsf{f}_R(\zeta q)\mathsf{f}_R(\zeta q^{-1})}\mathrm{Tr}_{a,A}\Big\{\theta_R\Big(T_{a,R}(\zeta q^{-1})C_{A,R}(\zeta)q^{S_R}T_{A,R}(\zeta q)^{-1}T_{a,R}(\zeta q)^{-1}\Big)\\
&\times\mathbb{T}_{a,L}(\zeta,\alpha)\mathbb{T}_{A,L}(\zeta,\alpha)\zeta^{\alpha-\mathbb{S}}(q^{-2S_L}X_L)\Big\}\\
&=\mathrm{Tr}_{a,A}\Big\{\theta_R(Z_{a,A,R})\mathbb{T}_{\{a,A\},R}(\zeta,\alpha)\zeta^{\alpha-\mathbb{S}}(q^{-2S_L}X_L)\Big\}
\end{aligned}$$

where

$$Z_{a,A,R}=\frac{\mathsf{f}_R(\zeta q^2)}{\mathsf{f}_R(\zeta q)\mathsf{f}_R(\zeta q^{-1})}F^{-1}_{a,A}T_{a,R}(\zeta q^{-1})C_{A,R}(\zeta)q^{S_R}F_{a,A}T_{\{a,A\},R}(\zeta q)^{-1}\,,$$

and we use (3.14). Let us write

$$T_{a,R}(\zeta)=\begin{pmatrix}a_R(\zeta)&b_R(\zeta)\\c_R(\zeta)&d_R(\zeta)\end{pmatrix}_a\,.$$

In what follows we shall drop the index a and work with 2 by 2 matrices with respect to this space.

We use several identities.

Fusion relation:

$$
\begin{pmatrix} 1 & \mathbf{a}_A \\ 0 & 1 \end{pmatrix} \begin{pmatrix} a_R(\zeta) & b_R(\zeta) \\ c_R(\zeta) & d_R(\zeta) \end{pmatrix} T_{A,R}(\zeta) \begin{pmatrix} 1 & -\mathbf{a}_A \\ 0 & 1 \end{pmatrix} = \begin{pmatrix} \mathsf{f}_R(\zeta) T_{A,R}(q\zeta) q^{-S_R} & 0 \\ C_{A,R}(\zeta) & \mathsf{f}_R(\zeta q) T_{A,R}(q^{-1}\zeta) q^{S_R} \end{pmatrix}, \tag{3.79}
$$

or equivalently,

$$
\begin{aligned}
\mathsf{f}_R(\zeta q)\mathsf{f}_R(\zeta q^{-1}) T_{A,R}(\zeta) &= \begin{pmatrix} d_R(q^{-1}\zeta) & -b_R(q^{-1}\zeta) \\ -c_R(q^{-1}\zeta) & a_R(q^{-1}\zeta) \end{pmatrix} \begin{pmatrix} 1 & -\mathbf{a}_A \\ 0 & 1 \end{pmatrix} \\
&\times \begin{pmatrix} \mathsf{f}_R(\zeta) T_{A,R}(q\zeta) q^{-S_R} & 0 \\ C_{A,R}(\zeta) & \mathsf{f}_R(\zeta q) T_{A,R}(q^{-1}\zeta) q^{S_R} \end{pmatrix} \begin{pmatrix} 1 & \mathbf{a}_A \\ 0 & 1 \end{pmatrix}.
\end{aligned} \tag{3.80}
$$

Crossing symmetry:

$$
\begin{pmatrix} d_R(q^{-1}\zeta) & -b_R(q^{-1}\zeta) \\ -c_R(q^{-1}\zeta) & a_R(q^{-1}\zeta) \end{pmatrix} \begin{pmatrix} a_R(\zeta) & b_R(\zeta) \\ c_R(\zeta) & d_R(\zeta) \end{pmatrix} = \mathsf{f}_R(\zeta q)\mathsf{f}_R(\zeta q^{-1}). \tag{3.81}
$$

We define $x^3_R(\zeta), x^\pm_R(\zeta)$ by

$$
\begin{aligned}
&\begin{pmatrix} a_R(q^{-1}\zeta) & b_R(q^{-1}\zeta) \\ c_R(q^{-1}\zeta) & d_R(q^{-1}\zeta) \end{pmatrix} \begin{pmatrix} d_R(\zeta) & -b_R(\zeta) \\ -c_R(\zeta) & a_R(\zeta) \end{pmatrix} \\
&\quad = \mathsf{f}_R(\zeta q)\mathsf{f}_R(\zeta q^{-1}) \Big(1 + x^3_R(\zeta)\sigma^3 + x^+_R(\zeta)\sigma^+ + x^-_R(\zeta)\sigma^- \Big).
\end{aligned} \tag{3.82}
$$

Let us proceed by matrix elements.

$(Z_{a,A,R})_{1,2}$: From the $(2,1)$ element in (3.79) we obtain $C_{A,R}(\zeta) = c_R(\zeta) T_{A,R}(\zeta)$. Using this we obtain

$$
\begin{aligned}
(Z_{a,A,R})_{1,2} &= \frac{1}{\mathsf{f}_R(\zeta q)\mathsf{f}_R(\zeta q^{-1})} \Big((b_R(q^{-1}\zeta) + \mathbf{a}_A d_R(q^{-1}\zeta)) c_R(\zeta) \\
&- (a_R(q^{-1}\zeta) + \mathbf{a}_A c_R(q^{-1}\zeta)) c_R(\zeta) T_{A,R}(\zeta) \mathbf{a}_A T_{A,R}(\zeta)^{-1} \Big).
\end{aligned}
$$

From the $(2,1)$ element in (3.81) we obtain $a_R(q^{-1}\zeta)c_R(\zeta) = c_R(q^{-1}\zeta)a_R(\zeta)$. From the $(1,2)$ element in (3.79) we obtain

$$
(a_R(\zeta) + \mathbf{a}_A c_R(\zeta))\, T_{A,R}(\zeta) \mathbf{a}_A T_{A,R}(\zeta)^{-1} = b_R(\zeta) + \mathbf{a}_A d_R(\zeta).
$$

Using these equalities we have

$$
\begin{aligned}
(Z_{a,A,R})_{1,2} &= \frac{1}{\mathsf{f}_R(\zeta q)\mathsf{f}_R(\zeta q^{-1})} \Big(b_R(q^{-1}\zeta) c_R(\zeta) - c_R(q^{-1}\zeta) b_R(\zeta) \\
&+ \mathbf{a}_A \left(d_R(q^{-1}\zeta) c_R(\zeta) - c_R(q^{-1}\zeta) d_R(\zeta) \right) \Big) \\
&= -x^3_R(\zeta) - \mathbf{a}_A x^-_R(\zeta).
\end{aligned}
$$

$(Z_{a,A,R})_{2,2}$: The calculation is similar, but we use the (2,2) element in (3.79). We obtain

$$
(Z_{a,A,R})_{2,2} = -x^-_R(\zeta) + \frac{1}{\mathsf{f}_R(\zeta q^{-1})} C_{A,R}(q^{-1}\zeta) q^{S_R} T_{A,R}(\zeta)^{-1}.
$$

$(Z_{a,A,R})_{1,1}$: Reducing a part of the computation to that for $(Z_{a,A,R})_{1,2}$, we obtain

$$(Z_{a,A,R})_{1,1} = \frac{\mathsf{f}_R(\zeta q^2)}{\mathsf{f}_R(\zeta q)^2 \mathsf{f}(\zeta q^{-1})} \left(a_R(q^{-1}\zeta) + \mathbf{a}_A c_R(q^{-1}\zeta)\right) C_{A,R}(\zeta) q^{2S_R} T_{A,R}(q^2\zeta)^{-1}$$
$$+ \frac{1}{\mathsf{f}_R(\zeta q)} \left(x_R^3(\zeta) + \mathbf{a}_A x_R^-(\zeta)\right) C_{A,R}(q\zeta) q^{S_R} T_{A,R}(q^2\zeta)^{-1}.$$

The $(2,1)$ element of (3.80) can be rewritten as

$$\left(a_R(q^{-1}\zeta) + \mathbf{a}_A c_R(q^{-1}\zeta)\right) C_{A,R}(\zeta) = \mathsf{f}_R(\zeta) c_R(q^{-1}\zeta) T_{A,R}(q\zeta) q^{-S_R}.$$

Multiply $\left(c_R(q^{-1}\zeta), d_R(q^{-1}\zeta)\right)$ from the left to (3.80) with ζ replaced by $q\zeta$. Use (3.82) and take the first component. The result is

$$\frac{\mathsf{f}_R(\zeta)\mathsf{f}_R(\zeta q^2)}{\mathsf{f}_R(\zeta q^{-1})\mathsf{f}_R(\zeta q)} c_R(q^{-1}\zeta) T_{A,R}(q\zeta)$$
$$= x_R^-(\zeta)\left(\mathsf{f}_R(\zeta q) T_{A,R}(q^2\zeta) q^{-S_R} - \mathbf{a}_A C_{A,R}(q\zeta)\right) + (1 - x_R^3(\zeta)) C_{A,R}(q\zeta).$$

Combining all these we obtain

$$(Z_{a,A,R})_{1,1} = x_R^-(\zeta) + \frac{1}{\mathsf{f}_R(\zeta q)} C_{A,R}(q\zeta) q^{S_R} T_{A,R}(q^2\zeta)^{-1}.$$

Now the proof is easy. If we substitute

(3.83)

$$Z_{a,A,R} = \begin{pmatrix} x_R^-(\zeta) & -x_R^3(\zeta) - \mathbf{a}_A x_R^-(\zeta) \\ * & -x_R^-(\zeta) \end{pmatrix}_a$$
$$+ \begin{pmatrix} \frac{1}{\mathsf{f}_R(\zeta q)} C_{A,R}(q\zeta) q^{S_R} T_{A,R}(q^2\zeta)^{-1} & 0 \\ 0 & \frac{1}{\mathsf{f}_R(\zeta q^{-1})} C_{A,R}(q^{-1}\zeta) q^{S_R} T_{A,R}(\zeta)^{-1} \end{pmatrix}_a,$$

we obtain III. The second, diagonal, term cancels with $I + II$. The last trick is introducing another auxiliary space in order to write the formula without using θ_R. In fact, we have

$$\mathrm{Tr}_c \left\{ \mathbb{T}_{c,R}(\zeta) \mathbf{u}_c(\zeta,\alpha)(X_L) \right\} = \mathrm{Tr}_{A,a,c} \theta_R (1 + x_R^3(\zeta)\sigma^3 + x_R^+(\zeta)\sigma^+ + x_R^-(\zeta)\sigma^-)$$
$$\times \left(-\frac{1}{2}\sigma_c^3 \sigma_a^+ + \sigma_c^+(\sigma_a^3 - \mathbf{a}_a \sigma_a^+) \right) \mathbb{T}_{\{a,A\},L}(\zeta,\alpha) \zeta^{\alpha - \mathbb{S}} (q^{-2S_L} X_L)$$
$$= \mathrm{Tr}_{A,a} \theta_R \left(x_R^-(\zeta)\sigma_a^3 - (x_R^3(\zeta) + \mathbf{a}_A x_R^-(\zeta))\sigma_a^+ \right) \mathbb{T}_{\{a,A\},L}(\zeta,\alpha) \zeta^{\alpha - \mathbb{S}} (q^{-2S_L} X_L).$$

The proof is over. □

This Lemma shows a deep resemblance between $\mathbf{b}_J^*(\zeta,\alpha)$ and $\mathbf{t}_J^*(\zeta,\alpha)$, but in order to be able to use the Taylor series we have to show that $\mathbf{g}_{c,J}(\zeta,\alpha)(X_J)$ is regular on (f_J); it is not at all obvious.

LEMMA 3.20. *The function* $\mathbf{g}_{c,J}(\zeta,\alpha)(X_J)$ *is regular on* (f_J).

PROOF. It is convenient to begin with the operator $\mathbf{b}_{\mathrm{rat},J}^*(\zeta,\alpha)$. It has the reduction property (3.75) with $\mathbf{g}_{\mathrm{rat},c,J}(\zeta,\alpha)(X_J)$, which is obtained from $\mathbf{g}_{c,J}(\zeta,\alpha)$ (X_J) by replacing all $\mathbf{f}_J(\zeta,\alpha)$ by $\mathbf{f}_{\mathrm{rat},J}(\zeta,\alpha)$. It suffices to consider the inhomogeneous case with inhomogeneity parameters ξ_j.

There are two dangerous terms, $\mathbb{T}_c(\zeta,\alpha)\mathbf{f}(\zeta,\alpha)$ and $\mathbf{u}_c(\zeta,\alpha)$. It is rather obvious that the remaining terms are regular on (f_J). We want to show that the residues vanish. Let $J=[k,m]$. Due to the R-matrix symmetry, it is sufficient to consider the residue at $\zeta^2=\xi_m^2$. So we want to show

$$\operatorname*{res}_{\zeta=\xi_m} \mathbb{T}_{c,J}(\zeta,\alpha)\mathbf{f}_{\mathrm{rat},J}(\zeta,\alpha)(X_J)\frac{d\zeta^2}{\zeta^2} = \operatorname*{res}_{\zeta=\xi_m} \mathbf{u}_{\mathrm{rat},c,J}(\zeta,\alpha)(X_J)\frac{d\zeta^2}{\zeta^2}\,. \tag{3.84}$$

Form Lemma 3.10 and the definition (3.69) it follows that

$$\operatorname*{res}_{\zeta^2=\xi_m^2} \mathbf{f}_{\mathrm{rat},J}(\zeta,\alpha)(X_J)\frac{d\zeta^2}{\zeta^2} = -\frac{1}{2}\operatorname*{res}_{\zeta^2=\xi_m^2}\left[\sigma_m^+\,,\ \mathbf{q}_J(\zeta,\alpha)(q^{-2S_J}X_J)\right]\frac{d\zeta^2}{\zeta^2}\,.$$

Noting that $\mathbb{T}_{c,J}(\xi_m,\alpha)=\mathbb{P}_{c,m}\mathbb{T}_{c,[k,m-1]}(\xi_m,\alpha)$, we can evaluate the left hand side of (3.84) as follows.

$$\begin{aligned}
(LHS) &= -\frac{1}{2}\mathbb{T}_{c,J}(\xi_m,\alpha)\operatorname*{res}_{\zeta^2=\xi_m^2}\left[\sigma_m^+\,,\ \mathbf{q}_J(\zeta,\alpha)(q^{-2S_J}X_J)\right]\frac{d\zeta^2}{\zeta^2}\\
&= -\frac{1}{2}\operatorname*{res}_{\zeta^2=\xi_m^2}\left[\sigma_c^+\,,\ \mathbb{T}_{c,J}(\zeta,\alpha)\mathbf{q}_J(\zeta,\alpha)(q^{-2S_J}X_J)\right]\frac{d\zeta^2}{\zeta^2}\\
&= -\frac{1}{2}\operatorname*{res}_{\zeta^2=\xi_m^2}\mathrm{Tr}_A\left[\sigma_c^+, F_{a,A}\mathbb{T}_{\{c,A\},J}(\zeta,\alpha)F_{a,A}^{-1}\zeta^{\alpha-\mathbb{S}}(q^{-2S_J}X_J)\right]\frac{d\zeta^2}{\zeta^2}\\
&= -\operatorname*{res}_{\zeta^2=\xi_m^2}\mathrm{Tr}_A\Big\{(\frac{1}{2}\sigma_c^3+\mathbf{a}_A\sigma_c^+)\mathbb{C}_A(\zeta,\alpha)\zeta^{\alpha-\mathbb{S}}(q^{-2S_J}X_J)\Big\}\frac{d\zeta^2}{\zeta^2}\\
&= (RHS)\,.
\end{aligned}$$

In the last equality we restore the term $\sigma_c^+\sigma_a^3$, because it does not contribute to the residue.

It remains to check that, when passing from $\mathbf{f}_{\mathrm{rat},J}(\zeta,\alpha)$ to $\mathbf{f}_J(\zeta,\alpha)$ by (3.109), we do not add singularities at (f_J). In the inhomogeneous case we compute the integral by residues, so we have to check that $\Delta_\zeta^{-1}\psi(\alpha,\zeta/\xi_j)$ and $\Delta_\zeta^{-1}\psi(\alpha,\zeta/\xi_j q)+\Delta_\zeta^{-1}\psi(\alpha,\zeta/\xi_j q^{-1})$ do not have poles at $\zeta^2=\xi_j^2$. For the first one it is trivial (3.72). For the second one it follows from the fact that the residue of $\Delta_\zeta^{-1}\psi(\zeta,\alpha)$ at $\zeta=q^{\pm1}$ have opposite signs. □

Now the situation is quite similar to the case of the operator $\mathbf{t}^*(\zeta,\alpha)$; we can develop $\mathbf{b}_J^*(\zeta,\alpha)(X_L)$ in power series in ζ^2-1, and the coefficients stabilise for sufficiently large J. This allows us to define an operator $\mathbf{b}^*(\zeta,\alpha)$ acting from $\mathcal{W}_{\alpha+1,s-1}$ to $\mathcal{W}_{\alpha,s}$ as follows.

DEFINITION 3.21. Consider an operator $q^{2(\alpha+1)S(k-1)}X_{[k,l]}\in W_{\alpha+1,s-1}$. We define the action of the operator $\mathbf{b}^*(\zeta,\alpha)$ by

$$\mathbf{b}^*(\zeta,\alpha)(q^{2(\alpha+1)S(k-1)}X_{[k,l]}) = q^{2\alpha S(k-1)}\mathbf{b}^*_{[k,\infty)}(\zeta,\alpha)(X_{[k,l]})\,. \tag{3.85}$$

This definition is consistent due to the left reduction relation. With this definition, the operator $\mathbf{b}^*(\zeta,\alpha)$ is given by a power series

$$\mathbf{b}^*(\zeta,\alpha)=\zeta^{\alpha+\mathbb{S}+2}\sum_{p=1}^{\infty}(\zeta^2-1)^{p-1}\mathbf{b}_p^*(\alpha)\,,$$

where the multiplier $\zeta^{\alpha+\mathbb{S}+2}$ is introduced for future convenience. It is easy to see that

$$\mathrm{lenght}\big(\mathbf{b}_p^*(\alpha)(q^{2(\alpha+1)S(0)}\mathcal{O})\big)\le \mathrm{lenght}(q^{2(\alpha+1)S(0)}\mathcal{O})+p\,. \tag{3.86}$$

We define further an operator $\mathbf{c}^*(\zeta,\alpha):\mathcal{W}_{\alpha-1,s+1}\to\mathcal{W}_{\alpha,s}$ by

$$\mathbf{c}^*(\zeta,\alpha)=-\phi(\mathbf{b}^*)(\zeta,\alpha)\,. \tag{3.87}$$

Notice the sign in (3.87). Certainly it is possible to define this operator directly starting with the representation W_ζ^- as the auxiliary space.

Collecting the blocks $\mathbf{b}^*(\zeta,\alpha),\mathbf{c}^*(\zeta,\alpha)$ we define the operators $\mathbf{b}^*(\zeta),\mathbf{c}^*(\zeta)$ on the entire space $\mathcal{W}$. Acting with several such operators we indicate the target space by writing, for example, as

$$\mathbf{b}^*(\zeta_5,\alpha)\mathbf{c}^*(\zeta_4)\mathbf{t}^*(\zeta_3)\mathbf{c}^*(\zeta_2)\mathbf{b}^*(\zeta_1)\,.$$

By fixing the leftmost block $\mathbf{b}^*(\zeta_5,\alpha)$ we automatically fix the blocks for the rest. Writing in full we have

$$\begin{aligned}&\mathbf{b}^*(\zeta_5,\alpha)\mathbf{c}^*(\zeta_4)\mathbf{t}^*(\zeta_3)\mathbf{c}^*(\zeta_2)\mathbf{b}^*(\zeta_1)\\&\quad=\mathbf{b}^*(\zeta_5,\alpha)\mathbf{c}^*(\zeta_4,\alpha+1)\mathbf{t}^*(\zeta_3,\alpha)\mathbf{c}^*(\zeta_2,\alpha)\mathbf{b}^*(\zeta_1,\alpha-1)\,.\end{aligned} \tag{3.88}$$

3.6. Homogeneous Versus Inhomogeneous Cases: Russian Doll Construction

Up to now we defined creation operators for an infinite homogeneous chain. Let us consider now an inhomogeneous model defined on a finite interval $[1,m]$ with inhomogeneities $\xi_1,\dots,\xi_m$. We denote $\mathfrak{H}_{[1,m]}=V_{\xi_1}\otimes\cdots\otimes V_{\xi_m}$. The action of the annihilation operators $\mathbf{b}(\zeta,\alpha)$, $\mathbf{c}(\zeta,\alpha)$ on $\mathfrak{H}_{[1,m]}$ is clear. What to do with the creation operators? As in the homogeneous case, let us begin with $\mathbf{b}^*_{[1,m]}(\zeta,\alpha)$. In the homogeneous case, the trick was to use the Taylor series in ζ^2-1. In the inhomogeneous case, this is replaced by evaluation of the operators at the inhomogeneities $\zeta=\xi_j$. This is not always possible because of singularities. Nevertheless, according to Lemma 3.19 and Lemma 3.20, the singularity at $\zeta=\xi_j$ is absent when $\mathbf{b}^*_{[1,m]}(\zeta,\alpha)$ acts on an operator located on the interval $[1,j-1]$. In this case we have

$$\begin{aligned}&\mathbf{b}^*_{[1,m]}(\xi_j,\alpha)(X_{[1,j-1]})\\&\quad=\mathrm{Tr}_c\Bigl\{\mathbb{R}_{c,m}(\xi_j/\xi_m)\cdots\mathbb{R}_{c,j+1}(\xi_j/\xi_{j+1})\mathbb{P}_{c,j}\mathbf{g}_{c,[1,j-1]}(\xi_j,\alpha)(X_{[1,j-1]})\Bigr\}\\&\quad=2\mathbf{g}_{j,[1,j-1]}(\xi_j,\alpha)(X_{[1,j-1]})\,.\end{aligned}$$

Note that the same computation gives

$$\mathbf{g}_{j,[1,j-1]}(\xi_j,\alpha)(X_{[1,j-2]})=\mathbb{R}_{j,j-1}(\xi_j/\xi_{j-1})\mathbf{g}_{j,[1,j-2]}(\xi_j,\alpha)(X_{[1,j-2]})\,. \tag{3.89}$$

In what follows we abbreviate $\mathbf{g}_{j,[1,j-1]}(\xi_j,\alpha)$ to $\mathbf{g}_j(\xi_j,\alpha)$. Since we want to use $\mathbf{b}^*$ and $\mathbf{t}^*$ as well, we introduce the symbols

(3.90)
$$\mathbf{g}_j^+(\xi_j,\alpha)=\mathbf{g}_j(\xi_j,\alpha)\,,\quad \mathbf{g}_j^-(\xi_j,\alpha)=\phi(\mathbf{g}_j)(\xi_j,\alpha)\,,\quad \mathbf{g}_j^0(\xi_j,\alpha)=\mathbb{T}_{j,[1,j-1]}(\xi_j,\alpha)\,.$$

Acting with several operators we follow the convention (3.88). We create operators from the unit operator on the interval by acting successively

$$\mathbf{g}_{j_p}^{\epsilon_p}(\xi_{j_p},\alpha)\mathbf{g}_{j_{p-1}}^{\epsilon_{p-1}}(\xi_{j_{p-1}})\cdots\mathbf{g}_{j_1}^{\epsilon_1}(\xi_{j_1})(I)\,,\quad j_p>j_{p-1}>\cdots>j_1\,,\quad \epsilon_k=\pm,0\,. \tag{3.91}$$

To be precise, $\mathbf{g}_{j_k}^{\epsilon_k}(\xi_{j_k})$ with $p > k \geq 1$ in the product (3.91) is understood as $\mathbf{g}_{j_k}^{\epsilon_k}(\xi_{j_k}, \alpha + \sum_{m=k+1}^{p} \epsilon_m)$. It is convenient to introduce also the identity operator

$$\mathbf{g}_j^{\bar{0}}(\xi_j, \alpha) = \mathbb{I}_{[1,j-1]}\,, \tag{3.92}$$

and to write (3.91) without gaps,

$$\mathbf{g}_m^{\epsilon_m}(\xi_m, \alpha)\mathbf{g}_{m-1}^{\epsilon_{m-1}}(\xi_{m-1}) \cdots \mathbf{g}_1^{\epsilon_1}(\xi_1)(I)\,, \quad \epsilon_k = \pm, 0, \bar{0}\,. \tag{3.93}$$

We shall see later that this is a complete set of operators (see Lemma 3.29 below). For an obvious reason we call this the Russian doll construction.

The domain of definition of the $\mathbf{g}_j^{\epsilon}(\xi_j)$ is not the whole space $\mathfrak{H}_{[1,m]}$. This does not look very democratic: the domain of definition of $\mathbf{g}_m^{\epsilon}(\xi_m)$ is $\mathfrak{H}_m = \mathfrak{H}_{[1,m-1]}$, which seems to be larger than the one for the others. This is certainly not the case because of R-matrix symmetry; the actual domain of definition of $\mathbf{g}_j^{\epsilon}(\xi_j)$ is

$$\mathfrak{H}_j = \mathbb{T}_{j,[j+1,m]}(\xi_j)\mathfrak{H}_j^{(0)}\,, \tag{3.94}$$

where $\mathfrak{H}_j^{(0)}$ is the space of operators which acts as the identity on the j-th component.

Now we come to a key relation between homogeneous and inhomogeneous cases. To match the notation with $\mathbf{g}^{\epsilon}(\zeta)$ we set

$$\mathbf{x}^{+*}(\zeta) = \mathbf{b}^*(\zeta), \quad \mathbf{x}^{-*}(\zeta) = \mathbf{c}^*(\zeta)\,, \quad \mathbf{x}^{0*}(\zeta) = \mathbf{t}^*(\zeta)\,. \tag{3.95}$$

THEOREM 3.22. *Take a set* $\epsilon_1, \ldots, \epsilon_n$, $\epsilon_j = \pm, 0$. *Fix* α, *set* $\alpha_0 = \alpha + \sum_{j=1}^{n} \epsilon_j$, *and consider a primary field* $q^{2\alpha_0 S(0)}$. *Then*

$$\begin{aligned} &\mathbf{x}^{\epsilon_n *}(\zeta_n, \alpha)\mathbf{x}^{\epsilon_{n-1} *}(\zeta_{n-1}) \ldots \mathbf{x}^{\epsilon_1 *}(\zeta_1)(q^{2\alpha_0 S(0)}) \\ &\qquad = q^{2\alpha S(0)} \mathrm{Tr}_{c_1,\ldots,c_n}\Big\{\mathbb{T}_{c_1,[1,\infty)}(\zeta_1) \cdots \mathbb{T}_{c_{n-1},[1,\infty)}(\zeta_{n-1})\mathbb{T}_{c_n,[1,\infty)}(\zeta_n) \\ &\qquad\qquad \times \mathbf{g}_{c_n}^{\epsilon_n}(\zeta_n, \alpha)\mathbf{g}_{c_{n-1}}^{\epsilon_{n-1}}(\zeta_{n-1}) \cdots \mathbf{g}_{c_1}^{\epsilon_1}(\zeta_1)(I)\Big\}. \end{aligned} \tag{3.96}$$

Operators $\mathbf{g}_{c_j}^{\epsilon_j}(\zeta_j)$ *in the right hand side are those for the inhomogeneous chain* $c_1, \ldots, c_n$ *with inhomogenieties* $\zeta_1, \ldots, \zeta_n$.

PROOF. We begin with the following statement

$$\begin{aligned} &\mathbf{x}_{[1,\infty)}^{\epsilon_2 *}(\zeta_2, \alpha)\mathbf{x}_{[1,\infty)}^{\epsilon_1 *}(\zeta_1, \alpha + \epsilon_2)(X_{[1,m]}) \\ &= \mathrm{Tr}_{c_1,c_2}\Big\{\mathbb{T}_{c_1,[m+1,\infty)}(\zeta_1)\mathbb{T}_{c_2,[m+1,\infty)}(\zeta_2)\mathbf{g}_{c_2,[1,m]\sqcup c_1}^{\epsilon_2}(\zeta_2, \alpha)\mathbf{g}_{c_1,[1,m]}^{\epsilon_1}(\zeta_1)(X_{[1,m]})\Big\}. \end{aligned} \tag{3.97}$$

In order to prove this we start with the finite interval $[1, m+2n]$, and use the entire collection of our tricks. First,

$$\begin{aligned} \mathbf{x}_{[1,m+2n]}^{\epsilon_1 *}(\zeta_1, \alpha + \epsilon_2)(X_{[1,m]}) &= \mathrm{Tr}_{c_1}\Big\{\mathbb{T}_{c_1,[m+1,m+2n]}(\zeta_1)\mathbf{g}_{c_1,[1,m]}^{\epsilon_1}(\zeta_1, \alpha + \epsilon_2)(X_{[1,m]})\Big\} \\ &\equiv \sum_{p=1}^{n-1} (\zeta_1^2 - 1)^{p-1} Y_{[1,m+p]} \qquad \mod \ (\zeta_1^2 - 1)^n\,, \end{aligned}$$

where the $Y_{[1,m+p]}$ are some operators located on the indicated intervals. Now,

$$\begin{aligned}(3.98)\quad & \mathbf{x}^{\epsilon_2 *}_{[1,m+2n]}(\zeta_2,\alpha)(Y_{[1,m+p]}) \\ &= \mathrm{Tr}_{c_2}\Big\{\mathbb{T}_{c_2,[m+p+1,m+2n.]}(\zeta_2)\mathbf{g}^{\epsilon_1}_{c_2,[1,m+p]}(\zeta_1,\alpha)(Y_{[1,m+p]})\Big\} \\ &\equiv \mathrm{Tr}_{c_2}\Big\{\mathbb{R}_{c_2,c_1}(\zeta_2/\zeta_1)\mathbb{T}_{c_2,[m+p+1,m+2n]}(\zeta_2)\mathbf{g}^{\epsilon_1}_{c_2,[1,m+p]}(\zeta_1,\alpha)(Y_{[1,m+p]})\Big\} \\ &\qquad\qquad \mathrm{mod}\quad (\zeta_2^2-1)^{n-1}.\end{aligned}$$

Inserting $\mathbb{R}_{c_2,c_1}(\zeta_2/\zeta_1)$ we follow the logic of Remark 3.15. The trace of this matrix with respect to c_2 equals 2 because the argument does not contain V_{c_1}. Now the proof is simple. In the right hand side of (3.98) we have nothing but $\mathbf{x}^{\epsilon_2 *}_{[1,m+2n]\sqcup c_1}(\zeta_2,\alpha)(Y_{[1,m+p]})$, so modulo $(\zeta_1^2-1)^{n-1}(\zeta_2^2-1)^{n-1}$ we have

$$\begin{aligned}&\mathbf{x}^{\epsilon_2 *}_{[1,m+2n]}(\zeta_2,\alpha)\mathbf{x}^{\epsilon_1 *}_{[1,m+2n]}(\zeta_1,\alpha+\epsilon_2)(X_{[1,m]}) \\ &\equiv \mathrm{Tr}_{c_1}\Big\{\mathbf{x}^{\epsilon_2 *}_{[1,m+2n]\sqcup c_1}(\zeta_2,\alpha)\mathbb{T}_{c_1,[m+1,m+2n]}(\zeta_1)\mathbf{g}^{\epsilon_1}_{c_1,[1,m]}(\zeta_1,\alpha+\epsilon_2)(X_{[1,m]})\Big\} \\ &= \mathrm{Tr}_{c_1}\Big\{\mathbb{T}_{c_1,[m+1,m+2n]}(\zeta_1)\mathbf{x}^{\epsilon_2 *}_{[1,m]\sqcup c_1\sqcup[m+1,m+2n]}(\zeta_2,\alpha)\mathbf{g}^{\epsilon_1}_{c_1,[1,m]}(\zeta_1,\alpha+\epsilon_2)(X_{[1,m]})\Big\} \\ &= \mathrm{Tr}_{c_1,c_2}\Big\{\mathbb{T}_{c_1,[m+1,m+2n]}(\zeta_1)\mathbb{T}_{c_2,[m+1,m+2n]}(\zeta_2) \\ &\qquad\qquad \times \mathbf{g}^{\epsilon_2}_{c_2,[1,m]\sqcup c_1}(\zeta_2,\alpha)\mathbf{g}^{\epsilon_1}_{c_1,[1,m]}(\zeta_1,\alpha+\epsilon_2)(X_{[1,m]})\Big\}.\end{aligned}$$

To prove the statement of the theorem one applies induction. □

The formula (3.96) is called dressed Russian doll. It allows us to reduce complicated problems with the infinite homogeneous case to a much simpler, finite inhomogeneous case. Due to Lemma 3.16 the annihilation operators act on the dressed Russian doll in the most natural way:

(3.99)

$$\begin{aligned}&q^{-2\alpha_0 S(0)}\mathbf{c}(\zeta,\alpha)\mathbf{x}^{\epsilon_n *}(\zeta_n)\cdots\mathbf{x}^{\epsilon_1 *}(\zeta_1)(q^{2\alpha S(0)}) \\ &= \mathrm{Tr}_{c_1,\dots,c_n}\Big\{\mathbb{T}_{c_1[1,\infty)}(\zeta_1)\cdots\mathbb{T}_{c_n,[1,\infty)}(\zeta_n)\mathbf{c}_{[c_1,\dots,c_n]}(\zeta,\alpha)\mathbf{g}^{\epsilon_n}_{c_n}(\zeta_n)\cdots\mathbf{g}^{\epsilon_1}_{c_1}(\zeta_1)(I)\Big\},\end{aligned}$$

where $\alpha_0=\alpha+1+\sum_{j=1}^n\epsilon_j$.

Let us give an application of the dressed Russian doll.

THEOREM 3.23. *The operator* $\bar{\mathbf{c}}$ *is obtained from* $\mathbf{c}$ *and* $\mathbf{t}^*$*:*

$$(3.100)\qquad \bar{\mathbf{c}}(\zeta,\alpha) = -\frac{1}{2\pi i}\oint_\Gamma \psi(\zeta/\xi,\alpha+\mathbb{S})\mathbf{t}^*(\xi,\alpha)\mathbf{c}(\xi,\alpha)\frac{d\xi^2}{\xi^2},$$

where Γ *goes around* $\xi^2=1$*, and* ζ^2 *is outside the contour.*

PROOF. The dressed Russian doll reduces the calculation into the inhomogeneous case. Consider an inhomogeneous chain with inhomogeneities $\xi_1,\dots,\xi_m$. The integral going now around $(\xi_1^2,\dots,\xi_m^2)$ is computed by residues

(3.101)

$$\frac{1}{2\pi i}\oint_\Gamma \psi(\zeta/\xi,\alpha+\mathbb{S})\mathbf{t}^*(\xi,\alpha)\mathbf{c}(\xi,\alpha)\frac{d\xi^2}{\xi^2} = \sum_{j=1}^m \psi(\zeta/\xi_j,\alpha+\mathbb{S})\mathbf{t}^*(\xi_j,\alpha)\mathbf{c}_j(\alpha),$$

where $\mathbf{c}_j(\alpha)$ is defined in (3.49). We have to make sure that the expression in the right hand side makes sense, which requires that $\mathbf{c}_j(\alpha)(X_{[1,m]})$ belongs to the domain of definition of $\mathbf{g}^0(\xi_j,\alpha)$. As usual, by symmetry it is sufficient to consider $j=m$. Using the definition of $\mathbf{c}$ (3.39) and the formulas for residues (3.10) we obtain

(3.102)
$$\operatorname*{res}_{\zeta^2=\xi_m^2}\mathbf{c}(\zeta,\alpha)(X_{[1,m]})\frac{d\zeta^2}{\zeta^2}=-\frac{1}{2}\operatorname*{res}_{\zeta^2=\xi_m^2}\left[\mathbf{q}_{[1,m]}(\zeta,\alpha)(q^{-2S_{[1,m]}}X_{[1,m]}),\sigma_m^+\right]_+\frac{d\zeta^2}{\zeta^2}\,.$$

Denote

$$Z_{A,[1,m]}(\zeta)=\mathbb{T}_{A,[1,m-1]}(\zeta,\alpha)\zeta^{\alpha-\mathbb{S}}(q^{-2S_{[1,m]}}X_{[1,m]})\,.$$

Recalling the formulas for the q-oscillator R-matrices (2.53) we find:

$$\begin{aligned}R_{A,m}(1)&=\begin{pmatrix}-\mathbf{a}_A\\1\end{pmatrix}_m\left(-\mathbf{a}_A^*,1\right)_m q^{-D_A\sigma_m^3}\,,\\ \operatorname*{res}_{\zeta^2=1}R_{A,m}(\zeta)^{-1}\frac{d\zeta^2}{\zeta^2}&=-q^{D_A\sigma_m^3}\begin{pmatrix}1\\\mathbf{a}_A^*\end{pmatrix}_m\left(1,\mathbf{a}_A\right)_m\,,\end{aligned}\tag{3.103}$$

hence

$$\begin{aligned}&\operatorname*{res}_{\zeta^2=\xi_m^2}\mathbf{c}(\zeta,\alpha)(X_{[1,m]})\frac{d\zeta^2}{\zeta^2}\\&=\frac{1}{2}\mathrm{Tr}_A\Big\{\begin{pmatrix}1\\0\end{pmatrix}_m\left(-\mathbf{a}_A^*,1\right)_m q^{-D_A\sigma_a^3}Z_{A,[1,m]}(\xi_m)q^{D_A\sigma_a^3}\begin{pmatrix}1\\\mathbf{a}_A^*\end{pmatrix}_m\left(1,\mathbf{a}_A\right)_m\\&+\begin{pmatrix}-\mathbf{a}_A\\1\end{pmatrix}_m\left(-\mathbf{a}_A^*,1\right)_m q^{-D_A\sigma_a^3}Z_{A,[1,m]}(\xi_m)q^{D_A\sigma_m^3}\begin{pmatrix}1\\\mathbf{a}_A^*\end{pmatrix}_m\left(0,1\right)_m\Big\}\\&=\frac{1}{2}\mathrm{Tr}_A\Big\{\left(-\mathbf{a}_A^*,1\right)_m q^{-D_A\sigma_a^3}Z_{A,[1,m]}(\xi_m)q^{D_A\sigma_a^3}\begin{pmatrix}1\\\mathbf{a}_A^*\end{pmatrix}_m\Big\}\,,\end{aligned}\tag{3.104}$$

which is a unit matrix in the m-th space. Notice a nice cancelation of terms containing $\mathbf{a}_A$. So we proved that $\mathrm{res}_{\zeta^2=\xi_m^2}\mathbf{c}(\zeta,\alpha)(X_{[1,m]})d\zeta^2/\zeta^2$ belongs to the domain of definition of $\mathbf{g}_m^0(\xi_m,\alpha)$. The rest is simple; we prove that

$$\operatorname*{res}_{\zeta^2=\xi_m^2}\bar{\mathbf{c}}_{[1,m]}(\zeta,\alpha)(X_{[1,m]})\frac{d\zeta^2}{\zeta^2}=-\mathbf{t}^*_{[1,m]}(\xi_m,\alpha)\operatorname*{res}_{\zeta^2=\xi_m^2}\mathbf{c}_{[1,m]}(\zeta,\alpha)(X_{[1,m]})\frac{d\zeta^2}{\zeta^2}\,,$$

using the right hand side of (3.104), the formulas (3.103) and the following way of writing $\bar{\mathbf{c}}_{[1,m]}(\zeta,\alpha)$:

$$\begin{aligned}&\bar{\mathbf{c}}_{[1,m]}(\zeta,\alpha)(X_{[1,m]})\\&\qquad=\frac{1}{2\pi i}\oint_\gamma\frac{d\xi^2}{\xi^2}\psi(\zeta/\xi,\alpha+\mathbb{S})\\&\qquad\times\mathrm{Tr}_{a,A}\Big\{\sigma_a^+\mathbb{T}_{a,[1,m]}(\xi,\alpha)\mathbb{T}_{A,[1,m]}(\xi,\alpha)\xi^{\alpha-\mathbb{S}}(q^{-2S_{[1,m]}}X_{[1,m]})\Big\}\,.\end{aligned}$$

□

Another immediate consequence of the dressed Russian doll construction is the commutativity of $\mathbf{b}^*(\zeta)$, $\mathbf{c}^*(\zeta)$ with $\mathbf{t}^*(\zeta)$.

THEOREM 3.24. *We have*

$$\mathbf{t}^*(\zeta_1)\mathbf{b}^*(\zeta_2)=\mathbf{b}^*(\zeta_2)\mathbf{t}^*(\zeta_1)\,,\quad\mathbf{t}^*(\zeta_1)\mathbf{c}^*(\zeta_2)=\mathbf{c}^*(\zeta_2)\mathbf{t}^*(\zeta_1)\,.\tag{3.105}$$

PROOF. We shall prove the case of $\mathbf{b}^*$. Let us consider, for example, the two operators on the left in the dressed Russian doll formula, assuming that $\epsilon_n = +$, $\epsilon_{n-1} = 0$. In the computation we indicate the dependance on all ζ_j's explicitly, without writing the indices for numbers of spaces implying that they follow ζ_j's:

(3.106)
$$\begin{aligned}
&\mathbf{g}^+(\zeta_n,\alpha|\zeta_1,\dots\zeta_{n-2},\zeta_{n-1})\mathbf{g}^0(\zeta_{n-1},\alpha+1|\zeta_1,\dots\zeta_{n-2})(X_{[n-2]})\\
&=\mathbf{g}^+(\zeta_n,\alpha|\zeta_1,\dots\zeta_{n-2},\zeta_{n-1})\mathbb{T}_{n-1,[1,n-2]}(\zeta_{n-1}|\zeta_1,\dots\zeta_{n-2})(q^{(\alpha+1)\sigma^3_{n-1}}X_{[n-2]})\\
&=\mathbb{T}_{n-1,[1,n-2]}(\zeta_{n-1}|\zeta_1,\dots\zeta_{n-2})\mathbf{g}^+(\zeta_n,\alpha|\zeta_{n-1},\zeta_1,\dots\zeta_{n-2})(q^{(\alpha+1)\sigma^3_{n-1}}X_{[n-2]})\\
&=\mathbb{T}_{n-1,[1,n-2]}(\zeta_{n-1}|\zeta_1,\dots\zeta_{n-2})q^{\alpha\sigma^3_{n-1}}\mathbf{g}^+(\zeta_n,\alpha|,\zeta_1,\dots\zeta_{n-2})(X_{[n-2]})\\
&=\mathbb{R}_{n-1,n}(\zeta_{n-1}/\zeta_n)^{-1}\mathbf{g}^0(\zeta_{n-1},\alpha|\zeta_1,\dots\zeta_{n-2},\zeta_n)\mathbf{g}^+(\zeta_n,\alpha|,\zeta_1,\dots\zeta_{n-2})(X_{[n-2]})\,,
\end{aligned}$$

where we used the R-matrix symmetry and the left reduction. Obviously, (3.106) proves the commutation relation for $\mathbf{b}^*(\zeta)$ and $\mathbf{t}^*(\zeta)$. □

REMARK 3.25. In the Russian doll construction we use the short-hand notation $\mathbf{g}_n^{\epsilon n}(\zeta_n)\mathbf{g}_{n-1}^{\epsilon n-1}(\zeta_{n-1})$ meaning that we 'add' the spaces $V_{\zeta_{n-1}}$ and V_{ζ_n} subsequently in this order. In what follows we shall allow ourselves to write them in the opposite way $\mathbf{g}_{n-1}^{\epsilon n-1}(\zeta_{n-1})\mathbf{g}_n^{\epsilon n}(\zeta_n)$ meaning that we add spaces in the opposite way. With this convention, the equation (3.106) takes the form of the relation in the Zamolodchikov algebra

$$\mathbf{g}_n^+(\zeta_n,\alpha)\mathbf{g}_{n-1}^0(\zeta_{n-1},\alpha+1)=\mathbb{R}_{n,n-1}(\zeta_n/\zeta_{n-1})\mathbf{g}_{n-1}^0(\zeta_{n-1},\alpha)\mathbf{g}_n^+(\zeta_n,\alpha)\,. \tag{3.107}$$

The operator $\mathbf{g}^0(\zeta)$ looks trivial, but it has a non-trivial commutation relation.

LEMMA 3.26. *In the Russian doll construction*

$$\mathbf{g}_n^+(\zeta_n,\alpha)\mathbf{g}_{n-1}^{\bar 0}(\zeta_{n-1},\alpha+1)=\mathbb{R}_{n,n-1}(\zeta_n/\zeta_{n-1})\mathbf{g}_{n-1}^{\bar 0}(\zeta_{n-1},\alpha)\mathbf{g}_n^+(\zeta_n,\alpha)\,. \tag{3.108}$$

PROOF. Recalling the definition (3.92)

$$\mathbf{g}_j^{\bar 0}(\xi_j,\alpha)=\mathbb{I}_{[1,j-1]}\,,$$

we observe that the stated relation

$$\begin{aligned}
&\mathbf{g}_n^+(\zeta_n,\alpha)\mathbf{g}_{n-1}^{\bar 0}(\zeta_{n-1},\alpha+1)(X_{[1,n-2]})\\
&\quad=\mathbb{R}_{n,n-1}(\zeta_n/\zeta_{n-1})\mathbf{g}_{n-1}^{\bar 0}(\zeta_{n-1},\alpha)\mathbf{g}_n^+(\zeta_n,\alpha)(X_{[1,n-2]})
\end{aligned}$$

is nothing but the right reduction relation (Lemma 3.19). □

Together with Theorem 3.18 proved previously, Theorem 3.24 shows that the commuting family of operators $\mathbf{t}^*(\zeta)$ is in the centre of the algebra created by $\mathbf{c}(\zeta),\mathbf{b}(\zeta),\mathbf{c}^*(\zeta),\mathbf{b}^*(\zeta),\mathbf{t}^*(\zeta)$. For that reason, in what follows we shall treat $\mathbf{t}^*(\zeta)$ as a constant. Due to Theorem 3.6.2, the relation between $\mathbf{b}^*_{\mathrm{rat}}(\zeta)$ and $\mathbf{b}^*(\zeta)$ may be written as the following Bogoliubov transformation which will be often used later

$$\mathbf{b}^*_{\mathrm{rat}}(\zeta)=\mathbf{b}^*(\zeta)-\frac{1}{2\pi i}\oint_\Gamma\Bigl(D_\zeta D_\xi\Delta^{-1}\psi(\zeta/\xi,\alpha+\mathbb{S})\Bigr)\mathbf{c}(\xi)\frac{d\xi^2}{\xi^2}\,, \tag{3.109}$$

where

$$D_\zeta F(\zeta):=F(\zeta q)+F(\zeta q^{-1})-\mathbf{t}^*(\zeta)F(\zeta)\,. \tag{3.110}$$

3.7. Commutation Relations Between Creation and Annihilation Operators

THEOREM 3.27. *We have the commutation relation*

$$[\mathbf{b}^*(\zeta), \mathbf{c}(\zeta')]_+ = 0\,.$$

PROOF. Due to the dressed Russian doll construction, it is sufficient to consider the inhomogeneous case and to prove the identity

$$\mathbf{c}_{J\sqcup c}(\zeta',\alpha)\mathbf{g}_{c,J}(\zeta,\alpha+1)(X_J)+\mathbf{g}_{c,J}(\zeta,\alpha)\mathbf{c}_J(\zeta',\alpha+1)(X_J)=0\,. \tag{3.111}$$

Recall that

$$\mathbf{g}_{c,J}(\zeta,\alpha)=\frac{1}{2}\mathbf{f}_J(q\zeta,\alpha)+\frac{1}{2}\mathbf{f}_J(q^{-1}\zeta,\alpha)-\mathbb{T}_{c,J}(\zeta,\alpha)\mathbf{f}_J(\zeta,\alpha)+\mathbf{u}_{c,J}(\zeta,\alpha)\,.$$

The annihilation operator $\mathbf{c}(\zeta')$ is given by the integral (3.39). Introduce

$$\mathcal{X}(\zeta,\xi)=\mathbf{k}_{J\sqcup c}(\xi,\alpha)\mathbf{g}_{c,J}(\zeta,\alpha+1)(X_J)+\mathbf{g}_{c,J}(\zeta,\alpha)\mathbf{k}_J(\xi,\alpha+1)(X_J)\,.$$

Then our goal is to prove that

$$\mathcal{X}(\zeta,\xi)\equiv 0 \mod q\text{-exact forms in } \xi\,. \tag{3.112}$$

By the very meaning of our construction, the q-exact forms must be regular on the divisor $(\mathrm{f}_J)\cup\zeta^2$. In this proof we shall write the equality modulo q-exact forms in ξ simply by $\equiv$.

The operator $\mathbf{g}_J(\zeta,\alpha)$ contains $\mathbf{f}_J(\zeta,\alpha)$, so, our first goal is to establish the commutation relations of the latter with $\mathbf{k}_J(\zeta,\alpha)$. From Lemma 3.8 we conclude that

$$\mathbf{f}_J(\zeta,\alpha)\mathbf{k}_J(\xi,\alpha+1)+\mathbf{k}_J(\xi,\alpha)\mathbf{f}_J(\zeta,\alpha+1)\equiv\mathbf{m}_J^{++}(\zeta,\xi,\alpha)\,.$$

This allows us to simplify the anticommutator for the first two terms in $\mathbf{g}_{c,J}(\zeta,\alpha)$. It applies also to the third term due to the identity

$$\mathbf{k}_{J\sqcup c}(\xi,\alpha)\mathbb{T}_{c,J}(\zeta,\alpha+1)\mathbf{f}_J(\zeta,\alpha+1)(X_J)=\mathbb{T}_{c,J}(\zeta,\alpha)\mathbf{k}_J(\xi,\alpha)\mathbf{f}_J(\zeta,\alpha+1)(X_J)\,,$$

which follows from the R-matrix symmetry and the left reduction. Altogether we obtain

$$\begin{aligned}\mathcal{X}(\zeta,\xi)=\Big\{\frac{1}{2}\mathbf{m}_J^{++}(\zeta q,\xi,\alpha)+\frac{1}{2}\mathbf{m}_J^{++}(\zeta q^{-1},\xi,\alpha)-\mathbb{T}_{c,J}(\zeta,\alpha)\mathbf{m}_J^{++}(\zeta,\xi,\alpha)\\+\mathbf{u}_{c,J}(\zeta,\alpha)\mathbf{k}_J(\xi,\alpha+1)+\mathbf{k}_{J\sqcup c}(\xi,\alpha)\mathbf{u}_J(\zeta,\alpha+1)\Big\}(X_J)\,.\end{aligned} \tag{3.113}$$

The first four terms from (3.113) can be written as

$$\mathcal{X}^{(1)}(\zeta,\xi)=\mathrm{Tr}_{a,b,A,B}\Big\{W^{(1)}_{a,b,c,A,B}(\zeta/\xi)\mathbb{T}_{\{a,A\}}(\zeta,\alpha)\mathbb{T}_{\{b,B\}}(\xi,\alpha)(\zeta\xi)^{\alpha-\mathbb{S}}(q^{-4S_J}X_J)\Big\}\,,$$

with

$$\begin{aligned}W^{(1)}_{a,b,c,A,B}(\eta)\equiv\frac{1}{2}\Big(\frac{\eta^2-1}{q^2\eta^2-1}\tau_a^+ +\frac{\eta^2-1}{q^{-2}\eta^2-1}\tau_a^-\Big)M^{(++)}_{b,A,B}(\eta)\\-F^{-1}_{a,A}P_{a,c}M^{(++)}_{b,A,B}(\eta)F_{a,A}+\sigma_b^+\eta^{-1}q^{2D_B-1}Y_{a,c,A}\,.\end{aligned} \tag{3.114}$$

The fifth term can be written as

$$\mathcal{X}^{(2)}(\zeta,\xi)=\mathrm{Tr}_{a,b,A,B}\Big\{W^{(2)}_{a,b,c,A,B}(\zeta/\xi)\mathbb{T}_{\{b,B\}}(\xi,\alpha)\mathbb{T}_{\{a,A\}}(\zeta,\alpha)(\zeta\xi)^{\alpha-\mathbb{S}}(q^{-4S_J}X_J)\Big\}\,,$$

where $W^{(2)}_{a,b,c,A,B}(\zeta/\xi)$ is computed using our usual trick of changing the order of multiplication applying the antiautomorphism θ:

$$W^{(2)}_{a,b,c,A,B}(\eta)=-\frac{\eta}{(1-\eta^{-2})(1-\eta^{-2}q^{2})(1-\eta^{-2}q^{-2})}$$
$$\times\,\theta_c\left(R_{\{b,B\},c}(\eta^{-1}q^{-1})\sigma_b^{+}Y_{a,c,A}q^{\sigma_c^3}\theta_c(R_{\{b,B\},c}(\eta^{-1}))\right)q^{2D_A+\sigma_a^3}\,,$$

where we did not use crossing symmetry in the last R-matrix in order to keep better track of singularities.

Now we are in a position to apply Yang-Baxter. After a very long but straightforward calculation, we obtain:

$$R_{\{a,A\},\{b,B\}}(\eta)W^{(1)}_{a,b,c,A,B}(\eta)R_{\{a,A\},\{b,B\}}(\eta)^{-1}=-W^{(2)}_{a,b,c,A,B}(\eta)$$
$$+\,q^{\sigma_b^3(D_A+\sigma_a^3/2)}\begin{pmatrix}S_{a,c,A,B}(\eta q^{-1}) & 0\\ 0 & -S_{a,c,A,B}(\eta q)\end{pmatrix}_b q^{-\sigma_b^3(D_A+\sigma_a^3/2)}+(\star)\sigma_a^-+(\star)\sigma_b^-\,,$$

where

$$S_{a,c,A,B}(\eta)=\frac{1}{\eta^{-2}-1}\Big[\frac{1}{2}\sigma_c^3\mathbf{a}_B^*\left(\sigma_a^3q^{-2D_B-1}\mathbf{a}_B^*-\sigma_a^+(\eta+q^{-2D_B-1}\mathbf{a}_B^*\mathbf{a}_A)\right)$$
$$-\,\sigma_c^+\sigma_a^+\left(\eta(q^{-2D_B}-\eta^{-2})\mathbf{a}_B^*\mathbf{a}_A-q^{2D_B+1}+\frac{1}{2}(q+q^{-1})\right)$$
$$+\,\sigma_c^+\sigma_a^3\eta(q^{-2D_B}-\eta^{-2})\mathbf{a}_B^*\Big]q^{\sigma_a^3+2D_A}\,.$$

The terms containing σ_a^-,σ_b^- do not count under the traces. So, we obtain

$$\mathbf{k}_{J\sqcup c}(\xi,\alpha)\mathbf{g}_{c,J}(\zeta,\alpha+1)(X_J)+\mathbf{g}_{c,J}(\zeta,\alpha)\mathbf{k}_J(\xi,\alpha+1)(X_J)$$
$$=\Delta_\xi\mathrm{Tr}_{a,A,B}\Big\{S_{a,c,A,B}(\zeta/\xi)\mathbb{T}_B(\xi,\alpha)\mathbb{T}_{\{a,A\}}(\zeta,\alpha)(\zeta\xi)^{\alpha-\mathbb{S}}(q^{-4S_J}X_J)\Big\}\equiv 0\,.$$

Indeed, the right hand side is an exact form in ξ since $S_{a,c,A,B}(\zeta/\xi)$ is singular only on the divisor $(\mathtt{f}_J)+(\zeta^2)$. □

THEOREM 3.28. *We have the commutation relation*

$$[\mathbf{b}^*(\zeta),\mathbf{b}(\zeta')]_+\big|_{\mathcal{W}_\alpha}=-\psi(\zeta/\zeta',\alpha+\mathbb{S})\,.$$

PROOF. We have to prove that

$$\mathbf{b}_{J\sqcup c}(\zeta',\alpha)\mathbf{g}_{c,J}(\zeta,\alpha-1)(X_J)+\mathbf{g}_{c,J}(\zeta,\alpha)\mathbf{b}_J(\zeta',\alpha+1)(X_J)$$
$$=-\psi(\zeta/\zeta',\alpha+\mathbb{S})\cdot X_J\,. \tag{3.115}$$

Once again, introduce

$$\mathcal{X}(\zeta,\xi)=\phi(\mathbf{k})_{J\sqcup c}(\xi,\alpha)\mathbf{g}_{c,J}(\zeta,\alpha-1)(X_J)+\mathbf{g}_{c,J}(\zeta,\alpha)\phi(\mathbf{k})_J(\xi,\alpha+1)(X_J)\,.$$

Similarly to the previous case, we have

$$\phi(\mathbf{k})_J(\xi,\alpha)\mathbf{f}_J(\zeta,\alpha-1)+\mathbf{f}_J(\zeta,\alpha)\phi(\mathbf{k})_J(\xi,\alpha+1)\equiv\mathbf{m}^{(+-)}(\zeta,\xi,\alpha)\,,$$

where $\mathbf{m}^{(+-)}(\zeta,\xi,\alpha)$ is given in Lemma 3.9. So, we obtain

$$\mathcal{X}(\zeta,\xi)\equiv\Big\{\frac{1}{2}\mathbf{m}_J^{(+-)}(\zeta q,\xi,\alpha)+\frac{1}{2}\mathbf{m}_J^{(+-)}(\zeta q^{-1},\xi,\alpha)-\mathbb{T}_{c,J}(\zeta,\alpha)\mathbf{m}_J^{(+-)}(\zeta,\xi,\alpha)$$
$$+\,\mathbf{u}_{c,J}(\zeta,\alpha)\phi(\mathbf{k})_J(\xi,\alpha+1)+\phi(\mathbf{k})_{J\sqcup c}(\xi,\alpha)\mathbf{u}_{c,J}(\zeta,\alpha-1)\Big\}(X_J)\,.$$

This time $\mathbf{m}^{(+-)}$ consists of two pieces which we separate

$$\mathcal{X}(\zeta,\xi)\equiv\mathcal{X}^{\mathrm{sing}}(\zeta,\xi)+\mathcal{X}^{\mathrm{reg}}(\zeta,\xi)\,.$$

The term $\mathcal{X}^{\mathrm{sing}}(\zeta,\xi)$ is simple:

(3.116) $$\Bigl(-\frac{1}{2}\psi(\zeta q/\xi,\alpha+\mathbb{S})-\frac{1}{2}\psi(\zeta q^{-1}/\xi,\alpha+\mathbb{S})+\mathbb{T}_{c,J}(\zeta,\alpha)\psi(\zeta/\xi,\alpha+\mathbb{S})\Bigr)(X_J)\,.$$

Recall $\mathbf{b}(\xi,\alpha)$:

$$\mathbf{b}_{J\sqcup c}(\zeta',\alpha)=\frac{1}{4\pi i}\oint_\Gamma\psi(\xi/\zeta',\alpha+\mathbb{S})\left(\phi(\mathbf{k})_{J\sqcup c}(\xi q,\alpha)+\phi(\mathbf{k})_{J\sqcup c}(\xi q,\alpha)\right)\frac{d\xi^2}{\xi^2}\,,$$

where Γ encircles inhomogeneities, and contains, in particular, ζ^2.

Obviously only the first two terms from (3.116) contribute to the integral, producing

$$-\frac{1}{2}\psi(\zeta/\zeta',\alpha+\mathbb{S})\,.$$

The $1/2$ disappears when we take trace over c. This yields the right hand side of (3.115). Notice that, if we did the same computation for $\bar{\mathbf{b}}$, the third term would contribute and produce $\mathbb{T}_c(\zeta,\alpha)\psi(\zeta/\zeta',\alpha+\mathbb{S})$, in agreement with Theorem 3.23.

Now our goal is to prove that

(3.117) $$\mathcal{X}^{\mathrm{reg}}(\zeta,\xi)\equiv 0$$

modulo exact forms singular on the divisor $(\mathsf{f}_J)+(\zeta^2)$. In what follows, it is convenient to use instead of $M^{(+-)}_{b,A,B}(\eta)$ its modification:

$$\tilde{M}^{(+-)}_{b,A,B}(\eta)=\frac{q}{1-\eta^2}q^{\sigma^3_bD_A}\left(\sigma^3_b+U(\eta)\sigma^-_b\right)q^{-\sigma^3_bD_A}\,.$$

This is possible because the difference is an exact form in ξ. We compute

$$\begin{aligned}\mathcal{X}^{\mathrm{reg}}(\zeta,\xi)=\mathrm{Tr}_{a,b,A,B}\Bigl\{&N(\alpha-\mathbb{S})W^{(1)}_{a,b,c,A,B}(\zeta/\xi)\mathbb{T}^+_{\{a,A\}}(\zeta,\alpha)\mathbb{T}^-_{\{b,B\}}(\xi,\alpha)\\&+N(\alpha-\mathbb{S}-1)W^{(2)}_{a,b,c,A,B}(\zeta/\xi)\mathbb{T}^-_{\{b,B\}}(\xi,\alpha)\mathbb{T}^+_{\{a,A\}}(\zeta,\alpha)\Bigr\}\eta^{\alpha-\mathbb{S}}(X_j)\,,\end{aligned}$$

where

(3.118) $$\begin{aligned}W^{(1)}_{a,b,c,A,B}(\eta)=&\frac{1}{2}\Bigl(\frac{\eta^2-1}{\eta^2q^2-1}\tau^+_a+\frac{\eta^2-1}{\eta^2q^{-2}-1}\tau^-_a\Bigr)\tilde{M}^{(+-)}_{b,A,B}(\eta)\\&-F^{-1}_{a,A}P_{a,c}\tilde{M}^{(+-)}_{b,A,B}(\eta)F_{a,A}+q^{-D_A}Y_{A,B}(\eta)q^{-D_A}Y_{a,c,A}\sigma^-_b\,.\end{aligned}$$

To evaluate the last term, we used (2.77) in order to change $N(\alpha-\mathbb{S}+1)$ into $N(\alpha-\mathbb{S})$, which is possible since the operator under the trace is left admissible. For the second part we compute

(3.119)
$$\begin{aligned}W^{(2)}_{a,b,c,A,B}(\eta)=&\frac{1}{(1-\eta^{-2})(1-\eta^{-2}q^2)(1-\eta^{-2}q^{-2})}\\&\times\theta_c\left(-\eta^{-1}R^-_{\{b,B\},c}(\eta^{-1}q^{-1})\sigma^-_bY_{a,c,A}q^{-\sigma^3_c}\theta_c(R^-_{\{b,B\},c}(\eta^{-1}))\right)q^{-2D_A-\sigma^3_a}\\=&\frac{1}{(1-\eta^{-2})(1-\eta^{-2}q^2)(1-\eta^{-2}q^{-2})}\\&\times\theta_c\left(\begin{pmatrix}-\eta^{-1}(1-q^{-2})\sigma^-_c\\1-\eta^{-2}q^{-2}\end{pmatrix}_bN_{a,c,A,B}(\eta)\left(1-\eta^{-2}q^2,-\eta^{-1}(1-q^2)\sigma^-_c\right)_b\right)\,,\end{aligned}$$

where

$$N_{a,c,A,B}(\eta)=-\eta^{-1}R^-_{B,c}(\eta^{-1})q^{\sigma^3_c/2}Y_{a,c,A}q^{-\sigma^3_c/2}\theta_c(R^-_{B,c}(\eta^{-1}q^{-1}))q^{-2D_A-\sigma^3_a}\,.$$

The presence of σ_b^- simplified the evaluation of (3.119). After a not very complicated computation, we obtain

$$\begin{aligned}
&N_{a,c,A,B}(\eta)\\
&= q^{-\sigma_a^3 D_B}(\eta^{-1}q^{2(D_B-D_A)+1})\Big\{-q^{-1}\overline{U}_{A,B}(\eta)\sigma_a^+\sigma_c^+\\
&+ (q^{\sigma_c^3+1}\eta^{-1}\overline{Y}_{A,B}(\eta) - \frac{1}{2}(1+\eta^{-2}))\sigma_a^+\sigma_c^3 + \mathbf{a}_B(\eta^{-1} - \eta^{-2}q^{-1}\overline{Y}_{A,B}(\eta))\sigma_a^+\sigma_c^-\\
&+ q^{-\sigma_a^3}\sigma_a^3\sigma_c^+ - \eta^{-2}\mathbf{a}_B^2 q^{2-3\sigma_a^3}\sigma_a^3\sigma_c^- + \eta^{-1}\mathbf{a}_B q^{\sigma_c^3-2\sigma_a^3}\sigma_a^3\sigma_c^3\Big\}q^{\sigma_a^3 D_B}\,.
\end{aligned}$$

The expression in braces is right admissible, so we can use (2.77) in order to match the argument of the N's, namely we change $W^{(2)}_{a,b,c,A,B}(\eta)$ to

$$\tilde{W}^{(2)}_{a,b,c,A,B}(\eta) = \frac{N(\alpha-\mathbb{S})}{N(\alpha-\mathbb{S}-1)}q^{-\sigma_a^3 D_B}\overline{Y}_{A,B}(\eta)(\eta q^{2(D_A-D_B)+1})q^{\sigma_a^3 D_B}W^{(2)}_{a,b,c,A,B}(\eta)\,.$$

Now we apply the quasi-intertwiner (Lemma 3.2) to $W^{(1)}_{a,b,c,A,B}(\eta)$. Then a straightforward, albeit extremely lengthy, calculation gives

(3.120)
$$\begin{aligned}
&q^{-D_B\sigma_a^3}\boldsymbol{\rho}\Big(\mathcal{F}_{\{a,A\},\{b,B\}}(\eta)\big(q^{-\sigma_b^3 D_A}W^{(1)}_{a,b,c,A,B}(\eta)q^{\sigma_b^3 D_A}\big)\mathcal{F}_{\{a,A\},\{b,B\}}(\eta)^{-1}\Big)q^{D_B\sigma_a^3}\\
&= -\tilde{W}^{(2)}_{a,b,c,A,B}(\eta)\\
&+ q^{\sigma_b^3(D_A+\sigma_a^3/2)}\begin{pmatrix} S_{a,c,A,B}(\eta q) & 0\\ 0 & -S_{a,c,A,B}(\eta q^{-1})\end{pmatrix}_b q^{-\sigma_b^3(D_A+\sigma_a^3/2)}\\
&+ (\star)\sigma_a^- + (\star)\sigma_b^+\,,
\end{aligned}$$

where

$$\begin{aligned}
S_{a,c,A,B}(\eta) &= \frac{1}{1-\eta^2}q^{-D_B\sigma_a^3}\Big\{-\frac{1}{2}(q - 2\eta q^{-\sigma_a^3}\overline{Y}_{A,B}(\eta))\sigma_a^3\sigma_c^3 - q\overline{Y}_{A,B}(\eta)\mathbf{a}_B\sigma_a^3\sigma_c^-\\
&- \eta^2 q\overline{U}_{A,B}(\eta)\sigma_a^3\sigma_c^+ + \frac{1}{2}q^{-1}\eta(q^2\eta - 2\overline{Y}_{A,B}(\eta))\sigma_a^+\sigma_c^3\\
&+ \frac{1}{2}q^{-1}(\eta(1+q^2) - 2\overline{Y}_{A,B}(\eta))\sigma_a^3\sigma_c^-\Big\}q^{D_B\sigma_a^3}\,.
\end{aligned}$$

Notice that, in spite of shifts in the diagonal matrix, the expression in the right hand side of (3.120) is right admissible due to

$$\begin{aligned}
q^{\pm D_A}\overline{U}_{A,B}(\eta q^{\pm 1})q^{\mp D_A} &= q^{\mp 1}\overline{U}_{A,B}(\eta),\\
q^{\pm D_A}\overline{Y}_{A,B}(\eta q^{\pm 1})q^{\mp D_A} &= q^{\mp 1}\overline{Y}_{A,B}(\eta)\,.
\end{aligned}$$

On the other hand, similarly to the previous case, the diagonal piece of (3.120) under the trace gives an exact form in ξ:

$$-\Delta_\xi \mathrm{Tr}_{B,a,A}\Big\{S_{a,c,A,B}(\eta)\mathbb{T}^-(\xi,\alpha)\mathbb{T}^+_{\{a,A\}}(\zeta,\alpha)\eta^{\alpha-\mathbb{S}}\Big\}\,,$$

hence

$$\mathcal{X}^{\mathrm{reg}}(\zeta,\xi) \equiv 0\,,$$

as was needed. □

With these commutation relations at hand, we prove the completeness of the Russian doll construction.

LEMMA 3.29. *The 4^l operators*

$$\mathbf{g}_l^{\epsilon_l}(\xi_l,\alpha)\mathbf{g}_{l-1}^{\epsilon_{l-1}}(\xi_{l-1})\cdots\mathbf{g}_1^{\epsilon_1}(\xi_1)(I)\,,\quad \epsilon_j=0,\bar{0},\pm\,,$$

are generically linearly independent.

PROOF. Acting by $\mathbf{b}(\zeta),\mathbf{c}(\zeta)$ one immediately proves that the linear dependence is possible if and only if the 2^l operators

$$\mathbf{g}_l^{\epsilon_l}(\xi_l,\alpha)\mathbf{g}_{l-1}^{\epsilon_{l-1}}(\xi_{l-1},\alpha)\cdots\mathbf{g}_1^{\epsilon_1}(\xi_1,\alpha)(I)\,,\quad \epsilon_j=0,\bar{0} \tag{3.121}$$

are linearly dependent. Here α does not change along the product because ϵ's are $0,\bar{0}$. Let us show that this is impossible in the generic case. We have two independent parameters: q and $x=q^\alpha$. By definition

$$\mathbf{g}_j^0(\xi_j,\alpha)(X_{[1,j-1]})=\mathbb{R}_{j,j-1}(\xi_j/\xi_{j-1})\cdots\mathbb{R}_{j,1}(\xi_j/\xi_1)\bigl(x^{\sigma_j^3}X_{[1,j-1]}\bigr)\,.$$

Set $q=1$. The R-matrix becomes the identity operator, and $\mathbf{g}_j^0(\xi_j,\alpha)$ is just an operator of multiplication by $x^{\sigma_j^3}$. So, (3.121) turns into

$$\prod_{j=1}^{l}x^{m_{\epsilon_j}\sigma_j^3}\,,\quad \epsilon_j=0,\bar{0}\,,\quad m_0=1,m_{\bar{0}}=0\,,$$

which are obviously linearly independent. □

The operators $\mathbf{g}_{c,[k,m]}^{\epsilon}$ will be very useful later. Let us summarise their reduction properties:

$$\mathbf{g}_{c,[k,m]}^{\epsilon}(\zeta,\alpha)(X_{[k,m-1]})=\mathbb{R}_{c,m}(\zeta)\mathbf{g}_{c,[k,m-1]}^{\epsilon}(\zeta,\alpha)(X_{[k,m-1]})\,, \tag{3.122}$$

$$\mathbf{g}_{c,[k,m]}^{\epsilon}(\zeta,\alpha)(q^{(\alpha+\epsilon)\sigma_k^3}X_{[k+1,m]})=q^{\alpha\sigma_k^3}\mathbf{g}_{c,[k+1,m]}^{\epsilon}(\zeta,\alpha)(X_{[k+1,m]})\,. \tag{3.123}$$

3.8. Summary

In this chapter, we introduced operators acting on the space of quasi-local operators $\mathcal{W}$ with the block structure

$$\begin{aligned}\mathbf{t}^*(\zeta):\ &\mathcal{W}_\alpha\to\mathcal{W}_\alpha\,,\\ \mathbf{b}^*(\zeta),\ \mathbf{c}(\zeta):\ &\mathcal{W}_{\alpha+1}\to\mathcal{W}_\alpha\,,\\ \mathbf{b}(\zeta),\ \mathbf{c}^*(\zeta):\ &\mathcal{W}_{\alpha-1}\to\mathcal{W}_\alpha\,.\end{aligned}$$

The annihilation operators are given by series

(3.124)

$$\mathbf{b}(\zeta)=\zeta^{-\alpha-\mathbb{S}}\bigl(\mathbf{b}_0+\sum_{p=1}^{\infty}(\zeta^2-1)^{-p}\mathbf{b}_p\bigr),\quad \mathbf{c}(\zeta)=\zeta^{\alpha+\mathbb{S}}\bigl(\mathbf{c}_0+\sum_{p=1}^{\infty}(\zeta^2-1)^{-p}\mathbf{c}_p\bigr)\,,$$

where the multipliers $\zeta^{\mp(\alpha+\mathbb{S})}$ are introduced for convenience, and the modes $\mathbf{b}_0$ and $\mathbf{c}_0$ are separated from the rest because actually they are not independent. For any $X\in\mathcal{W}$ we have

$$\mathbf{b}(\zeta_1)\cdots\mathbf{b}(\zeta_L)(X)=0\quad\text{if}\quad L>\mathrm{length}(X)\,,$$

and similarly for $\mathbf{c}(\zeta)$. The creation operators are given by series

$$\mathbf{b}^*(\zeta) = \zeta^{2+\alpha+\mathbb{S}} \sum_{p=1}^{\infty} (\zeta^2 - 1)^{p-1} \mathbf{b}_p^* , \tag{3.125}$$

$$\mathbf{c}^*(\zeta) = \zeta^{2-\alpha-\mathbb{S}} \sum_{p=1}^{\infty} (\zeta^2 - 1)^{p-1} \mathbf{c}_p^* , \tag{3.126}$$

$$\mathbf{t}^*(\zeta) = \sum_{p=1}^{\infty} (\zeta^2 - 1)^{p-1} \mathbf{t}_p^* . \tag{3.127}$$

For any $X \in \mathcal{W}$

$$\mathrm{length}(\mathbf{t}_p^*(X)) \leq \mathrm{length}(X) + p ,$$

and similarly for $\mathbf{b}_p^*$, $\mathbf{c}_p^*$.

The fermionic operators $\mathbf{b}, \mathbf{c}, \mathbf{b}^*, \mathbf{c}^*$ have the following commutation relations:

$$\begin{aligned}
&[\mathbf{c}(\zeta_1), \mathbf{c}(\zeta_2)]_+ = [\mathbf{c}(\zeta_1), \mathbf{b}(\zeta_2)]_+ = [\mathbf{b}(\zeta_1), \mathbf{b}(\zeta_2)]_+ = 0 , \\
&[\mathbf{c}(\zeta_1), \mathbf{b}^*(\zeta_2)]_+ = [\mathbf{b}(\zeta_1), \mathbf{c}^*(\zeta_2)]_+ = 0 , \\
&[\mathbf{b}(\zeta_1), \mathbf{b}^*(\zeta_2)]_+ \big|_{\mathcal{W}_\alpha} = -\psi(\zeta_2/\zeta_1, \alpha + \mathbb{S}) , \\
&[\mathbf{c}(\zeta_1), \mathbf{c}^*(\zeta_2)]_+ \big|_{\mathcal{W}_\alpha} = \psi(\zeta_1/\zeta_2, \alpha + \mathbb{S}) .
\end{aligned} \tag{3.128}$$

This implies the canonical commutation relations for the coefficients

$$[\mathbf{c}_p, \mathbf{b}_r^*]_+ = [\mathbf{b}_p, \mathbf{c}_r^*]_+ = 0 , \quad [\mathbf{b}_p, \mathbf{b}_r^*]_+ = \delta_{p,r} , \quad [\mathbf{c}_p, \mathbf{c}_r^*]_+ = \delta_{p,r} , \quad p, r \geq 1 .$$

The operator $\mathbf{t}^*(\zeta)$ is in the centre:

$$\begin{aligned}
&[\mathbf{t}^*(\zeta_1), \mathbf{t}^*(\zeta_2)] = [\mathbf{t}^*(\zeta_1), \mathbf{c}(\zeta_2)] = [\mathbf{t}^*(\zeta_1), \mathbf{b}(\zeta_2)] \\
&= [\mathbf{t}^*(\zeta_1), \mathbf{c}^*(\zeta_2)] = [\mathbf{t}^*(\zeta_1), \mathbf{b}^*(\zeta_2)] = 0 .
\end{aligned}$$

Still missing are the anti-commutation relations for the creation operators. An algebraic proof, which was hard even for those between annihilation and creation operators, would be extremely complicated in this case. Fortunately we do not need to do this, because in the next section a simple indirect proof will be given.

Another problem which has not been discussed is the completeness: do the operators $\mathbf{t}_p^*$, $\mathbf{b}_p^*$, $\mathbf{c}_p^*$ together with the left shift $(\frac{1}{2}\mathbf{t}_1^*)^{-1}$ create the entire space $\mathcal{W}_\alpha$ acting on the primary field $q^{2\alpha S(0)}$? The answer is positive, but the proof requires the anti-commutation relations between the creation operators, so, we postpone it to Chapter 4.

CHAPTER 4

Main Theorem

In the previous two chapters, we did preparations in the *Matsubara* and the *Space* directions separately. The goal of the present chapter is to bring them together and to compute the functional $\mathscr{Z}^{\kappa}$ (see Chapter 1) on operators created by the action of $\mathbf{t}^*(\zeta)$, $\mathbf{b}^*(\zeta)$, $\mathbf{c}^*(\zeta)$, i.e., on the descendants. This will require further preparation from both the *Space* and the *Matsubara* directions.

We begin with the *Space* direction, explaining the relation between our creation operators and the dual rqKZ equations. This result is interesting by itself. Then we pass to the *Matsubara* direction, introducing deformed Abelian integrals and deformed Riemann bilinear identity. After that we prove our main theorem, which states that the functional $\mathscr{Z}^{\kappa}$ is evaluated on the descendants with the help of only two functions. In Appendix, we discuss the relation of our construction to cohomologies of deformed affine Jacobi varieties. Readers who are interested only in applications to physics can skip Appendix, but it is strongly recommended to anyone interested in the mathematical meaning of our construction.

4.1. Fermionic Basis and Difference Equations

In Section 1.5, we introduced the difference equations satisfied by the inhomogeneous density matrix in infinite volume. The inhomogeneity parameters are denoted by $\xi_1,\dots,\xi_l$. The dual equations are

$$g(\xi_1,\dots,\xi_{l-1},\xi_l,\alpha)=\mathbf{A}(\xi_1,\dots,\xi_l,\alpha)g(\xi_1,\dots,\xi_{l-1},\xi_lq^{-1},\alpha) \tag{4.1}$$

where

$$\mathbf{A}(\xi_1,\dots,\xi_l,\alpha)(X_{[1,l]})=T_{l,[1,l-1]}(\xi_l)q^{\alpha\sigma_l^3}\theta_l\Bigl(X_{[1,l]}\ \theta_l(T_{l,[1,l-1]}(\xi_l)^{-1})\Bigr)\,. \tag{4.2}$$

We want to show that the equations (4.1) are solved by

$$\mathbf{g}_m^{\epsilon_m}(\xi_m,\alpha)(X_{[1,l-1]})\,.$$

Then one can apply the Russian doll (3.93) to compute $\mathscr{Z}^0_\infty$ (see Section 1.5).

We begin with a simple statement.

LEMMA 4.1. *The following two identities hold:*

$$\begin{aligned}\mathbf{A}(\xi_1,\dots,\xi_l,\alpha)(X_{[1,l-1]}x_l)&=\mathbb{T}_{l,[1,l-1]}(\xi_l,\alpha)(X_{[1,l-1]})\theta_l(x_l)\,,\\ \mathbf{A}(\xi_1,\dots,\xi_l,\alpha)\mathbb{T}_{l,[1,l-1]}(\xi_lq^{-1},\alpha)(X_{[1,l-1]})&=X_{[1,l-1]}\,.\end{aligned} \tag{4.3}$$

PROOF. The first relation is obvious, the second follows from the crossing symmetry. □

In particular, the relations (4.3) for $x_l=I$ can be rewritten as

$$\mathbf{A}(\xi_1,\dots,\xi_l,\alpha)\mathbf{g}_l^{\epsilon}(\xi_lq^{-1},\alpha)(X_{[1,l-1]})=\mathbf{g}_l^{\bar\epsilon}(\xi_l,\alpha)(X_{[1,l-1]})\,,\quad \epsilon=0,\bar 0\,. \tag{4.4}$$

This is the first, modest, step towards our goal. Now we prove a more complicated relation.

LEMMA 4.2. *We have*

$$\mathbf{g}_l^+(\xi_l,\alpha)(X_{[1,l-1]}) = -\mathbf{A}(\xi_1,\dots,\xi_l,\alpha)\mathbf{g}_l^+(\xi_l q^{-1},\alpha)(X_{[1,l-1]})\,. \tag{4.5}$$

PROOF. Denote $J=[1,l-1]$, and write c for the space index l. After some algebra using the definition of $\mathbf{g}_l^+(\zeta,\alpha)$ (Lemma 3.19), the identity $\Delta_\zeta\mathbf{f}(\zeta,\alpha)=\mathbf{k}(\zeta,\alpha)$ and the identities (4.3), we reduce (4.5) to

$$\begin{aligned}\mathrm{Tr}_{a,A}\Bigl\{\mathbb{T}_{c,J}(\xi_l,\alpha)\mathbb{T}_{\{a,A\},J}(\xi_l q^{-1},\alpha)(\xi_l q^{-1})^{\alpha-\mathbb{S}}(q^{-2S_J}X_J)\left(\frac{1}{2}\sigma_a^+ + Y_{a,c,A}\right)\Bigr\}\\ = \mathrm{Tr}_{a,A}\Bigl\{\mathbb{T}_{\{a,A\},J}(\xi_l,\alpha)\xi_l^{\alpha-\mathbb{S}}(q^{-2S_J}X_J)\left(\frac{1}{2}\sigma_a^+ + Y_{a,c,A}\right)\Bigr\}.\end{aligned} \tag{4.6}$$

Let us transform the left hand side by rewriting

$$\frac{1}{2}\sigma_a^+ + Y_{a,c,A} = \tau_a^+\sigma_c^+ - F_{a,A}^{-1}(\mathcal{P}_{a,c}^-\tau_a^-\sigma_c^+)F_{a,A}\,. \tag{4.7}$$

Here $\mathcal{P}_{a,c}^- = \frac{1}{2}(I-P_{a,c})$ is the anti-symmetriser, for which we have

$$\mathbb{T}_{c,J}(\zeta)\mathbb{T}_{a,J}(\zeta q^{-1})\mathcal{P}_{a,c}^- = \mathcal{P}_{a,c}^-\,. \tag{4.8}$$

Here one can refer to (2.89), or to the crossing symmetry.

Now we proceed, using the definition of the fusion matrix (2.78) for the first term in (4.7), and undoing fusion for the second term followed by the use of (4.8),

$$\begin{aligned}(LHS) &= \mathrm{Tr}_A\Bigl\{\mathbb{T}_{c,J}(\xi_l,\alpha)\mathbb{T}_{A,J}(\xi_l,\alpha)\sigma_c^+\xi_l^{\alpha-\mathbb{S}}(q^{-2S_J}X_J)\Bigr\}\\ &\quad - \mathrm{Tr}_A\Bigl\{\mathbb{T}_{A,J}(\xi_l q^{-1},\alpha)(\xi_l q^{-1})^{\alpha-\mathbb{S}}(q^{-2S_J}X_J)\Bigr\}\sigma_c^+\\ &= \mathrm{Tr}_A\Bigl\{F_{c,A}\mathbb{T}_{\{c,A\},J}(\xi_l,\alpha)\sigma_c^+\xi_l^{\alpha-\mathbb{S}}(q^{-2S_J}X_J)\Bigr\}\\ &\quad - \mathrm{Tr}_A\Bigl\{\mathbb{T}_A(\xi_l q^{-1},\alpha)(\xi_l q^{-1})^{\alpha-\mathbb{S}}(q^{-2S_J}X_J)\Bigr\}\sigma_c^+\\ &= \mathrm{Tr}_{a,A}\Bigl\{P_{a,c}F_{a,A}\mathbb{T}_{\{a,A\},J}(\xi_l,\alpha)\sigma_c^+\xi_l^{\alpha-\mathbb{S}}(q^{-2S_J}X_J)\Bigr\}\\ &\quad -\mathrm{Tr}_{a,A}\Bigl\{\tau_a^-\mathbb{T}_{\{a,A\}}(\xi_l,\alpha)\xi_l^{\alpha-\mathbb{S}}(q^{-2S_J}X_J)\Bigr\}\sigma_c^+\\ &= \mathrm{Tr}_{a,A}\Bigl\{(\sigma_a^+\tau_c^- + \sigma_a^3\sigma_c^+ - \mathbf{a}_A\sigma_a^+\sigma_c^+)\mathbb{T}_{\{a,A\},J}(\xi_l,\alpha)\xi_l^{\alpha-\mathbb{S}}(q^{-2S_J}X_J)\Bigr\},\end{aligned}$$

which is the right hand side of (4.6). □

As usual, we obtain the same equation for $\mathbf{g}_l^-(\xi_l,\alpha)$ by reflection. Together with (4.4) we obtain

(4.9)

$$\mathbf{g}_l^{\bar\epsilon}(\xi_l,\alpha)(X_{[1,l-1]}) = (-1)^\epsilon\mathbf{A}(\xi_1,\dots,\xi_l,\alpha)\mathbf{g}_l^\epsilon(\xi_l q^{-1},\alpha)(X_{[1,l-1]})\,,\quad \epsilon=\pm,0,\bar 0\,,$$

where we set $\bar\epsilon=\epsilon$ for $\epsilon=\pm$.

Already at this point, we can relate our results with the original approach [**JM**] to correlation functions on the plane based on q-difference equations (1.20). This is an equation for the function $h(\xi_1,\dots,\xi_l,\alpha)$ introduced in Section 1.5 as density matrix. Consider the pairing

$$U^\epsilon(\xi_l) = h(\xi_1,\dots,\xi_l,\alpha)\ \mathbf{g}^\epsilon(\xi_l,\alpha)(X_{[1,l-1]})\,,$$

where we do not exhibit the dependence on parameters $\xi_1, \dots, \xi_{l-1}$ explicitly. We have

$$U^\epsilon(\xi_l q^{-1}) = (-1)^\epsilon U^\epsilon(\xi_l)\,.$$

The function $U^\epsilon(\xi_l)$ is regular for $\arg \xi_l \le \pi\nu/2$, hence we can determine it by studying its behaviour at $\xi_l = 0, \infty$. It can be shown that

$$U^{\pm}(\xi_l) \to 0, \quad \xi_l \to 0, \infty\,. \tag{4.10}$$

These estimates follow from (1.27) and from the case $\mathbf{m} = 0$ of Lemma 4.20 proved below, whose proof in this case does not require the results of the present section.

Assuming (4.10) we conclude by induction for the Russian doll:

$$\begin{aligned} &h(\xi_1, \dots, \xi_l, \alpha)\mathbf{g}_m^{\epsilon_m}(\xi_m, \alpha)\mathbf{g}_{m-1}^{\epsilon_{m-1}}(\xi_{m-1}) \cdots \mathbf{g}_1^{\epsilon_1}(\xi_1)(I) \\ &\qquad = \begin{cases} 1 & \text{if } \epsilon_j = 0, \bar{0} \text{ for all } j, \\ 0 & \text{otherwise.} \end{cases} \end{aligned} \tag{4.11}$$

So, we found a basis of operators which have very simple expectation values on the plane. Our goal is to show that for this basis the expectation values on a cylinder are also not very complicated.

4.2. Deformed Abelian Differentials

According to the discussion of Section 1.8, our main goal is to learn the best way of computing the matrix elements $\langle \kappa + \alpha | Y_{\mathbf{M}} | \kappa \rangle$ between two eigenvectors corresponding to maximal eigenvalues of $T_{\mathbf{M}}(1, \kappa + \alpha)$ and $T_{\mathbf{M}}(1, \kappa)$. This requires some mathematical apparatus which we introduce in the present section. The fact that the eigenvalues are maximal will not be used; $\langle \kappa + \alpha |$, $| \kappa \rangle$ can be considered here as any eigenvectors of corresponding transfer matrices. *However, we require that they have equal spins, so that the scalar product does not vanish identically.* On the other hand, in the spirit of Section 2.5, we shall consider the *Matsubara* chain with representations of arbitrary spins and arbitrary spectral parameters: $s_{\mathbf{1}}, \tau_{\mathbf{1}}, \dots, s_{\mathbf{n}}, \tau_{\mathbf{n}}$.

Recall the functions $a(\zeta), d(\zeta)$ (2.93). We introduce a function $\varphi(\zeta)$ satisfying

$$d(\zeta q)\varphi(\zeta q) = a(\zeta)\varphi(\zeta)\,. \tag{4.12}$$

Explicitly we have

$$\varphi(\zeta) = \prod_{\mathbf{m}=\mathbf{1}}^{\mathbf{n}} \varphi_{s_{\mathbf{m}}}(\zeta/\tau_{\mathbf{m}}), \quad \varphi_s = \prod_{k=0}^{2s} \frac{1}{1 - \zeta^2 q^{2(s-k)}}\,. \tag{4.13}$$

We shall denote by $\Gamma_{\mathbf{m}}$ a simple closed contour going counterclockwise around the poles of $\varphi_{s_{\mathbf{m}}}(\zeta/\tau_{\mathbf{m}})$. In addition, we shall consider a contour $\Gamma_{\mathbf{0}}$ going around $\zeta = 0$. We shall call the totality of the poles of $\varphi(\zeta)$ in the ζ^2-plane *φ-divisor.*

Let us define two measures:

$$d\mu^{\pm}(\zeta) = Q^{\mp}(\zeta, \kappa + \alpha)Q^{\pm}(\zeta, \kappa)\varphi(\zeta)\frac{d\zeta^2}{\zeta^2}\,. \tag{4.14}$$

We introduce two kinds of deformed Abelian differentials as follows.

DEFINITION 4.3. A deformed Abelian differential with singularities at $\zeta^2 = 0, \infty$ is an expression of the form

$$f^{\pm}(\zeta)d\mu^{\pm}(\zeta)\,, \tag{4.15}$$

where $\zeta^{\mp\alpha} f^{\pm}(\zeta)$ is a Laurent polynomial of ζ^2.

The differential form (4.15) is single-valued on the ζ^2-plane. Deformed Abelian integrals are integrals of deformed Abelian differentials over the cycles $\Gamma_{\mathbf{m}}$, $\mathbf{m} = \mathbf{0}, \mathbf{1}, \cdots, \mathbf{n}$.

In Appendix, we shall explain that (4.15) provides a deformation of hyper-elliptic Abelian differentials with two parameters: α which can be introduced even on the classical level, and q which is the parameter of quantisation.

It is possible to consider rational functions $\zeta^{\mp\alpha} f^{\pm}(\zeta)$ with singularities which do not overlap the φ-divisor and the point $\zeta^2 = 0$. However, we shall need yet another generalisation for our goals.

Let us introduce two definitions which will be often used later:

$$\rho(\zeta|\alpha) = \frac{T(\zeta, \kappa + \alpha)}{T(\zeta, \kappa)}, \tag{4.16}$$

$$(\delta_\zeta^- f)(\zeta) = f(\zeta q) - \rho(\zeta|\alpha) f(\zeta). \tag{4.17}$$

REMARK 4.4. The function $\rho(\zeta|\alpha)$ depends on α, as well as on κ and on all the *Matsubara* spectral parameters. We indicate explicitly only its dependance on α. The same applies to the function $\omega(\zeta, \xi|\alpha)$ which will be introduced later.

We start our study of deformed Abelian integrals with the following technical Lemma.

LEMMA 4.5. *Let $\zeta^{\mp\alpha} f^{\pm}(\zeta)$ be a Laurent polynomial in ζ^2. Then, for $\mathbf{m} = \mathbf{0}, \ldots, \mathbf{n}$, the following identities hold.*

$$\begin{aligned} &\int_{\Gamma_{\mathbf{m}}} T(\zeta, \kappa) \delta_\zeta^- \left(\Delta_\zeta^{-1} (f^{\pm}(\zeta))\right) d\mu^{\pm}(\zeta) \\ &= \int_{\Gamma_{\mathbf{m}}} f^{\pm}(\zeta) d(\zeta) Q^{\mp}(\zeta, \kappa + \alpha) Q^{\pm}(\zeta q^{-1}, \kappa) \varphi(\zeta) \frac{d\zeta^2}{\zeta^2}. \end{aligned} \tag{4.18}$$

PROOF. This can be verified directly by applying the Baxter equations (2.12), (2.16) to $T(\zeta, \kappa) Q^{\pm}(\zeta, \kappa)$ or $T(\zeta, \kappa + \alpha) Q^{\mp}(\zeta, \kappa + \alpha)$ and moving contours of integration. □

This Lemma suggests the following definition.

DEFINITION 4.6. Consider a non-zero ξ such that ξ^2, $\xi^2 q^{\pm 2}$ do not overlap the φ-divisor. A deformed second kind Abelian differential with singularity at $\zeta^2 = \xi^2$ is an expression of the form (4.15) with

$$f^{\pm}(\zeta) = T(\zeta, \kappa) \delta_\zeta^- \left(\Delta_\zeta^{-1} (\psi(\zeta/\xi, \alpha))\right).$$

For these differentials we *define* their periods by the formula (4.18).

DEFINITION 4.7. Consider a function $g^{\pm}(\zeta)$ such that $\zeta^{\mp\alpha} g^{\pm}(\zeta)$ is a Laurent polynomial. A q-exact form $E_\zeta(g^{\pm}(\zeta))$ is the following deformed Abelian differential

$$\begin{aligned} &E_\zeta(g^{\pm}(\zeta)) \\ &= \Big\{ T(\zeta, \kappa) \Big(\delta_\zeta^- \left(\Delta_\zeta^{-1} \left(g^{\pm}(\zeta q^{-1}) T(\zeta q^{-1}, \kappa) \right) \right) - \delta_\zeta^- \left(\Delta_\zeta^{-1} \left(g^{\pm}(\zeta) T(\zeta, \kappa + \alpha) \right) \right) \Big) \\ &+ a(\zeta) d(\zeta q) g^{\pm}(\zeta q) - a(\zeta q^{-1}) d(\zeta) g^{\pm}(\zeta q^{-1}) \Big\} d\mu^{\pm}(\zeta). \end{aligned} \tag{4.19}$$

This definition is justified by the following simple statement.

LEMMA 4.8. *We have*

$$E_\zeta\big(g^\pm(\zeta)\big) = F(\zeta) - F(\zeta q^{-1}),$$

where

$$\begin{aligned}F(\zeta) = \Big\{&\Delta_\zeta^{-1}\big(g^\pm(\zeta)T(\zeta,\kappa)\big)Q^\mp(\zeta,\kappa+\alpha)Q^\pm(\zeta q,\kappa)\\&+\Delta_\zeta^{-1}\big(g^\pm(\zeta)T(\zeta,\kappa+\alpha)\big)Q^\mp(\zeta q,\kappa+\alpha)Q^\pm(\zeta,\kappa)\\&-\Delta_\zeta^{-1}\big(g^\pm(\zeta q)T(\zeta q,\kappa)\big)Q^\mp(\zeta q,\kappa+\alpha)Q^\pm(\zeta,\kappa)\\&-\Delta_\zeta^{-1}\big(g^\pm(\zeta q)T(\zeta q,\kappa+\alpha)\big)Q^\mp(\zeta,\kappa+\alpha)Q^\pm(\zeta q,\kappa)\\&+g^\pm(\zeta q)d(\zeta q)Q^\mp(\zeta,\kappa+\alpha)Q^\pm(\zeta,\kappa)\\&+g^\pm(\zeta)a(\zeta)Q^\mp(\zeta q,\kappa+\alpha)Q^\pm(\zeta q,\kappa)\Big\}a(\zeta)\varphi(\zeta)\frac{d\zeta^2}{\zeta^2}\,.\end{aligned}\tag{4.20}$$

PROOF. Proof consists in direct computation using the Baxter equation. ☐

This Lemma implies that periods of q-exact forms vanish. We have $2\mathbf{n}$ different deformed Abelian differentials with singularities at $\zeta^2=0,\infty$ modulo these q-exact forms.

Using Definition 4.6, one defines the q-exact forms $E_\zeta\big(\psi(\zeta/\xi,\pm\alpha)\big)$. Substitute formally $\psi(\zeta/\xi,\pm\alpha)$ into (4.19), and rewrite the result as

$$\begin{aligned}&E_\zeta(\psi(\zeta/\xi,\pm\alpha)) = \widetilde{E}_\zeta(\psi(\zeta/\xi,\pm\alpha))\\&+T(\zeta,\kappa)\Big(T(\xi,\kappa)\delta^-_\zeta\big(\Delta_\zeta^{-1}\psi(\zeta/(q\xi,\alpha))\big)-T(\xi,\kappa+\alpha)\delta^-_\zeta\big(\Delta_\zeta^{-1}\psi(\zeta/\xi,\alpha)\big)\Big)d\mu^\pm(\zeta)\,,\end{aligned}\tag{4.21}$$

where $\zeta^{-\alpha}\widetilde{E}_\zeta(\psi(\zeta/\xi,\pm\alpha))$ is a rational form of ζ^2 with poles at $\zeta^2=\xi^2q^{\pm2}$ for which the periods are defined, while the remainder is a sum of two q-second kind differentials whose periods are defined through (4.18).

Since the periods are defined using (4.18), let us check that they vanish.

LEMMA 4.9. *We have*

$$\int_{\Gamma_\mathbf{m}} E_\zeta(\psi(\zeta/\xi,\pm\alpha)) = 0\,,$$

for $\mathbf{m}=\mathbf{0},\dots,\mathbf{n}$, *and in addition for* $\mathbf{m}=\infty$ *which corresponds to the contour going around* $\zeta^2=\infty$.

PROOF. For $\mathbf{m}=\mathbf{1},\dots,\mathbf{n}$, and generic ξ the proof follows from Lemma 4.8: we can use the original definition of $E_\zeta(\psi(\zeta/\xi,\pm\alpha))$ because the transcendental functions entering it are analytic in the vicinity of $\Gamma_\mathbf{m}$.

Let us consider $\mathbf{m}=\mathbf{0}$.

According to our definition of the period of the transcendental part, we have

$$\int_{\Gamma_\mathbf{0}} E_\zeta(\psi(\zeta/\xi,\pm\alpha)) = \int_{\Gamma_\mathbf{0}} \mathcal{E}(\zeta)\,,$$

where

$$\begin{aligned}\mathcal{E}(\zeta) &= \widetilde{E}_\zeta(\psi(\zeta/\xi, \pm\alpha)) \\ &+ \big(T(\xi,\kappa)\psi(q^{-1}\zeta/\xi, \pm\alpha) - T(\xi,\kappa+\alpha)\psi(\zeta/\xi,\pm\alpha)\big)d(\zeta) \\ &\times Q^{\mp}(\zeta,\kappa+\alpha)Q^{\pm}(\zeta q^{-1},\kappa)\varphi(\zeta)\frac{d\zeta^2}{\zeta^2}\,.\end{aligned}$$

We have to check that the residue at $\zeta^2 = 0$ of $\mathcal{E}(\zeta)$ vanishes. This is easy to do using

$$Q^{\pm}(\zeta,\kappa) = \zeta^{\pm(\kappa-s)}(1+O(\zeta^2)), \quad T(\zeta,\kappa) = q^{\kappa-s} + q^{-\kappa+s} + O(\zeta^2)\,,$$

where s is the spin of $|\kappa\rangle$.

In order to prove that the integral over Γ_∞ vanishes, we have to compute the sum of residues of $\mathcal{E}(\zeta)$ at finite points lying outside the contours $\Gamma_{\mathbf{m}}$, $\mathbf{m} = \mathbf{0},\dots,\mathbf{n}$, which are $\zeta^2 = \xi^2, \zeta^2 = \xi^2 q^{\pm 2}$. We have for this sum

$$\begin{aligned}&T(\xi,\kappa)d(\xi q)Q^{\mp}(\xi q,\kappa+\alpha)Q^{\pm}(\xi,\kappa)\varphi(\xi q) \\ &- T(\xi,\kappa+\alpha)d(\xi)Q^{\mp}(\xi,\kappa+\alpha)Q^{\pm}(\xi q^{-1},\kappa)\varphi(\xi) \\ &+ a(\xi q^{-1})d(\xi)Q^{\mp}(\xi q^{-1},\kappa+\alpha)Q^{\pm}(\xi q^{-1},\kappa)\varphi(\xi q^{-1}) \\ &- a(\xi)d(\xi q)Q^{\mp}(\xi q,\kappa+\alpha)Q^{\pm}(\xi q,\kappa)\varphi(\xi q) = 0\,,\end{aligned}$$

where we used the Baxter equation and (4.12). □

The most important property of the deformed Abelian differentials is given by the following Lemma.

LEMMA 4.10. *Consider the following function in two variables*

$$r(\zeta,\xi) = r^+(\zeta,\xi) - r^-(\xi,\zeta)\,,$$

where

$$\begin{aligned}r^{\pm}(\zeta,\xi) &= T(\zeta,\kappa)\Delta_\zeta^{-1}\left(\psi(\zeta/\xi,\pm\alpha)(T(\zeta,\kappa) - T(\xi,\kappa))\right) \\ &+ T(\zeta,\kappa+\alpha)\Delta_\zeta^{-1}\left(\psi(\zeta/\xi,\pm\alpha)(T(\zeta,\kappa+\alpha) - T(\xi,\kappa+\alpha))\right) \\ &- T(\zeta,\kappa)\Delta_\zeta^{-1}\left(\psi(q\zeta/\xi,\pm\alpha)(T(\zeta q,\kappa+\alpha) - T(\xi,\kappa+\alpha))\right) \\ &- T(\zeta,\kappa+\alpha)\Delta_\zeta^{-1}\left(\psi(q^{-1}\zeta/\xi,\pm\alpha)(T(\zeta q^{-1},\kappa) - T(\xi,\kappa))\right) \\ &+ \big(d(\zeta q) - d(\xi)\big)a(\zeta)\psi(q\zeta/\xi,\pm\alpha) - \big(a(\zeta q^{-1}) - a(\xi)\big)d(\zeta)\psi(q^{-1}\zeta/\xi,\pm\alpha)\,,\end{aligned} \tag{4.22}$$

so that $(\xi/\zeta)^{\pm\alpha}r^{\pm}(\zeta,\xi)$ are polynomials of ζ^2, ξ^2. Then for $\mathbf{i},\mathbf{j} = \mathbf{0},\dots,\mathbf{n}$ we have

$$\int_{\Gamma_{\mathbf{i}}}\int_{\Gamma_{\mathbf{j}}} r(\zeta,\xi)d\mu^+(\zeta)d\mu^-(\xi) = 0\,. \tag{4.23}$$

PROOF. Denote the integrand in (4.23) by $\mathcal{I}(\zeta,\xi)$. Due to

$$\psi(\xi/\zeta,-\alpha) = -\psi(\zeta/\xi,\alpha)\,, \quad \Delta_\xi^{-1}\psi(\xi/\zeta,-\alpha) = \Delta_\zeta^{-1}\psi(\zeta/\xi,\alpha)\,,$$

we have

$$\mathcal{I}(\zeta,\xi) = E_\zeta(\psi(\zeta/\xi,\alpha))d\mu^-(\xi) - d\mu^+(\zeta)E_\xi(\psi(\xi/\zeta,-\alpha))\,. \tag{4.24}$$

The proof seems to be finished, however, one has to be careful at this point. Lemma 4.9 allows us to conclude (4.23) only for

$$\int_{\Gamma_{\mathbf{i}}}\int_{\Gamma_{\mathbf{j}}} \mathcal{I}(\zeta,\xi), \quad \mathbf{i}\neq\mathbf{j}\,.$$

The trouble is with the case $\mathbf{i}=\mathbf{j}$, $\mathbf{i}=\mathbf{1},\dots\mathbf{n}$, where the singularities in ξ and ζ overlap and Lemma 4.9 is not applicable. Let us write in this case

$$\int_{\Gamma_{\mathbf{i}}}\int_{\Gamma_{\mathbf{i}}}\mathcal{I}(\zeta,\xi)=-\sum_{\mathbf{j}\neq\mathbf{i}}\int_{\Gamma_{\mathbf{j}}}\int_{\Gamma_{\mathbf{i}}}\mathcal{I}(\zeta,\xi)-\int_{\Gamma_{\infty}}\int_{\Gamma_{\mathbf{i}}}\mathcal{I}(\zeta,\xi)\,. \tag{4.25}$$

The sum over $\mathbf{j}$ vanishes. We have

$$d\mu^{\pm}(\eta)=O\bigl(\eta^{\mp\alpha-2\mathbf{n}}\bigr)\frac{d\eta^2}{\eta^2}\,,\qquad \eta\to\infty\,. \tag{4.26}$$

Using (4.26) and Lemma 4.9 one shows that the integral over Γ_∞ in the right hand side of (4.25) vanishes. □

Clearly $\xi^{\alpha}r^{+}(\zeta,\xi)$ is a polynomial in ξ^2 and $\zeta^{-\alpha}r^{-}(\xi,\zeta)$ is a polynomial in ζ^2, both of degree $\mathbf{n}$. This allows us to define the polynomials $p^{\pm}_{\mathbf{m}}(\zeta^2)$ by

$$r^{+}(\zeta,\xi)=\sum_{\mathbf{m}=0}^{\mathbf{n}}\zeta^{\alpha}p^{+}_{\mathbf{m}}(\zeta^2)\xi^{-\alpha+2\mathbf{m}}\,,\qquad r^{-}(\xi,\zeta)=\sum_{\mathbf{m}=0}^{\mathbf{n}}\xi^{-\alpha}p^{-}_{\mathbf{m}}(\xi^2)\zeta^{\alpha+2\mathbf{m}}\,.$$

We define the deformed Abelian differentials of the first and the second kind by

$$\Omega^{\pm}_{\mathbf{m}}(\zeta)=\zeta^{\pm\alpha+2\mathbf{m}}d\mu^{\pm}(\zeta)\,,\quad \widetilde{\Omega}^{\pm}_{\mathbf{m}}(\zeta)=\zeta^{\pm\alpha}p^{\pm}_{\mathbf{m}}(\zeta^2)d\mu^{\pm}(\zeta)\,. \tag{4.27}$$

Introduce the $(\mathbf{n}+\mathbf{1})\times(\mathbf{n}+\mathbf{1})$ period matrices

$$\mathcal{A}^{\pm}_{\mathbf{i},\mathbf{j}}=\int_{\Gamma_{\mathbf{i}}}\Omega^{\pm}_{\mathbf{j}}(\zeta)\,,\quad \mathcal{B}^{\pm}_{\mathbf{i},\mathbf{j}}=\int_{\Gamma_{\mathbf{i}}}\widetilde{\Omega}^{\pm}_{\mathbf{j}}(\zeta)\,. \tag{4.28}$$

Then (4.23) reads as

$$\mathcal{B}^{+}(\mathcal{A}^{-})^{t}=\mathcal{A}^{+}(\mathcal{B}^{-})^{t}\,. \tag{4.29}$$

Let us impose the genericity condition

$$\det\bigl(\mathcal{A}^{\pm}\bigr)\neq 0\,. \tag{4.30}$$

We shall soon show that this is equivalent to the genericity condition (1.55) introduced previously. The condition (4.30) corresponds to the well-known fact in the classical theory of Abelian integrals that the matrix of a-periods of holomorphic differentials is non-degenerate. Assuming (4.30), one can rewrite the identity (4.29) as

$$\mathcal{X}^{+}=(\mathcal{X}^{-})^{t},\qquad \mathcal{X}^{\pm}=(\mathcal{A}^{\pm})^{-1}\mathcal{B}^{\pm}\,. \tag{4.31}$$

Let us show that the genericity condition (1.55) holds generally. Recall that we assume $\mathbf{n}$ is even.

LEMMA 4.11. *Consider all the eigenvectors of $T_{\mathrm{M}}(\zeta,\kappa)$ with a given spin s. Denote them by $|\kappa,I,s\rangle$, $I=1,\dots,\binom{\mathbf{n}}{\mathbf{n}/2-s}$. For sufficiently small α they are continued analytically to give $|\kappa+\alpha,I,s\rangle$. We claim that for generic α and all I,J*

$$\langle\kappa+\alpha,I,s|\kappa,J,s\rangle\neq 0\,.$$

PROOF. Our first claim is that for all I,J the scalar product $\langle\kappa+\alpha,I,s|\kappa,J,s\rangle$ does not vanish identically as function of α. Indeed, a simple analysis of the free fermion case $q=i$ shows that $|\kappa,J,s\rangle$ for all I are just different branches of the same analytic function defined on a Riemann surface over the plane of $x=q^{\kappa}$. Then by genericity argument this is generally true, and identical vanishing of any scalar product implies by analytic continuation identical vanishing of all of them which is absurd. Now we have a finite number of analytic functions of α neither of which is identically zero. Obviously for generic α all of them are not zero. $\square$

Now we are ready to introduce an object which will play a fundamental role in what follows: the deformed normalised differential of the second kind. Consider the form of two variables $E_\zeta(\psi(\zeta/\xi,\alpha))d\mu^-(\xi)$. It has singularity at $\zeta^2=\infty$, being at the same time regular as $\xi^2\to\infty$. Let us eliminate the singularity at $\zeta^2=\infty$. As usual in algebraic geometry, we collect the first and second kind differentials into vectors $\Omega(\zeta)$ and $\widetilde{\Omega}(\zeta)$. Then we define

$$\Omega(\zeta,\xi)=E_\zeta(\psi(\zeta/\xi,\alpha))d\mu^-(\xi)-\widetilde{\Omega}^+(\zeta)^t\,\Omega^-(\xi)+\Omega^+(\zeta)^t\,\mathcal{M}\,\Omega^-(\xi)\,.$$

The last term is regular for $\zeta^2,\xi^2\to\infty$ for any constant matrix $\mathcal{M}$. We fix this arbitrariness by imposing the requirement that the periods vanish:

$$\int_{\Gamma_{\mathbf{m}}}\Omega(\zeta,\xi)=0\,,\quad \mathbf{m}=\mathbf{0},\dots,\mathbf{n}\,, \tag{4.32}$$

where the integrals are taken with respect to ζ. Clearly this requires

$$\mathcal{M}=\mathcal{X}^+\,,$$

so

$$\Omega(\zeta,\xi)=E_\zeta(\psi(\zeta/\xi,\alpha))d\mu^-(\xi)-\widetilde{\Omega}^+(\zeta)^t\,\Omega^-(\xi)+\Omega^+(\zeta)^t\,\mathcal{X}^+\,\Omega^-(\xi)\,. \tag{4.33}$$

On the other hand from (4.24) and (4.31) we have an alternative representation

$$\Omega(\zeta,\xi)=d\mu^+(\zeta)E_\xi(\psi(\xi/\zeta,-\alpha))-\Omega^+(\zeta)^t\,\widetilde{\Omega}^-(\xi)+\Omega^-(\xi)^t\,\mathcal{X}^-\,\Omega^+(\zeta)\,,$$

which means that the periods in ξ also vanish. This is a remarkable property of the normalised second kind q-differential which is known in the classical case due to Riemann and Roch.

Later we shall use the function $\omega(\zeta,\xi|\alpha)$ which is related to $\Omega(\zeta,\xi)$ as

$$\Omega(\zeta,\xi)=\frac{1}{4}\omega(\zeta,\xi|\alpha)T(\zeta,\kappa)T(\xi,\kappa)d\mu^+(\zeta)d\mu^-(\xi)\,. \tag{4.34}$$

The function $\omega(\zeta,\xi|\alpha)$ depends on all parameters $\kappa,\mathbf{n},\tau_{\mathbf{m}}$, and we shall write them explicitly when necessary. But we shall mostly indicate only α as the most important one. Let us write an explicit formula for $\omega(\zeta,\xi|\alpha)$ introducing the vector $v(\zeta,\alpha)$ with components $v(\zeta,\alpha)_{\mathbf{m}}=\zeta^{\alpha+2\mathbf{m}}$,

(4.35)

$$\begin{aligned}\omega(\zeta,\xi|\alpha)\quad&=\omega_{\mathrm{sing}}(\zeta,\xi|\alpha)+\frac{4}{T(\zeta,\kappa)T(\xi,\kappa)}\,v(\zeta,\alpha)^t\,\mathcal{X}^+\,v(\xi,-\alpha)\,,\\ \omega_{\mathrm{sing}}(\zeta,\xi|\alpha)&=4\Bigl(\delta^-_\zeta\delta^-_\xi\Delta^{-1}_\zeta\psi(\zeta/\xi,\alpha)\\ &\quad+\frac{1}{T(\zeta,\kappa)T(\xi,\kappa)}\Bigl(a(\zeta)d(\xi)\psi(q\zeta/\xi,\alpha)-a(\xi)d(\zeta)\psi(q^{-1}\zeta/\xi,\alpha)\Bigr)\Bigr)\,.\end{aligned}$$

Let us return to the genericity condition (4.30).

LEMMA 4.12. *Consider an inhomogeneous Matsubara chain with parameters* $s_{\mathbf{m}},\tau_{\mathbf{m}}$, $\mathbf{m}=\mathbf{1},\dots,\mathbf{n}$. *As before, let* $\mathbf{N}=2\sum_{\mathbf{m}=\mathbf{1}}^{\mathbf{n}}s_{\mathbf{m}}$. *Following the notation of Section* 2.1 *we set*

$$Q^-(\zeta,\kappa+\alpha)Q^+(\zeta,\kappa)=\zeta^{-\alpha}\prod_{j=1}^{\mathbf{N}}\bigl(1-\zeta^2/\lambda_j^2\bigr)\,.$$

Genericity conditions (4.30) *and* (1.55) *are equivalent due to the identity*

$$\mathcal{N}\cdot\det\mathcal{A}^{+}=\langle\kappa+\alpha|\kappa\rangle\,, \tag{4.36}$$

where

$$\mathcal{N}=-\frac{(-1)^{\frac{1}{2}\mathbf{N}(\mathbf{N}+2\mathbf{n}+1)}}{(2\pi i)^{\mathbf{n}}}\prod_{j=1}^{\mathbf{N}}\lambda_j^{2\mathbf{n}+1}\prod_{\mathbf{m}=1}^{\mathbf{n}}\prod_{p=1}^{2s_{\mathbf{m}}}(q^p-q^{-p})\prod_{\mathbf{m}=1}^{\mathbf{n}}\tau_{\mathbf{m}}^{-2(\mathbf{N}+\mathbf{n})-2s_{\mathbf{m}}(2\mathbf{n}-1)}$$
$$\times\prod_{\mathbf{j}<\mathbf{k}}\Big\{\prod_{p=0}^{2(s_{\mathbf{j}}+s_{\mathbf{k}})}(q^{p-s_{\mathbf{j}}-s_{\mathbf{k}}}\tau_{\mathbf{k}}^2-q^{s_{\mathbf{j}}+s_{\mathbf{k}}-p}\tau_{\mathbf{j}}^2)\Big\}.$$

PROOF. First of all, notice that the contours of integration entering $\mathcal{A}^{+}$ get clutched between two poles when $\tau_{\mathbf{j}}=\tau_{\mathbf{k}}q^{s_{\mathbf{j}}+s_{\mathbf{k}}-p}$, $p=0,\dots,2(s_{\mathbf{j}}+s_{\mathbf{k}})$. The resulting singularities are cancelled by the multiplier $\mathcal{N}$. So the left hand side is regular.

Consider first the case when all the *Matsubara* spins equal 1/2. In this case the product $\zeta^{\alpha}Q^{-}(\zeta,\kappa+\alpha)Q^{+}(\zeta,\kappa)$ is a polynomial of degree $\mathbf{n}$. Obviously,

$$\int_{\Gamma_0}\Omega_{\mathbf{j}}(\zeta)=2\pi i\delta_{\mathbf{j},\mathbf{0}}\,,$$

hence

$$\det\mathcal{A}^{+}=2\pi i\det\Big(\int_{\Gamma_{\mathbf{i}}}\prod_{\mathbf{m}=1}^{\mathbf{n}}\Big(1-\frac{\zeta^2}{\lambda_{\mathbf{m}}^2}\Big)\zeta^{2\mathbf{j}}\varphi(\zeta)\frac{d\zeta^2}{\zeta^2}\Big)_{\mathbf{i},\mathbf{j}=\mathbf{1},\dots,\mathbf{n}}.$$

The right hand side of (4.36) satisfies the recurrence relation (2.19). Let us show the same for the left hand side. We set $\lambda_{\mathbf{1}}=\tau_{\mathbf{1}}q^{1/2}$. The pole of $\varphi_{s_{\mathbf{1}}}(\zeta/\tau_{\mathbf{1}})$ at $\zeta^2=\tau_{\mathbf{1}}q^{1/2}$ cancels. Now we multiply $\mathcal{A}^{+}$ from the right by $1-\tau_{\mathbf{1}}^{-2}qE$ where E is a matrix with entries $E_{i,j}=\delta_{i-1,j}$. In the first $\mathbf{n}-1$ columns $\varphi_{s_{\mathbf{1}}}(\zeta/\tau_{\mathbf{1}})$ cancels completely. That is why the first $\mathbf{n}-1$ elements of the first row vanish. It remains to compute the $\mathbf{n}$-th element of the first row which is sitting on one residue. This proves the recurrence relation. The initial condition for $\mathbf{n}=1$ is easy to check.

It remains to prove that the left hand side reproduces correctly the multipliers $P(\lambda_j/\tau_{\mathbf{m}},1/2,s_{\mathbf{m}})$ in (2.94). To this end we make the dependence on parameters transparent, denoting the left hand side of (4.36) by $X_{\mathbf{n}}(s_{\mathbf{1}},\tau_{\mathbf{1}},\dots,s_{\mathbf{n}},\tau_{\mathbf{n}}|\lambda_1,\dots\lambda_{\mathbf{N}})$. It is easy to see that we have to check the relation

$$\begin{aligned}&X_{\mathbf{n}}(1/2,\tau_{\mathbf{1}},s_{\mathbf{2}},\tau_{\mathbf{2}},\dots,s_{\mathbf{n}},\tau_{\mathbf{n}}|\lambda_1,\dots\lambda_{\mathbf{N}})|_{\tau_{\mathbf{1}}=\tau_{\mathbf{2}}q^{-s_{\mathbf{2}}-1/2}}\\&=\prod_{j=1}^{\mathbf{N}}(1-(\lambda_j/\tau_{\mathbf{2}})^2q^{2s_{\mathbf{2}}})X_{\mathbf{n}-1}(s_{\mathbf{2}}+1/2,\tau_{\mathbf{2}}q^{-1/2},\dots,s_{\mathbf{n}},\tau_{\mathbf{n}}|\lambda_1,\dots\lambda_{\mathbf{N}}).\end{aligned} \tag{4.37}$$

Take the determinant in $X_{\mathbf{n}}(1/2,\tau_{\mathbf{1}},s_{\mathbf{2}},\tau_{\mathbf{2}},\dots,s_{\mathbf{n}},\tau_{\mathbf{n}}|\lambda_1,\dots\lambda_{\mathbf{N}})$, and add the first row to the second. With this all the rows except the first one become regular at $\tau_{\mathbf{1}}=\tau_{\mathbf{2}}q^{-s_{\mathbf{2}}-1/2}$. Multiply the matrix under the determinant by $1-\tau_{\mathbf{2}}^{-2}q^{s}E$ from the right. Then multiply the first row by $\tau_{\mathbf{1}}^2q^{2s_{\mathbf{2}}+1}-\tau_{\mathbf{2}}^2$ from $\mathcal{N}$. Now we are ready to send $\tau_{\mathbf{1}}\to\tau_{\mathbf{2}}q^{-s_{\mathbf{2}}-1/2}$, in the first row only the $\mathbf{n}$-th element has a non-vanishing limit. We compute it and obtain (4.37). □

Finally, we show that $\Omega(\zeta,\xi)$ behaves nicely under the fusion.

LEMMA 4.13. *Consider the function* $\omega(\zeta,\xi|\alpha,\frac{1}{2},\tau_{\mathbf{1}},s_{\mathbf{2}},\tau_{\mathbf{2}},\dots,s_{\mathbf{n}},\tau_{\mathbf{n}})$, *and assume that the corresponding Matsubara eigenvectors* $|\kappa\rangle$, $|\kappa+\alpha\rangle$ *satisfy*

$$\mathcal{P}^{2s_{\mathbf{2}}+1}_{2s_{\mathbf{2}},1}|\kappa\rangle\neq 0,\qquad \mathcal{P}^{2s_{\mathbf{2}}+1}_{2s_{\mathbf{2}},1}|\kappa+\alpha\rangle\neq 0$$

for $\tau_{\mathbf{1}}=\tau_{\mathbf{2}}q^{-s_{\mathbf{2}}-1/2}$. *Then we have*

$$\omega(\zeta,\xi|\alpha,\frac{1}{2},\tau_{\mathbf{1}},s_{\mathbf{2}},\tau_{\mathbf{2}},\dots,s_{\mathbf{n}},\tau_{\mathbf{n}})\Big|_{\tau_{\mathbf{1}}=\tau_{\mathbf{2}}q^{-s_{\mathbf{2}}-1/2}} = \omega(\zeta,\xi|\alpha,s_{\mathbf{2}}+\frac{1}{2},\tau_{\mathbf{2}}q^{-1/2},\dots,s_{\mathbf{n}},\tau_{\mathbf{n}})\,. \tag{4.38}$$

PROOF. We shall denote the right hand side of (4.38) by $\omega_{\mathrm{fus}}(\zeta,\xi|\alpha)$, and similarly for all other fused objects. The fusion procedure (2.90) implies that the vectors $\mathcal{P}^{2s_{\mathbf{2}}+1}_{2s_{\mathbf{2}},1}|\kappa\rangle$ and $\mathcal{P}^{2s_{\mathbf{2}}+1}_{2s_{\mathbf{2}},1}|\kappa+\alpha\rangle$ are eigenvectors for the fused transfer matrices. For the eigenvalues we have

$$\begin{aligned}T(\zeta,\kappa)&=(1-q^{2s_{\mathbf{2}}}(\zeta/\tau_{\mathbf{2}})^2)T_{\mathrm{fus}}(\zeta,\kappa)\,,\\ T(\zeta,\kappa+\alpha)&=(1-q^{2s_{\mathbf{2}}}(\zeta/\tau_{\mathbf{2}})^2)T_{\mathrm{fus}}(\zeta,\kappa+\alpha)\,.\end{aligned}\tag{4.39}$$

The function $\omega(\zeta,\xi|\alpha)$ in (4.35) consists of two pieces. For $\omega_{\mathrm{sing}}(\zeta,\xi|\alpha)$ the equation (4.38) obviously follows from (4.39). So the non-trivial part concerns the term with $\mathcal{X}^+$.

Let us rewrite this term as

$$\begin{aligned}&v(\zeta,\alpha)^t(\mathcal{A}^+)^{-1}\mathcal{B}^+v(\xi,-\alpha)=v(\zeta,\alpha)^tM(N\mathcal{A}^+M)^{-1}(N\mathcal{B}^+)v(\xi,-\alpha)\,,\\ &M=U-q^{2s_{\mathbf{2}}}\tau_{\mathbf{2}}^{-2}I\,.\end{aligned}$$

Here U is the matrix whose non-zero entries are $U_{\mathbf{m},\mathbf{m}+1}=1$ ($\mathbf{m}=\mathbf{0},\dots,\mathbf{n}-\mathbf{1}$), and the matrix N rearranges the contours $\Gamma_{\mathbf{m}}$ in such a way that only the first one is unnatural after the fusion:

$$\Gamma_{\mathbf{0}},\Gamma_{\mathbf{1}},\Gamma_{\mathbf{2}},\Gamma_{\mathbf{3}},\cdots,\Gamma_{\mathbf{n}}\ \to\ \Gamma_{\mathbf{1}},\Gamma_{\mathbf{0}},\Gamma_{\mathbf{2}}+\Gamma_{\mathbf{1}},\Gamma_{\mathbf{3}},\cdots,\Gamma_{\mathbf{n}}\,.$$

In the limit $\tau_{\mathbf{1}}\to\tau_{\mathbf{2}}q^{-s_{\mathbf{2}}-1/2}$, all matrix elements of $N\mathcal{A}^+M$ remain finite except for the one at the top left corner. It means in turn that the first row and the first column of the inverse matrix $(N\mathcal{A}^+M)^{-1}$ vanishes in this limit. It is then easy to see that as $\tau_{\mathbf{1}}\to\tau_{\mathbf{2}}q^{-s_{\mathbf{2}}-1/2}$

$$v(\zeta,\alpha)^t(N\mathcal{A}^+M)^{-1}\to(1-q^{2s_{\mathbf{2}}}(\zeta/\tau_{\mathbf{2}})^2)\cdot\big(0,\quad v_{\mathrm{fus}}(\zeta,\alpha)^t(\mathcal{A}^+_{\mathrm{fus}})^{-1}\big)\,,\tag{4.40}$$

where we use a tautological notation $v_{\mathrm{fus}}(\zeta,\alpha)$ for the $\mathbf{n}$-dimensional vector.

In order to evaluate $(N\mathcal{B}^+)v(\xi,-\alpha)$ we use the identity

$$\widetilde{\Omega}^+(\eta)^tv(\xi,-\alpha)\Big|_{\tau_{\mathbf{1}}=\tau_{\mathbf{2}}q^{-s_{\mathbf{2}}-1/2}}=\widetilde{\Omega}^+_{\mathrm{fus}}(\eta)^tv_{\mathrm{fus}}(\xi,-\alpha)-\frac{1}{2}E_{\mathrm{fus},\eta}\big((\eta/\xi)^\alpha(\eta^2+\xi^2)\big)\,.$$

The integrals over η of $E_{\mathrm{fus},\eta}\big((\eta/\xi)^\alpha(\eta^2+\xi^2)\big)$ vanishes for all cycles except for $\Gamma_{\mathbf{1}}$, which does not count due to (4.40). The statement of Lemma follows from this. $\square$

We finish this section with a remark which will be useful in the future.

REMARK 4.14. Consider a function $g^{\pm}(\zeta)$ which may have singularities away from the φ-divisor. Suppose that

$$g^{\pm}(\zeta)=\zeta^{\pm\alpha}(C+O(\zeta^2)),\quad \zeta\to 0\,.$$

Then using the definition (4.18) we have

$$\int_{\Gamma_0} T(\zeta,\kappa)\delta^-_\zeta \Delta^{-1}_\zeta (g^\pm(\zeta))d\mu^\pm(\zeta) = 2\pi i q^{\mp(\kappa-s)}C\,.$$

4.3. Main Theorem

Now we are ready to put together the *Space* and the *Matsubara* directions; namely, to evaluate the functional $\mathcal{Z}^\kappa_{\mathbf{n}}$ on quasi-local operators created by $\mathbf{b}^*(\zeta)$, $\mathbf{c}^*(\zeta)$, $\mathbf{t}^*(\zeta)$.

We begin with a simple Lemma which shows that under $\mathcal{Z}^\kappa_{\mathbf{n}}$ the dressed Russian doll changes its dress from adjoint in the *Space* to *Matsubara* . In what follows we set

$$\mathbf{b}^*(\zeta)=\mathbf{x}^{+*}(\zeta)\,,\quad \mathbf{c}^*(\zeta)=\mathbf{x}^{-*}(\zeta)\,,\quad \mathbf{t}^*(\zeta)=\mathbf{x}^{0*}(\zeta)\,.$$

LEMMA 4.15. *Let $\epsilon_1,\dots,\epsilon_l\in\{+,-,0\}$ such that $\sum_{j=1}^l\epsilon_j=0$. Then we have*

(4.41)

$$\mathcal{Z}^\kappa_{\mathbf{n}}\Big\{\mathbf{x}^{\epsilon_l*}(\zeta_l)\cdots\mathbf{x}^{\epsilon_1*}(\zeta_1)(q^{2\alpha S(0)})\Big\}$$
$$=2^l\frac{\langle\kappa+\alpha|\mathrm{Tr}_{c_1,\dots,c_l}\Big\{T_{c_1,\mathbf{M}}(\zeta_1,\kappa)\cdots T_{c_l,\mathbf{M}}(\zeta_l,\kappa)\mathbf{g}^{\epsilon_l}_{c_l}(\zeta_l,\alpha)\cdots\mathbf{g}^{\epsilon_1}_{c_1}(\zeta_1)(I)\Big\}|\kappa\rangle}{\prod_{j=1}^l T(\zeta_j,\kappa)\ \langle\kappa+\alpha|\kappa\rangle}\,.$$

As before, $\langle\kappa+\alpha|$ and $|\kappa\rangle$ are the eigenvectors corresponding to the maximal eigenvalues of $T_{\mathbf{M}}(1,\kappa+\alpha)$ and $T_{\mathbf{M}}(1,\kappa)$, respectively.

PROOF. Recall the formula for the dressed Russian doll

(4.42) $$\mathbf{x}^{\epsilon_l*}(\zeta_l,\alpha)\cdots\mathbf{x}^{\epsilon_1*}(\zeta_1)(q^{2\alpha S(0)})$$
$$=q^{2\alpha S(0)}\mathrm{Tr}_{c_1,\dots,c_l}\Big\{\mathbb{T}_{c_1[1,\infty)}(\zeta_1)\cdots\mathbb{T}_{c_l,[1,\infty)}(\zeta_l)\mathbf{g}^{\epsilon_l}_{c_l}(\zeta_l,\alpha)\cdots\mathbf{g}^{\epsilon_1}_{c_1}(\zeta_1)(I)\Big\}\,.$$

Due to the condition $\sum_{j=1}^l\epsilon_j=0$, the initial and the final α's coincide.

On the other hand we can rewrite the formula (1.54) for arbitrary m as

(4.43) $$\mathcal{Z}^\kappa_{\mathbf{n}}\Big\{q^{2S(0)}X_{[1,m]}\Big\}=\frac{\langle\kappa+\alpha|\mathrm{Tr}_{[1,\infty)}\big(T^{\mathrm{norm}}_{[1,\infty),\mathbf{M}}(1,\kappa)X_{[1,m]}\big)|\kappa\rangle}{\langle\kappa+\alpha|\kappa\rangle}\,,$$

where we introduced the normalised monodromy matrix:

$$T^{\mathrm{norm}}_{j,\mathbf{M}}(1,\kappa)=\frac{1}{T(1,\kappa)}T_{j,\mathbf{M}}(1,\kappa)\,.$$

Due to Remark 3.15 we have

(4.44) $$\mathrm{Tr}_{c_1,\dots,c_l}\Big\{\mathbb{T}_{c_1[1,\infty)}(\zeta_1)\cdots\mathbb{T}_{c_l,[1,\infty)}(\zeta_l)\mathbf{g}^{\epsilon_l}_{c_l}(\zeta_l,\alpha)\cdots\mathbf{g}^{\epsilon_1}_{c_1}(\zeta_1)(1)\Big\}|\kappa\rangle$$
$$=\prod_{j=1}^l\frac{2}{T(\zeta_j,\kappa)}\ \mathrm{Tr}_{c_1,\dots,c_l}\Big\{T_{c_1,\mathbf{M}}(\zeta_1,\kappa)\mathbb{T}_{c_1[1,\infty)}(\zeta_1)\cdots$$
$$\times T_{c_l,\mathbf{M}}(\zeta_l,\kappa)\mathbb{T}_{c_l,[1,\infty)}(\zeta_l)\mathbf{g}^{\epsilon_l}_{c_l}(\zeta_l,\alpha)\cdots\mathbf{g}^{\epsilon_1}_{c_1}(\zeta_1)(1)\Big\}|\kappa\rangle\,.$$

Though $T_{c_j,\mathbf{M}}(\zeta_j,\kappa)$ are operators acting in the *Matsubara* space, we can apply the argument of Remark 3.15 noting that $|\kappa\rangle$ is an eigenvector of the transfer matrix

$T_{\mathbf{M}}(\zeta,\kappa)$. The rest is simple: we plug (4.44) into (4.43), use the Yang-Baxter equation and

$$T^{\mathrm{norm}}_{[1,\infty),\mathbf{M}}(1,\kappa)|\kappa\rangle = |\kappa\rangle\,.$$

□

This Lemma suggests us to introduce a functional

$$Z_{\zeta_1,\dots,\zeta_l}\{X_{[1,l]}\} = \frac{\langle\kappa+\alpha|\mathrm{Tr}_{1,\dots,l}\{T_{1,\mathbf{M}}(\zeta_1,\kappa)\cdots T_{l,\mathbf{M}}(\zeta_l,\kappa)X_{[1,l]}\}|\kappa\rangle}{\prod_{j=1}^{l}T(\zeta_j,\kappa)\ \ \langle\kappa+\alpha|\kappa\rangle}\,, \tag{4.45}$$

which plays the role of $\mathcal{Z}^{\kappa}_{\mathbf{n}}$ in the inhomogeneous case. This is an intermediate object, so we do not indicate other parameters such as $\mathbf{n},\kappa$, etc. Obviously,

$$Z_{\zeta_1,\dots,\zeta_l}\{X_{[1,l-1]}\} = Z_{\zeta_1,\dots,\zeta_{l-1}}\{X_{[1,l-1]}\}\,. \tag{4.46}$$

REMARK 4.16. We are concerned in the end with the maximal eigenvectors $|\kappa\rangle$ and $|\kappa+\alpha\rangle$ of $T_{\mathbf{M}}(1,\kappa)$ and $T_{\mathbf{M}}(1,\kappa+\alpha)$. Nevertheless, in what follows we shall not use this property; all we need is that they are eigenvectors satisfying the genericity condition

$$\langle\kappa+\alpha|\kappa\rangle\neq 0\,,$$

which stipulates in particular that

$$\mathbb{S}(X_{[1,l]}) = 0\,. \tag{4.47}$$

The genericity condition is true for generic values of parameters due to Lemma 4.11.

In what follows by *Matsubara* data we understand length $\mathbf{n}$, spins, inhomogeneities, twist κ and two eigenvectors $|\kappa\rangle$, $|\kappa+\alpha\rangle$.

We shall need the following simple property.

LEMMA 4.17. *Let $X_{[1,l]}$ be an element of* $\mathrm{End}\,V^{\otimes l}$ *of spin* 0*, possibly depending on $\zeta_1,\dots,\zeta_l$. If the condition*

$$Z_{\zeta_1,\dots,\zeta_l}\{X_{[1,l]}\} = 0 \tag{4.48}$$

holds for general Matsubara data, then

$$X_{[1,l]} = 0\,.$$

PROOF. The statement is easy to believe in view of infinitely many possible choices for the *Matsubara* data. So, readers who are not particularly concerned with mathematical rigour can skip the proof.

As in Lemma (4.11), consider all eigenvectors $\langle\kappa{+}\alpha,I,s|$, $|\kappa,J,s\rangle$ for all possible spins s. Since the scalar product does not vanish generically, (4.48) implies that

$$\langle\kappa+\alpha,I,s|\mathrm{Tr}_{1,\dots,l}\{T_{1,\mathbf{M}}(\zeta_1,\kappa)\cdots T_{l,\mathbf{M}}(\zeta_l,\kappa)X_{[1,l]}\}|\kappa,J,s\rangle = 0 \tag{4.49}$$

holds for all s, I, J. Set $\mathbf{n}=l$, $\tau_j=\zeta_j q^{-1/2}$. Then the product of the monodromy matrices turns into the permutation of *Space* and *Matsubara* , and (4.49) becomes

$$\langle\kappa+\alpha,I,s|X_{\mathbf{M}}|\kappa,J,s\rangle = 0\,.$$

This implies $X=0$ by the completeness of the eigenvectors. □

We compute the right hand side of (4.41) inductively, evaluating first

$$Z_{\zeta_1,\dots,\zeta_l}\{\mathbf{g}^{\epsilon}_l(\zeta_l,\alpha)\bigl(X_{[1,l-1]}\bigr)\}\,,\quad \epsilon=0,\pm\,.$$

Here we require $\mathbb{S}(X_{[1,l-1]})=-\epsilon$ in view of (4.47). It is probably worth recalling that $\mathbf{g}^{\epsilon}_l(\zeta_l,\alpha)$ depends on the parameters $\zeta_1,\dots,\zeta_{l-1}$ as well. We begin with an almost trivial statement.

LEMMA 4.18. *For $\epsilon=0$ we have*

$$Z_{\zeta_1,\dots,\zeta_l}\{\mathbf{g}^{0}_l(\zeta_l,\alpha)\bigl(X_{[1,l-1]}\bigr)\}=\rho(\zeta_l|\alpha)Z_{\zeta_1,\dots,\zeta_{l-1}}\{X_{[1,l-1]}\}\,,$$

where

$$\rho(\zeta|\alpha)=\frac{T(\zeta,\kappa+\alpha)}{T(\zeta,\kappa)}\,,$$

as we already introduced, see (4.16).

PROOF. By definition

$$\mathbf{g}^{0}_l(\zeta_l,\alpha)\bigl(X_{[1,l-1]}\bigr)=\mathbb{T}_{l,[1,l-1]}(\zeta_l)\bigl(q^{\alpha\sigma^3_l}X_{[1,l-1]}\bigr)\,.$$

Using the Yang-Baxter equation, we derive

$$\begin{aligned}
&Z_{\zeta_1,\dots,\zeta_l}\{\mathbf{g}^{0}_l(\zeta_l,\alpha)\bigl(X_{[1,l-1]}\bigr)\}\\
&=\frac{1}{\prod_{j=1}^{l}T(\zeta_j,\kappa)\langle\kappa+\alpha|\kappa\rangle}\langle\kappa+\alpha|\mathrm{Tr}_{1,\dots,l}\{T_{1,\mathbf{M}}(\zeta_1,\kappa)\cdots T_{l-1,\mathbf{M}}(\zeta_{l-1},\kappa)T_{l,\mathbf{M}}(\zeta_l,\kappa)\\
&\times\mathbb{T}_{l,[1,l-1]}(\zeta_l)\bigl(q^{\alpha\sigma^3_l}X_{[1,l-1]}\bigr)\}|\kappa\rangle\\
&=\frac{1}{\prod_{j=1}^{l}T(\zeta_j,\kappa)\ \ \langle\kappa+\alpha|\kappa\rangle}\\
&\times\langle\kappa+\alpha|\mathrm{Tr}_l(T_{l,\mathbf{M}}(\zeta_l,\kappa+\alpha))\mathrm{Tr}_{1,\dots,l-1}\{T_{1,\mathbf{M}}(\zeta_1,\kappa)\cdots T_{l-1,\mathbf{M}}(\zeta_{l-1},\kappa)X_{[1,l-1]}\}|\kappa\rangle\\
&=\rho(\zeta_l|\alpha)Z_{\zeta_1,\dots,\zeta_{l-1}}\{X_{[1,l-1]}\}\,.
\end{aligned}$$

□

Now we consider the case $\epsilon=+$ which is much more complicated. We do it in two steps. We skip the superscript $+$ for $\mathbf{g}$ in accordance with our previous convention.

LEMMA 4.19. *The expression*

(4.50)

$$\zeta_l^{-\alpha}T(\zeta_l,\kappa)Z_{\zeta_1,\dots,\zeta_l}\Bigl\{\Bigl(\mathbf{g}_l(\zeta_l,\alpha)-\frac{1}{4\pi i}\oint_{\Gamma}\omega_{\mathrm{sing}}(\zeta_l,\xi|\alpha)\mathbf{c}(\xi,\alpha)\frac{d\xi^2}{\xi^2}\Bigr)\bigl(X_{[1,l-1]}\bigr)\Bigr\}\,,$$

is a polynomial in ζ_l^2 of degree $\mathbf{n}$. Here the contour Γ encircles the singularities $\zeta_1,\dots,\zeta_{l-1}$, and the function $\omega_{\mathrm{sing}}(\zeta_l,\xi|\alpha)$ is defined in (4.35).

PROOF. We introduced $\mathbf{f}_{\mathrm{rat}}(\zeta,\alpha)$ in (3.69) such that $\zeta^{-\alpha}\mathbf{f}_{\mathrm{rat}}(\zeta,\alpha)$ is a rational function of ζ^2. Replacing $\mathbf{f}(\zeta,\alpha)$ with $\mathbf{f}_{\mathrm{rat}}(\zeta,\alpha)$ in the definition of $\mathbf{g}(\zeta,\alpha)$, we define $\mathbf{g}_{\mathrm{rat}}(\zeta,\alpha)$ with the same rationality property. From Theorem 3.23 we obtain (cf. (3.109))

$$\mathbf{g}_{\mathrm{rat}}(\zeta,\alpha)=\mathbf{g}(\zeta,\alpha)-\frac{1}{4\pi i}\oint_{\Gamma}D_\zeta D_\xi\Delta_\zeta^{-1}\psi(\zeta/\xi,\alpha)\mathbf{c}(\xi,\alpha)\frac{d\xi^2}{\xi^2}\,,\tag{4.51}$$

where D_ζ is defined in (3.110). Now using the definition of $\omega_{\rm sing}(\zeta,\xi|\alpha)$ and Lemma 4.18 we see that (4.50) is equal to

(4.52)

$$\zeta_l^{-\alpha}T(\zeta_l,\kappa)Z_{\zeta_1,\dots,\zeta_l}\Big\{\Big(\mathbf{g}_{{\rm rat},l}(\zeta_l,\alpha)-\frac{1}{4\pi i}\oint_\Gamma\tilde{\omega}_{\rm sing}(\zeta_l,\xi|\alpha)\mathbf{c}(\xi,\alpha)\frac{d\xi^2}{\xi^2}\Big)(X_{[1,l-1]})\Big\},$$

where

$$\begin{aligned}\tilde{\omega}_{\rm sing}(\zeta,\xi|\alpha)&=\big(2\rho(\zeta|\alpha)-2\rho(\xi|\alpha)-\Delta_\zeta\big)\psi(\zeta/\xi,\alpha)\\&\quad+\frac{4}{T(\zeta,\kappa)T(\xi,\kappa)}\Big(a(\zeta)d(\xi)\psi(q\zeta/\xi,\alpha)-a(\xi)d(\zeta)\psi(q^{-1}\zeta/\xi,\alpha)\Big).\end{aligned}$$

The expression (4.52) is a rational function of ζ_l^2 with poles at $\zeta_l^2=\zeta_j^2q^{\pm2}$, $j=1,\dots,l-1$. We shall consider its behaviour at $\zeta^2=0,\infty$ later. For the moment we concentrate on the residues.

Let us compute the residues of $\mathbf{g}_{{\rm rat},l}(\zeta_l,\alpha)$ at $\zeta_l^2=\zeta_{l-1}^2q^{\pm2}$, the other residues are obtained by R-matrix symmetry. Denote $J=[1,l-1]$. Recall

(4.53)

$$\begin{aligned}&\mathbf{g}_{{\rm rat},l,J}(\zeta_l,\alpha)(X_J)\\&=\Big(\frac{1}{2}\mathbf{f}_{{\rm rat},J}(q\zeta_l,\alpha)+\frac{1}{2}\mathbf{f}_{{\rm rat},J}(q^{-1}\zeta_l,\alpha)\Big)(X_J)\\&-\mathbb{T}_{l,J}(\zeta_l,\alpha)\mathbf{f}_{{\rm rat},J}(\zeta_l,\alpha)(X_J)\\&+{\rm Tr}_{A,a}\Big\{\Big(-\frac{1}{2}\sigma_l^3\sigma_a^++\sigma_l^+\sigma_a^3-\mathbf{a}_A\sigma_l^+\sigma_a^+\Big)\mathbb{T}_{\{a,A\},J}(\zeta_l,\alpha)\zeta_l^{\alpha-\mathbb{S}}\left(q^{-2S_J}X_J\right)\Big\}.\end{aligned}$$

The second line in the right hand side of (4.53) does not contain the residues at $\zeta_l=\zeta_jq^{\pm1}$ when inserted into (4.50), because $\mathbb{T}_{l,J}(\zeta_l,\alpha)$ can be removed using Lemma 4.18, $\rho(\zeta_l|\alpha)T(\zeta_l,\kappa)$ is a polynomial and $\mathbf{f}_{{\rm rat},J}(\zeta_l,\alpha)(X_J)$ has poles only at $\zeta_l^2=\zeta_j^2$. Using the definitions and Lemma 3.10 we compute:

$$\begin{aligned}&{\rm res}_{\zeta_l=\zeta_{l-1}q^{\pm1}}\Big(\frac{1}{2}\mathbf{f}_{{\rm rat},J}(q\zeta_l,\alpha)+\frac{1}{2}\mathbf{f}_{{\rm rat},J}(q^{-1}\zeta_l,\alpha)\Big)(X_J)\frac{d\zeta_l^2}{\zeta_l^2}\\&=-\frac{1}{4}{\rm res}_{\zeta_l=\zeta_{l-1}}\Big[\sigma_{l-1}^+,\mathbf{q}_J(\zeta_l,\alpha)(X_J)\Big]\frac{d\zeta_l^2}{\zeta_l^2},.\end{aligned}$$

Now we compute the residues of the last line of (4.53) which consists of three terms. Recall that

$${\rm Tr}_{A,a}\left\{\sigma_a^+\mathbb{T}_{\{a,A\},J}(\zeta_l,\alpha)\zeta^{\alpha-\mathbb{S}}\left(q^{-2S_J}X_J\right)\right\}=\mathbf{k}_J(\zeta_l,\alpha)\left(X_J\right),$$

so, for the first term we can use directly Lemma 3.10. In the second term

$${\rm Tr}_{A,a}\left\{\sigma_a^3\mathbb{T}_{\{a,A\},J}(\zeta_l,\alpha)\zeta^{\alpha-\mathbb{S}}\left(q^{-2S_J}X_J\right)\right\}=\left(\mathbf{q}_J(\zeta_lq,\alpha)-\mathbf{q}_J(\zeta_lq^{-1},\alpha)\right)\left(X_J\right),$$

which allows us to express the residues in terms of the residues of $\mathbf{q}_J(\zeta_l,\alpha)\left(X_J\right)$. The only unusual term is the third one. For the residue at $\zeta_l=\zeta_{l-1}q^{-1}$ we perform exactly the same computation as in Lemma 3.10 using at the end the identity

$$\mathbf{a}_A\sigma_{l-1}^+R_{A,l-1}(1)=-\tau_{l-1}^+R_{A,l-1}(1),$$

obtaining

$$\operatorname*{res}_{\zeta_l=\zeta_{l-1}q^{-1}} \mathrm{Tr}_{A,a}\left\{\mathbf{a}_A\sigma_a^+\mathbb{T}_{\{a,A\},J}(\zeta_l,\alpha)\zeta^{\alpha-\mathbb{S}}\left(q^{-2S_J}X_J\right)\right\}\frac{d\zeta_l^2}{\zeta_l^2}$$
$$=\operatorname*{res}_{\zeta_l=\zeta_{l-1}}\tau_{l-1}^+\mathbf{q}_J(\zeta_l,\alpha)(X_J)\frac{d\zeta_l^2}{\zeta_l^2}\,.$$

Similarly

$$\operatorname*{res}_{\zeta_l=\zeta_{l-1}q} \mathrm{Tr}_{A,a}\left\{\mathbf{a}_A\sigma_a^+\mathbb{T}_{\{a,A\},J}(\zeta_l,\alpha)\zeta_l^{\alpha-\mathbb{S}}\left(q^{-2S_J}X_J\right)\right\}\frac{d\zeta_l^2}{\zeta_l^2}$$
$$=-\operatorname*{res}_{\zeta_l=\zeta_{l-1}}\mathbf{q}_J(\zeta_l,\alpha)(X_J)\tau_{l-1}^-\frac{d\zeta_l^2}{\zeta_l^2}\,.$$

Putting all these together we obtain

$$\operatorname*{res}_{\zeta_l=q^\varepsilon\zeta_{l-1}}\left(\mathbf{g}_{\mathrm{rat},l,J}(\zeta_l,\alpha)(X_J)+\mathbb{T}_{l,J}(\zeta_l,\alpha)\mathbf{f}_{\mathrm{rat},J}(\zeta_l,\alpha)(X_J)\right)\frac{d\zeta_l^2}{\zeta_l^2}$$
$$=\begin{cases}-\frac{1}{4}[\sigma_{l-1}^+,U_J]_+-U_J(\tau_{l-1}^+\sigma_l^+-\sigma_{l-1}^+\tau_l^+) & (\varepsilon=+),\\ \frac{1}{4}[\sigma_{l-1}^+,U_J]_++(\tau_{l-1}^-\sigma_l^+-\sigma_{l-1}^+\tau_l^-)U_J & (\varepsilon=-),\end{cases}\tag{4.54}$$

where

$$U_J=\operatorname*{res}_{\zeta_l=\zeta_{l-1}}\mathbf{q}_J(\zeta_l,\alpha)(X_J)\frac{d\zeta_l^2}{\zeta_l^2}\,.$$

In the proof of Theorem 3.23 we had

$$\operatorname*{res}_{\zeta_l=\zeta_{l-1}}\mathbf{c}_J(\zeta_l,\alpha)(X_J)\frac{d\zeta_l^2}{\zeta_l^2}=-\frac{1}{2}[\sigma_{l-1}^+,U_J]_+\,.$$

It was also shown that $[\sigma_{l-1}^+,U_J]_+$ acts as unit operator in $(l-1)$-th space:

$$[\sigma_{l-1}^+,U_J]_+=x_{[1,l-2]}I_{l-1},\quad\text{implying}\quad x_{[1,l-2]}=\mathrm{Tr}_{l-1}(\sigma_{l-1}^+U_J)\,.\tag{4.55}$$

Notice that

$$\tau_{l-1}^+\sigma_l^+-\sigma_{l-1}^+\tau_l^+=(\tau_{l-1}^+\sigma_l^+-\sigma_{l-1}^+\tau_l^+)\mathcal{P}_{l-1,l}^-\,,$$
$$\tau_{l-1}^-\sigma_l^+-\sigma_{l-1}^+\tau_l^-=\mathcal{P}_{l-1,l}^-(\tau_{l-1}^-\sigma_l^+-\sigma_{l-1}^+\tau_l^-)\,.$$

So, plugging the residues into the trace we can use the quantum determinant:

$$T_{l-1,\mathbf{M}}(\zeta_{l-1},\kappa)T_{l,\mathbf{M}}(\zeta_{l-1}q^{-1},\kappa)\mathcal{P}_{l-1,l}^-=a(\zeta_{l-1}q^{-1})d(\zeta_{l-1})\mathcal{P}_{l-1,l}^-\,,$$
$$\mathcal{P}_{l-1,l}^-T_{l-1,\mathbf{M}}(\zeta_{l-1},\kappa)T_{l,\mathbf{M}}(\zeta_{l-1}q,\kappa)=a(\zeta_{l-1})d(\zeta_{l-1}q)\mathcal{P}_{l-1,l}^-\,.$$

Now we compute

$$\operatorname*{res}_{\zeta_l=\zeta_{l-1}q}Z_{\zeta_1,\dots,\zeta_l}\left\{\mathbf{g}_l(\zeta_l,\alpha)(X_J)\right\}\frac{d\zeta_l^2}{\zeta_l^2}=\frac{1}{2}\operatorname*{res}_{\zeta_l=\zeta_{l-1}}Z_{\zeta_1,\dots,\zeta_l}\left\{\mathbf{c}(\zeta_l,\alpha)(X_J)\right\}\frac{d\zeta_l^2}{\zeta_l^2}$$
$$-Z_{\zeta_1,\dots,\zeta_{l-1},\zeta_{l-1}q}\left\{U_J(\tau_{l-1}^+\sigma_l^+-\sigma_{l-1}^+\tau_l^+)\right\}.$$

The first term is already as wanted, so let us evaluate the second one. We need to compute

$$\begin{aligned}
&\mathrm{Tr}_{l-1,l}\big(T_{l-1,\mathbf{M}}(\zeta_{l-1},\kappa)T_{l,\mathbf{M}}(\zeta_{l-1}q,\kappa)U_J(\tau^+_{l-1}\sigma^+_l-\sigma^+_{l-1}\tau^+_l)\big)\\
&\quad= a(\zeta_{l-1})d(\zeta_{l-1}q)\mathrm{Tr}_{l-1,l}\big(U_J(\tau^+_{l-1}\sigma^+_l-\sigma^+_{l-1}\tau^+_l)\big)\\
&\quad= -\frac{1}{2}a(\zeta_{l-1})d(\zeta_{l-1}q)\mathrm{Tr}_{l-1}\big([U_J,\sigma^+_{l-1}]_+\big)\\
&\quad= \frac{2a(\zeta_{l-1})d(\zeta_{l-1}q)}{T(\zeta_{l-1},\kappa)T(\zeta_{l-1}q,\kappa)}\mathrm{Tr}_{l-1,l}\big(T_{l-1,\mathbf{M}}(\zeta_{l-1},\kappa)T_{l,\mathbf{M}}(\zeta_{l-1}q,\kappa)\\
&\quad\times \operatorname*{res}_{\zeta_l=\zeta_{l-1}}\big(\mathbf{c}(\zeta_l,\alpha)(X_J)\big)\big)\frac{d\zeta_l^2}{\zeta_l^2}\,.
\end{aligned}$$

Thus the residues at $\zeta_l=\zeta_{l-1}q$ of (4.50) vanishes. A similar computation shows that the residue at $\zeta_l=\zeta_{l-1}q^{-1}$ also vanishes. By R-matrix symmetry, it follows that (4.50) does not have other singularities than those at $\zeta^2=0,\infty$.

It is easy to see from the definition that

$$\begin{aligned}
\mathbf{g}_{\mathrm{rat},l}(\zeta,\alpha)(X_{[1,l-1]})&=O(\zeta^\alpha),\ \zeta\to 0\,,\\
\mathbf{g}_{\mathrm{rat},l}(\zeta,\alpha)(X_{[1,l-1]})&=O(\zeta^\alpha),\ \zeta\to \infty\,,
\end{aligned}$$

if the spin of $X_{[1,l-1]}$ equals -1. This implies that (4.50) is a polynomial of ζ^2 of degree $\mathbf{n}$.

The proof is finished. □

The formula (4.50) is nice because it expresses the action of a creation operator in terms of an annihilation operator. On the other hand it is not exact: a polynomial of degree $\mathbf{n}$ remains undetermined. We need $\mathbf{n}+1$ conditions to fix it. This is done in the spirit of the previous section.

LEMMA 4.20. *The following integrals vanish:*

$$\int_{\Gamma_{\mathbf{m}}} T(\zeta_l,\kappa)Z_{\zeta_1,\dots,\zeta_l}\{\mathbf{g}_l(\zeta_l,\alpha)(X_{[1,l-1]})\}d\mu^+(\zeta_l)=0\,, \tag{4.56}$$

for $\mathbf{m}=\mathbf{0},\mathbf{1},\dots,\mathbf{n}$.

PROOF. We consider first the case $\mathbf{m}\neq\mathbf{0}$. According to Section 2.4 the fusion in the *Matsubara* direction is a simple procedure. So we shall assume that $s_{\mathbf{m}}=1/2$, $\mathbf{m}=\mathbf{1},\dots,\mathbf{n}$. We prove the identity with $\mathbf{m}=\mathbf{n}$; the other cases follow by symmetry. It has been said that in this case the integrals can be taken literally even in the presence of transcendental functions.

Since

$$\langle\kappa+\alpha|Q^-_{\mathbf{M}}(\zeta,\kappa+\alpha)=\langle\kappa+\alpha|Q^-(\zeta,\kappa+\alpha)\,,\quad Q^+_{\mathbf{M}}(\zeta,\kappa)|\kappa\rangle=Q^+(\zeta,\kappa)|\kappa\rangle\,,$$

it is convenient to consider the following object:

$$\begin{aligned}
I_{\mathbf{M}}(\zeta_l)=Q^-_{\mathbf{M}}(\zeta,\kappa+\alpha)\mathrm{Tr}_{[1,l]}\Big(&T_{1,\mathbf{M}}(\zeta_1,\kappa)\cdots T_{l-1,\mathbf{M}}(\zeta_{l-1},\kappa)T_{l,\mathbf{M}}(\zeta_l,\kappa)\\
&\times\mathbf{g}_l(\zeta_l,\alpha)(X_{[1,l-1]})\Big)Q^+_{\mathbf{M}}(\zeta,\kappa)\,.
\end{aligned}$$

Then using

$$\mathrm{res}_{\zeta=q^{-1/2}\tau_{\mathbf{n}}}\varphi(\zeta)\frac{d\zeta^2}{\zeta^2}=\frac{d(\tau_{\mathbf{n}}q^{1/2})}{a(\tau_{\mathbf{n}}q^{-1/2})}\mathrm{res}_{\zeta=q^{1/2}\tau_{\mathbf{n}}}\varphi(\zeta)\frac{d\zeta^2}{\zeta^2}\,,$$

we rewrite the identity to be proved for $\mathbf{m}=\mathbf{n}$, $s_{\mathbf{n}}=1/2$, as

$$a(\tau_{\mathbf{n}}q^{-1/2})I_{\mathbf{M}}(\tau_{\mathbf{n}}q^{1/2})+d(\tau_{\mathbf{n}}q^{1/2})I_{\mathbf{M}}(\tau_{\mathbf{n}}q^{-1/2})=0\,. \tag{4.57}$$

We show that this identity follows from the difference equation (4.5).

Let us evaluate $I_{\mathbf{M}}(\tau_{\mathbf{n}}q^{1/2})$. First, use the Yang-Baxter equation under the trace:

$$\begin{aligned}&\mathrm{Tr}_{[1,l]}\Bigl(T_{1,\mathbf{M}}(\zeta_1,\kappa)\cdots T_{l-1,\mathbf{M}}(\zeta_{l-1},\kappa)T_{l,\mathbf{M}}(\tau_{\mathbf{n}}q^{1/2},\kappa)\mathbf{g}_l(\tau_{\mathbf{n}}q^{1/2},\alpha)(X_{[1,l-1]})\Bigr)\\&=\mathrm{Tr}_{[1,l]}\Bigl(T_{l,\mathbf{M}}(\tau_{\mathbf{n}}q^{1/2},\kappa)T_{1,\mathbf{M}}(\zeta_1,\kappa)\cdots T_{l-1,\mathbf{M}}(\zeta_{l-1},\kappa)\\&\times T_{l,[1,l-1]}(\tau_{\mathbf{n}}q^{1/2})^{-1}\mathbf{g}_l(\tau_{\mathbf{n}}q^{1/2},\alpha)(X_{[1,l-1]})T_{l,[1,l-1]}(\tau_{\mathbf{n}}q^{1/2})\Bigr)\,.\end{aligned}$$

Recalling the definition of the *Matsubara* monodromy matrix and the relation between the L-operator and the R-matrix (1.17), (1.15) we find

$$T_{l,\mathbf{M}}(\tau_{\mathbf{n}}q^{1/2},\kappa)=\eta P_{l,\mathbf{n}}T_{l,\mathbf{M}'}(\tau_{\mathbf{n}}q^{1/2},\kappa)=\eta T_{\mathbf{n},\mathbf{M}'}(\tau_{\mathbf{n}}q^{1/2},\kappa)P_{l,\mathbf{n}}\,,$$

where $\mathbf{M}'=[\mathbf{1},\mathbf{n}-1]$, $\eta=q^{1/2}(q^{-1}-q)$. Further,

$$\eta T_{\mathbf{n},\mathbf{M}'}(\tau_{\mathbf{n}}q^{1/2},\kappa)=T_{\mathbf{M}}(\tau_{\mathbf{n}}q^{1/2},\kappa+\alpha)q^{-\alpha\sigma^3_{\mathbf{n}}}\,.$$

At the special point $\zeta=\tau_{\mathbf{n}}q^{1/2}$ the Baxter equation (2.105) contains only one term in the right hand side. Hence,

$$\begin{aligned}&I_{\mathbf{M}}(\tau_{\mathbf{n}}q^{1/2})\\&=d(\tau_{\mathbf{n}}q^{1/2})Q^-_{\mathbf{M}}(\tau_{\mathbf{n}}q^{-1/2},\kappa+\alpha)\mathrm{Tr}_{[1,l]}\Bigl(q^{-\alpha\sigma^3_{\mathbf{n}}}P_{l,\mathbf{n}}T_{1,\mathbf{M}}(\zeta_1,\kappa)\cdots T_{l-1,\mathbf{M}}(\zeta_{l-1},\kappa)\\&\times T_{l,[1,l-1]}(\tau_{\mathbf{n}}q^{1/2})^{-1}\mathbf{g}_l(\tau_{\mathbf{n}}q^{1/2},\alpha)(X_{[1,l-1]})T_{l,[1,l-1]}(\tau_{\mathbf{n}}q^{1/2})\Bigr)Q^+_{\mathbf{M}}(\tau_{\mathbf{n}}q^{1/2},\kappa)\\&=\mu(\tau_{\mathbf{n}})d(\tau_{\mathbf{n}}q^{1/2})Q^-_{\mathbf{M}}(\tau_{\mathbf{n}}q^{-1/2},\kappa+\alpha)\mathrm{Tr}_{[1,l]}\Bigl(T_{1,\mathbf{M}'}(\zeta_1,\kappa)\cdots T_{l-1,\mathbf{M}'}(\zeta_{l-1},\kappa)P_{l,\mathbf{n}}\\&\times q^{-\alpha\sigma^3_l}T_{l,[1,l-1]}(\tau_{\mathbf{n}}q^{1/2})^{-1}\mathbf{g}_l(\tau_{\mathbf{n}}q^{1/2},\alpha)(X_{[1,l-1]})\Bigr)Q^+_{\mathbf{M}}(\tau_{\mathbf{n}}q^{1/2},\kappa)\,,\end{aligned} \tag{4.58}$$

where $\mu(\tau_{\mathbf{n}})$ is an irrelevant constant appearing in the identity

$$T_{l,[1,l-1]}(\tau_{\mathbf{n}}q^{1/2})L_{1,l}(\zeta_1/\tau_{\mathbf{n}})\cdots L_{l-1,l}(\zeta_{l-1}/\tau_{\mathbf{n}})=\mu(\tau_{\mathbf{n}})\,.$$

Now let us evaluate $I_{\mathbf{M}}(\tau_{\mathbf{n}}q^{-1/2})$ in a similar fashion. Using

$$\begin{aligned}&T_{l,\mathbf{M}}(\tau_{\mathbf{n}}q^{-1/2},\kappa)=-2\eta q^{-1}\mathcal{P}^-_{l,\mathbf{n}}T_{l,\mathbf{M}'}(\tau_{\mathbf{n}}q^{-1/2},\kappa)=2\mathcal{P}^-_{l,\mathbf{n}}T_{\mathbf{M}}(\tau_{\mathbf{n}}q^{-1/2},\kappa)\,,\\&T_{\mathbf{M}}(\tau_{\mathbf{n}}q^{-1/2},\kappa)Q_{\mathbf{M}}(\tau_{\mathbf{n}}q^{-1/2},\kappa)=a(\tau_{\mathbf{n}}q^{-1/2})Q_{\mathbf{M}}(\tau_{\mathbf{n}}q^{1/2},\kappa)\,,\\&L_{1,\mathbf{n}}(\zeta_1/\tau_{\mathbf{n}})\cdots L_{l-1,\mathbf{n}}(\zeta_{l-1}/\tau_{\mathbf{n}})\mathcal{P}^-_{l,\mathbf{n}}=\mu(\tau_{\mathbf{n}})\theta_l\bigl(T_{l,[1,l-1]}(\tau_{\mathbf{n}}q^{1/2})^{-1}\bigr)\mathcal{P}^-_{l,\mathbf{n}}\,,\end{aligned}$$

we compute

(4.59)

$$
\begin{aligned}
&I_{\mathbf{M}}(\tau_{\mathbf{n}}q^{-1/2})\\
&=\mu(\tau_{\mathbf{n}})a(\tau_{\mathbf{n}}q^{-1/2})Q^-_{\mathbf{M}}(\tau_{\mathbf{n}}q^{-1/2},\kappa+\alpha)\mathrm{Tr}_{[1,l]}\Bigl(T_{1,\mathbf{M}'}(\zeta_1,\kappa)\cdots T_{l-1,\mathbf{M}'}(\zeta_{l-1},\kappa)\\
&\times\mathbf{g}_l(\tau_{\mathbf{n}}q^{-1/2},\alpha)\theta_l\bigl(T_{l,[1,l-1]}(\tau_{\mathbf{n}}q^{1/2})^{-1}\bigr)2\mathcal{P}^-_{l,\mathbf{n}}\Bigr)Q^+_{\mathbf{M}}(\tau_{\mathbf{n}}q^{1/2},\kappa)\\
&=\mu(\tau_{\mathbf{n}})a(\tau_{\mathbf{n}}q^{-1/2})Q^-_{\mathbf{M}}(\tau_{\mathbf{n}}q^{-1/2},\kappa+\alpha)\mathrm{Tr}_{[1,l]}\Bigl(T_{1,\mathbf{M}'}(\zeta_1,\kappa)\cdots T_{l-1,\mathbf{M}'}(\zeta_{l-1},\kappa)P_{l,\mathbf{n}}\\
&\times\theta_l\bigl(\mathbf{g}_l(\tau_{\mathbf{n}}q^{-1/2},\alpha)\theta_l\bigl(T_{l,[1,l-1]}(\tau_{\mathbf{n}}q^{1/2})^{-1}\bigr)\bigr)Q^+_{\mathbf{M}}(\tau_{\mathbf{n}}q^{1/2},\kappa)\,.
\end{aligned}
$$

Going to the last line we used $\mathrm{Tr}_l(\theta_l(x))=\mathrm{Tr}_l(x)$, $\theta_l(2\mathcal{P}^-_{l,\mathbf{n}})=P_{l,\mathbf{n}}$.

Comparing (4.58) and (4.59) we see that (4.57) indeed follows from the difference equation (4.5).

It remains to consider the integral over $\Gamma_{\mathbf{0}}$. Recall that $\mathbf{g}_l(\zeta_l,\alpha)$ consists of two parts:

$$
\begin{aligned}
\mathbf{g}_l(\zeta_l,\alpha)=&\Bigl(\frac{1}{2}\mathbf{f}_{[1,l-1]}(q\zeta_l,\alpha)+\frac{1}{2}\mathbf{f}_{[1,l-1]}(q^{-1}\zeta_l,\alpha)-\mathbb{T}_{l,[1,l-1]}(\zeta_l,\alpha)\mathbf{f}_{[1,l-1]}(\zeta_l,\alpha)\Bigr)\\
&+\mathbf{u}_{l,[1,l-1]}(\zeta_l,\alpha)\,.
\end{aligned}
$$

The second term $\mathbf{u}_{l,[1,l-1]}(\zeta_l,\alpha)$ does not contain transcendental functions, and it is easy to see that

$$
\mathrm{Tr}_l\{T_{l,\mathbf{M}}(\zeta_l,\kappa)\mathbf{u}_{l,[1,l-1]}(\zeta_l,\alpha)(X_{[1,l-1]})\}=O(\zeta_l^{\alpha+2}),\quad \zeta_l\to 0\,,
$$

if spin of $X_{[1,l-1]}$ equals -1. So, we have to consider

$$
\begin{aligned}
&\frac{1}{2}\mathbf{f}_{[1,l-1]}(q\zeta_l,\alpha)+\frac{1}{2}\mathbf{f}_{[1,l-1]}(q^{-1}\zeta_l,\alpha)-\mathbb{T}_{l,[1,l-1]}(\zeta_l,\alpha)\mathbf{f}_{[1,l-1]}(\zeta_l,\alpha)\\
&=\mathbf{f}_{[1,l-1]}(q\zeta_l,\alpha)-\mathbb{T}_{l,[1,l-1]}(\zeta_l,\alpha)\mathbf{f}_{[1,l-1]}(\zeta_l,\alpha)-\frac{1}{2}\mathbf{k}_{[1,l-1]}(\zeta_l,\alpha)\,,
\end{aligned}
$$

where we use $\Delta_\zeta\mathbf{f}_{[1,l-1]}(\zeta,\alpha)=\mathbf{k}_{[1,l-1]}(\zeta,\alpha)$. Under the functional $Z_{\zeta_1,\cdots,\zeta_l}\{\cdot\}$, we can replace $\mathbb{T}_{l,[1,l-1]}(\zeta_l,\alpha)$ by $\rho(\zeta_l|\alpha)$. Hence Remark 4.14 applies, and the integral over $\Gamma_{\mathbf{0}}$ in the right hand side vanishes because

$$
\mathbf{k}_{[1,l-1]}(\zeta_l,\alpha)(X_{[1,l-1]})\}=O(\zeta_l^{\alpha+2})\,,\quad \zeta_l\to 0\,,
$$

if spin of $X_{[1,l-1]}$ equals -1 according to (3.36). □

Now we are ready to prove our main theorem which will be formulated in several equivalent ways. The first one is as follows.

THEOREM 4.21. *We have*

(4.60)

$$
\begin{aligned}
Z_{\zeta_1,\dots,\zeta_l}\bigl\{\mathbf{g}^+_l(\zeta_l,\alpha)(X_{[1,l-1]})\bigr\}&=\frac{1}{4\pi i}\oint_\Gamma \omega(\zeta_l,\xi|\alpha)Z_{\zeta_1,\dots,\zeta_l}\bigl\{\mathbf{c}(\xi,\alpha)(X_{[1,l-1]})\bigr\}\frac{d\xi^2}{\xi^2}\,,\\
Z_{\zeta_1,\dots,\zeta_l}\bigl\{\mathbf{g}^-_l(\zeta_l,\alpha)(X_{[1,l-1]})\bigr\}&=-\frac{1}{4\pi i}\oint_\Gamma \omega(\xi,\zeta_l|\alpha)Z_{\zeta_1,\dots,\zeta_l}\bigl\{\mathbf{b}(\xi,\alpha)(X_{[1,l-1]})\bigr\}\frac{d\xi^2}{\xi^2}\,,
\end{aligned}
$$

where $\omega(\zeta,\xi|\alpha)$ *is defined in* (4.34), (4.35).

PROOF. We begin with the first identity. Lemma 4.19 shows that the difference between the left hand side and the right hand side is of the form $\zeta_l^{\alpha} Pol_{\mathbf{n}}(\zeta_l^2)$, where $Pol_{\mathbf{n}}(\zeta_l^2)$ is a polynomial in ζ^2 of degree $\mathbf{n}$. Lemma 4.20 together with the genericity condition (4.30) imply that $Pol_{\mathbf{n}}(\zeta_l^2)$ vanishes.

The second identity in (4.60) is obtained similarly. □

This theorem has an important corollary.

THEOREM 4.22. *The operators $\mathbf{g}^{\epsilon}(\zeta_j,\alpha)$ satisfy the Zamolodchikov algebra,*

$$\begin{aligned}\mathbf{g}_l^{\epsilon_l}(\zeta_l,\alpha)&\mathbf{g}_{l-1}^{\epsilon_{l-1}}(\zeta_{l-1},\alpha+\epsilon_l)\\&=(-1)^{\epsilon_l\epsilon_{l-1}}\mathbb{R}_{l,l-1}(\zeta_l/\zeta_{l-1})\mathbf{g}_{l-1}^{\epsilon_{l-1}}(\zeta_{l-1},\alpha)\mathbf{g}_l^{\epsilon_l}(\zeta_l,\alpha+\epsilon_{l-1})\,.\end{aligned}\tag{4.61}$$

PROOF. Denote

$$\begin{aligned}U_{l,l-1}^{\epsilon_l,\epsilon_{l-1}}(\alpha)&=\mathbf{g}_l^{\epsilon_l}(\zeta_l,\alpha)\mathbf{g}_{l-1}^{\epsilon_{l-1}}(\zeta_{l-1},\alpha+\epsilon_l)\\&-(-1)^{\epsilon_l\epsilon_{l-1}}\mathbb{R}_{l,l-1}(\zeta_l/\zeta_{l-1})\mathbf{g}_{l-1}^{\epsilon_{l-1}}(\zeta_{l-1},\alpha)\mathbf{g}_l^{\epsilon_l}(\zeta_l,\alpha+\epsilon_{l-1})\,.\end{aligned}$$

We show

$$U_{l,l-1}^{\epsilon_l,\epsilon_{l-1}}(\alpha)(X_{[1,l-2]})=0,\quad \forall X_{[1,l-2]}\,.$$

It is obvious if the ϵ's are either 0 or $\bar{0}$. The case when one of ϵ's equals 0 or $\bar{0}$ is covered by (3.107), (3.108).

Consider the case $\epsilon_l=+$, $\epsilon_{l-1}=-$. Denoting by $\mathfrak{H}_s$ the subspace of operators of spin s, we assume first that

$$U^{+,-}(\alpha)(X_{[1,l-2]})\in\mathfrak{H}_0\,.$$

Applying twice (4.60) we have

$$\begin{aligned}Z_{\zeta_1,\dots,\zeta_{l-1},\zeta_l}\{\mathbf{g}_l^{+}(\zeta_l,\alpha)\mathbf{g}_{n-1}^{-}(\zeta_{l-1},\alpha+1)(X_{[1,l-2]})\}=Z_{\zeta_1,\dots,\zeta_{l-1},\zeta_l}\Big\{\Big(\frac{1}{2}\omega(\zeta_l,\zeta_{l-1}|\alpha)\\+\frac{1}{(4\pi i)^2}\oint_\Gamma\oint_\Gamma\frac{d\xi_1^2}{\xi_1^2}\frac{d\xi_2^2}{\xi_2^2}\omega(\zeta_l,\xi_1|\alpha)\omega(\xi_2,\zeta_{l-1}|\alpha)\mathbf{b}(\xi_2,\alpha)\mathbf{c}(\xi_1,\alpha-1)\Big)(X_{[l-2]})\Big\}\,.\end{aligned}$$

Computing

$$Z_{\zeta_1,\dots,\zeta_l,\zeta_{l-1}}\{\mathbf{g}_{n-1}^{-}(\zeta_{l-1},\alpha)\mathbf{g}_l^{+}(\zeta_l,\alpha-1)(X_{[1,l-2]})\}$$

we obtain $Z_{\zeta_1,\dots,\zeta_l,\zeta_{l-1}}\{\cdots\}$ where the expression $\{\cdots\}$ inside the braces is the same as above but with the opposite sign. The obvious relation

$$Z_{\zeta_1,\dots,\zeta_l,\zeta_{l-1}}\{X\}=Z_{\zeta_1,\dots,\zeta_{l-1},\zeta_l}\{\mathbb{R}_{l.l-1}(\zeta_l/\zeta_{l-1})X\}$$

then shows that

$$Z_{\zeta_1,\dots,\zeta_{l-1},\zeta_l}\{U_{l,l-1}^{+,-}(\alpha)(X_{[1,l-2]})\}=0\,.\tag{4.62}$$

From Lemma 4.17, we obtain

$$U_{l,l-1}^{+,-}(\alpha)(X_{[1,l-2]})=0\,.$$

The case $\epsilon_l=-$, $\epsilon_{l-1}=+$ is similar, so we conclude that

$$U_{l,l-1}^{\epsilon_l,\epsilon_{l-1}}(\alpha)(X_{[1,l-2]})=0\,,\quad \forall\epsilon_l,\epsilon_{l-1}\,.\tag{4.63}$$

In order to consider other spins we proceed inductively. For example, using the fact that the kernel of $\mathbf{b}(\zeta,\alpha)$ in $\mathfrak{H}_1$ is trivial, we obtain:

$$\mathbf{b}(\zeta,\alpha+1)U_{l,l-1}^{\epsilon_l,\epsilon_{l-1}}(\alpha)(X_{[1,l-2]})=0\iff U_{l,l-1}^{\epsilon_l,\epsilon_l-1}(\alpha)(X_{[1,l-2]})=0\,.$$

Moving $\mathbf{b}(\zeta,\alpha+1)$ through $U_{l,l-1}^{\epsilon_l,\epsilon_{l-1}}(\alpha)$ (we have seen that they commute) we obtain (4.63) for $X_{[1,l-2]}\in\mathfrak{H}_{-(\epsilon_l+\epsilon_{l-1})+1}$. Then we continue using either $\mathbf{b}(\zeta)$ or $\mathbf{c}(\zeta)$. □

This theorem has a corollary which can be considered as a complement to Chapter 3.

THEOREM 4.23. *In the homogeneous case, the fermionic creation operators anti-commute.*

PROOF. The statement follows from the previous theorem and the dressed Russian doll construction. □

Now in the homogeneous case we can order the creation operators in an arbitrary way; for example $\mathbf{t}^*(\zeta)$ to the left, then $\mathbf{b}^*(\zeta)$, then $\mathbf{c}^*(\zeta)$. The main theorem is

THEOREM 4.24. *We have*

$$\begin{aligned}(4.64)\qquad &\mathcal{Z}^{\kappa}_{\mathbf{n}}\{\mathbf{t}^*(\eta_1)\cdots\mathbf{t}^*(\eta_m)\mathbf{b}^*(\zeta_1)\cdots\mathbf{b}^*(\zeta_n)\mathbf{c}^*(\xi_1)\cdots\mathbf{c}^*(\xi_n)(q^{2\alpha S(0)})\}\\ &=\prod_{j=1}^{m}2\rho(\eta_j|\alpha)\cdot\det\bigl(\omega(\zeta_i,\xi_j|\alpha)\bigr)_{i,j=1,\dots,n}\,.\end{aligned}$$

PROOF. The proof immediately follows from Lemma 4.41, Lemma 4.18 and Theorem 4.21. □

The equation (4.64) should be understood in terms of generating functions. According to our general procedure, quasi-local operators are found in the power series $\mathbf{t}^*(\eta_1)\cdots\mathbf{t}^*(\eta_m)\mathbf{b}^*(\zeta_1)\cdots\mathbf{b}^*(\zeta_n)\mathbf{c}^*(\xi_1)\cdots\mathbf{c}^*(\xi_n)(q^{2\alpha S(0)})$ in η_j^2-1, ζ_j^2-1, ξ_j^2-1. So, the expectation values of these quasi-local operators are given by the Taylor coefficients of the right hand side of (4.64).

It remains to show that the fermionic basis is complete.

4.4. Completeness in Homogeneous Case

4.4.1. Linear Independence. In the homogeneous case, we have the creation operators

$$\mathbf{b}^*(\zeta)=\zeta^{2+\alpha+\mathbb{S}}\sum_{p=1}^{\infty}(\zeta^2-1)^{p-1}\mathbf{b}^*_p\,,\quad \mathbf{c}^*(\zeta)=\zeta^{2-\alpha-\mathbb{S}}\sum_{p=1}^{\infty}(\zeta^2-1)^{p-1}\mathbf{c}^*_p\,,$$

$$\mathbf{t}^*(\zeta)=\sum_{p=1}^{\infty}(\zeta^2-1)^{p-1}\mathbf{t}^*_p\,.$$

The operator $\frac{1}{2}\mathbf{t}^*_1=\boldsymbol{\tau}$ plays a special role: it is the right shift by one site on the lattice. These operators create a number of local operators:

$$\begin{aligned}(4.65)\qquad &\boldsymbol{\tau}^p\mathbf{t}^*_{i_1}\cdots\mathbf{t}^*_{i_r}\mathbf{b}^*_{j_1}\cdots\mathbf{b}^*_{j_s}\mathbf{c}^*_{k_1}\cdots\mathbf{c}^*_{k_t}(q^{2\alpha S(0)})\,,\\ &p\in\mathbb{Z},\ \ i_r\geq\cdots\geq i_1\geq 2\,,\ \ j_s>\cdots>j_1\geq 1\,,\ \ k_t>\cdots>k_1\geq 1\,.\end{aligned}$$

Recall that

(4.66)

$$\mathrm{length}\bigl(\boldsymbol{\tau}^p\mathbf{t}^*_{i_1}\cdots\mathbf{t}^*_{i_r}\mathbf{b}^*_{j_1}\cdots\mathbf{b}^*_{j_s}\mathbf{c}^*_{k_1}\cdots\mathbf{c}^*_{k_t}(q^{2\alpha S(0)})\bigr)\leq\sum_{a=1}^{r}i_a+\sum_{a=1}^{s}j_a+\sum_{a=1}^{t}k_a\,.$$

Our first concern is to show the linear independence of the vectors (4.65). Obviously, it amounts to proving the linear independence of the joint kernel of the annihilation operators spanned by the vectors

$$\boldsymbol{\tau}^p \mathbf{t}^*_{i_1}\cdots\mathbf{t}^*_{i_r}(q^{2\alpha S(0)})\,,\quad p\in\mathbb{Z},\quad i_r\geq\cdots\geq i_1\geq 2\,. \tag{4.67}$$

We begin with the following proposition.

LEMMA 4.25. *For generic $\xi_1,\dots,\xi_l$, varying the Matsubara data we can consider*

$$\eta_j=\rho(\xi_j|\alpha)$$

as independent variables.

PROOF. We consider the Mastubara chain consisting of spins 1/2 with length $\mathbf{n}=l$ and inhomogeneities τ_j. We set $\kappa=-l/2$, and consider the Bethe vector of lowest spin. Then

$$\rho(\zeta|\alpha)=\frac{q^\alpha F(\zeta)+q^{-\alpha}}{F(\zeta)+1}\,,$$

where

$$F(\zeta)=\frac{a(\zeta)}{d(\zeta)}=\prod_{j=1}^{l}\frac{\tau_j^2-q^{-1}\zeta^2}{\tau_j^2-q\zeta^2}\,.$$

Obviously it is sufficient to prove that

$$\det\Bigl(\frac{\partial}{\partial\tau_i^2}\rho(\xi_j|\alpha)\Bigr)_{i,j=1,\dots,l}\neq 0\,.$$

We have

$$\det\Bigl(\frac{\partial}{\partial\tau_i^2}\rho(\xi_j|\alpha)\Bigr)_{i,j=1,\dots,l}=\prod_{j=1}^{l}\frac{(q^\alpha-q^{-\alpha})\xi_j^2F(\xi_j)}{(1+F(\xi_j))^2}\cdot\mathcal{D}\,,$$

where

$$\mathcal{D}=\det\Bigl(\frac{q^{-1}-q}{(\tau_i^2-q^{-1}\xi_j^2)(\tau_i^2-q\xi_j^2)}\Bigr)_{i,j=1,\dots,l}\,.$$

This determinant does not vanish identically. Let us give a formal proof of this fact which will be used later. Consider the case $|q|=1$, Notice that

$$\prod_{j=1}^{l}(\tau_j^{-2l+1}\xi_j)\frac{\prod_{i,j}(\tau_i^2-q\xi_j^2)(\tau_i^2-q^{-1}\xi_j^2)}{\prod_{i<j}(\tau_i^2-\tau_j^2)\prod_{i<j}(\xi_i^2-\xi_j^2)}\cdot\mathcal{D}=M_l(\xi_1,\dots,\xi_l|\tau_1,\dots,\tau_l) \tag{4.68}$$

is the domain wall partition function (2.20). At the point P_0 of the parameter space $\xi_1=\cdots=\xi_l=\tau_1=\cdots=\tau_l=1$, and for ξ_j and τ_j being unimodular and sufficiently close to P_0, all the Boltzmann weights are real positive, so, the same is true for the domain wall partition function. □

The linear independence follows from this Lemma.

COROLLARY 4.26. *A finite linear combination of operators* (4.67) *does not vanish unless all the coefficients vanish.*

PROOF. Consider some linear combination assuming that $\sum i_k < l$ for all summands. By the very meaning of the construction of the operator $\mathbf{t}^*(\zeta)$ (Section 3.4.1) and Lemma 4.18, it follows that the corollary holds if the first l Taylor coefficients of the function $\rho(\zeta|\alpha)$ in $\zeta^2 - 1$ can be considered as independent variables. This is equivalent to the statement that generically

$$\frac{\mathcal{D}}{\prod_{i<j}(\xi_i^2 - \xi_j^2)}\bigg|_{\xi_1=1,\dots,\xi_l=1} \tag{4.69}$$

does not vanish. This follows from (4.68) since in (4.69) we need the vicinity of the point P_0 (see proof of Lemma 4.25) in the parameter space. □

4.4.2. Operators $\bar{\mathbf{t}}^*, \bar{\mathbf{b}}^*, \bar{\mathbf{c}}^*$. The main property of the operators $\mathbf{b}_p^*, \mathbf{c}_p^*, \mathbf{t}_p^*$ is that acting on a local operator they enlarge its support by at most p sites to the right. In order to control the length of operators precisely, it is convenient to have also operators which enlarge the support to the left. In this section we describe a simple construction of such operators.

We begin with the counterpart of $\mathbf{t}^*(\zeta)$. Introduce the operator

$$\bar{\mathbf{t}}^*(\zeta) = \mathbf{t}^*(\zeta)^{-1}, \tag{4.70}$$

which is well-defined as power series in $\zeta^2 - 1$. It is easy to see that

$$\bar{\mathbf{t}}^*(\zeta)(q^{2\alpha S(0)}X_{[1,l]}) = \mathrm{Tr}_a\{\mathbb{T}_{a,(-\infty,l]}(\zeta)^{-1}(q^{2\alpha S(0)}X_{[1,l]})\}, \tag{4.71}$$

where $\mathbb{T}_{a,(-\infty,l]}(\zeta)^{-1}$ is defined as power series in $\zeta^2 - 1$ in exactly the same way as $\mathbb{T}_{a,[k,\infty)}(\zeta)$ discussed before. As opposed to $\mathbf{t}^*(\zeta)$, $\bar{\mathbf{t}}^*(\zeta)$ enlarges the support to the *left* and has the *right* reduction property. This is exactly what we want.

We proceed with defining the operators

$$\bar{\mathbf{b}}^*(\zeta) = -\bar{\mathbf{t}}^*(\zeta)\mathbf{b}^*(\zeta), \quad \bar{\mathbf{c}}^*(\zeta) = -\bar{\mathbf{t}}^*(\zeta)\mathbf{c}^*(\zeta). \tag{4.72}$$

Recalling the formula

(4.73)
$$\mathbf{b}^*(\zeta,\alpha)(q^{2(\alpha+1)S(k-1)}X_{[k,m]}) = q^{2\alpha S(k-1)}\mathrm{Tr}_c\{\mathbb{T}_{c,[m+1,\infty)}(\zeta)\mathbf{g}_{c,[k,m]}(\zeta,\alpha)(X_{[k,m]})\},$$

one easily derives

$$\begin{aligned}\bar{\mathbf{b}}^*(\zeta,\alpha)&(q^{2(\alpha+1)S(k-1)}X_{[k,m]})\\ &= \mathrm{Tr}_c\{\mathbb{T}_{c,(-\infty,k-1]}(\zeta)^{-1}(q^{2\alpha S(k-1)}\bar{\mathbf{g}}_{c,[k,m]}(\zeta,\alpha)(X_{[k,m]}))\},\end{aligned} \tag{4.74}$$

where

$$\bar{\mathbf{g}}_{c,J}(\zeta,\alpha) = -\mathbb{T}_{c,J}(\zeta)^{-1}\mathbf{g}_{c,J}(\zeta,\alpha).$$

The reduction properties of $\mathbf{g}_{c,J}(\zeta,\alpha)$ (3.122), (3.123) ensure the consistency of (4.73). Likewise the properties of $\bar{\mathbf{g}}_{c,J}(\zeta,\alpha)$

$$\bar{\mathbf{g}}_{c,[k,m]}(\zeta,\alpha)(X_{[k,m-1]}) = \bar{\mathbf{g}}_{c,[k,m-1]}(\zeta,\alpha)(X_{[k,m-1]}), \tag{4.75}$$

$$\bar{\mathbf{g}}_{c,[k,m]}(\zeta,\alpha)(q^{(\alpha+1)\sigma_k^3}X_{[k+1,m]}) = \mathbb{R}_{c,k}(\zeta)^{-1}q^{\alpha\sigma_k^3}\bar{\mathbf{g}}_{c,[k+1,m]}(\zeta,\alpha)(X_{[k+1,m]}), \tag{4.76}$$

ensure the consistency of (4.74).

Operators $\bar{\mathbf{b}}^*(\zeta), \bar{\mathbf{c}}^*(\zeta), \bar{\mathbf{t}}^*(\zeta)$ are given by the Russian doll construction of the form (4.74) via

$$\bar{\mathbf{g}}^+_{c,J}(\zeta,\alpha) = \bar{\mathbf{g}}_{c,J}(\zeta,\alpha), \quad \bar{\mathbf{g}}^-_{c,J}(\zeta,\alpha) = \phi(\bar{\mathbf{g}})_{c,J}(\zeta,\alpha), \quad \bar{\mathbf{g}}^0_{c,J}(\zeta,\alpha) = \mathbb{T}_{c,J}(\zeta,\alpha)^{-1}.$$

From the definition (4.72) and Theorem 3.23, it follows that the operators $\bar{\mathbf{b}}^*(\zeta)$, $\bar{\mathbf{c}}^*(\zeta)$ satisfy the anti-commutation relations with $\bar{\mathbf{b}}(\zeta)$, $\bar{\mathbf{c}}(\zeta)$:

$$[\bar{\mathbf{b}}(\zeta),\bar{\mathbf{c}}^*(\zeta')]_+ = [\bar{\mathbf{c}}(\zeta),\bar{\mathbf{b}}^*(\zeta')]_+ = 0\,,$$
$$[\bar{\mathbf{b}}(\zeta),\bar{\mathbf{b}}^*(\zeta')]_+ = -\psi(\zeta'/\zeta,\alpha+\mathbb{S})\,,\quad [\bar{\mathbf{c}}(\zeta),\bar{\mathbf{c}}^*(\zeta')]_+ = \psi(\zeta/\zeta',\alpha+\mathbb{S})\,.$$

Hence, defining

$$\bar{\mathbf{b}}(\zeta) = \zeta^{-\alpha-\mathbb{S}}\big(\bar{\mathbf{b}}_0 + \sum_{p=1}^{\infty}(\zeta^2-1)^{-p}\bar{\mathbf{b}}_p\big)\,,\quad \bar{\mathbf{c}}(\zeta) = \zeta^{\alpha+\mathbb{S}}\big(\bar{\mathbf{c}}_0 + \sum_{p=1}^{\infty}(\zeta^2-1)^{-p}\bar{\mathbf{c}}_p\big)\,,$$
$$\bar{\mathbf{b}}^*(\zeta) = \zeta^{2+\alpha+\mathbb{S}}\sum_{p=1}^{\infty}(\zeta^2-1)^{p-1}\bar{\mathbf{b}}^*_p\,,\qquad \bar{\mathbf{c}}^*(\zeta) = \zeta^{2-\alpha-\mathbb{S}}\sum_{p=1}^{\infty}(\zeta^2-1)^{p-1}\bar{\mathbf{c}}^*_p\,,$$
$$\bar{\mathbf{t}}^*(\zeta) = \sum_{p=1}^{\infty}(\zeta^2-1)^{p-1}\bar{\mathbf{t}}^*_p\,,$$

we see that $\bar{\mathbf{b}}_p, \bar{\mathbf{b}}^*_p$ and $\bar{\mathbf{c}}_p, \bar{\mathbf{c}}^*_p$, for $p \geq 1$ have canonical anti-commutation relations.

Consider the singular part of the product $\mathbf{c}^*(\zeta)\mathbf{b}(\zeta)$. As usual we define this operator block-wise

$$\mathbf{s}(\zeta,\alpha) = \frac{1}{2\pi i}\int_\Gamma \psi(\zeta/\xi, 1-2\alpha-2\mathbb{S})\mathbf{c}^*(\xi,\alpha)\mathbf{b}(\xi,\alpha+1)\frac{d\xi^2}{\xi^2}\,, \tag{4.77}$$

where 1 is inside Γ and ζ^2 is outside. We claim that $\mathbf{s}(\zeta)$ is a bosonic annihilation operator which lowers spin by 2. This is a corollary of the two facts discussed above.

REMARK 4.27. Definition (4.72) and Theorem 3.23 imply an alternative formula

$$\mathbf{s}(\zeta,\alpha) = \frac{1}{2\pi i}\int_\Gamma \psi(\zeta/\xi, 1-2\alpha-2\mathbb{S})\bar{\mathbf{c}}^*(\xi,\alpha)\bar{\mathbf{b}}(\xi,\alpha+1)\frac{d\xi^2}{\xi^2}\,.$$

Together with (4.77), this means that this operator does not enlarge the support neither to the right nor to the left.

4.4.3. Basis. Here we shall construct a basis of operators $q^{2\alpha S(0)}\mathcal{O}$ supported on an interval $[1,n]$ for any given n. Consider first the operator

$$B_n(\alpha) = \mathbf{b}^*_{[1,n]}q^{2(\alpha+n)S(0)} \in \mathcal{W}_{\alpha,n}\,. \tag{4.78}$$

Since $\bar{\mathbf{b}}^*_p = -\sum_{r+s=p+1}\bar{\mathbf{t}}^*_r\mathbf{b}^*_s$, we can write it also as

$$B_n(\alpha) = (-1)^n\boldsymbol{\tau}^n\bar{\mathbf{b}}^*_{[1,n]}q^{2(\alpha+n)S(0)}\,, \tag{4.79}$$

which implies that this operator is supported on $[1,n]$. Applying annihilation operators on $B_n(\alpha)$ we get various other operators with support on the same interval $[1,n]$.

We begin with the spin 2 annihilation operator $\mathbf{s}(\zeta)$ defined in the previous subsection. Combine the blocks $\mathbf{s}(\zeta,\alpha)$ into $\mathbf{s}(\zeta)$, and define the modes

$$\mathbf{s}(\zeta) = \zeta^{1-2\alpha-2\mathbb{S}}\Big(\mathbf{s}_0 + \sum_{p=1}^{\infty}(\zeta^2-1)^{-p}\mathbf{s}_p\Big)\,. \tag{4.80}$$

As in the case of fermionic annihilation operators, $\mathbf{s}_0$ is not independent. So we shall not use it. For other modes we have

$$\mathbf{s}_a = \sum_{j=1}^{\infty} \mathbf{c}_j^* \mathbf{b}_{a+j-1} \,.$$

Given a partition $\lambda = \{\lambda_1, \cdots, \lambda_l\}$ $(\lambda_1 \geq \cdots \geq \lambda_l \geq 1)$, define

$$\mathbf{s}_\lambda = \mathbf{s}_{\lambda_1} \mathbf{s}_{\lambda_2} \cdots \mathbf{s}_{\lambda_l} \,.$$

For $k = 0, 1, \cdots, n$, let Λ_k be the set of all partitions $\lambda = \{\lambda_1, \ldots, \lambda_k\}$ such that

$$n - k + 1 \geq \lambda_1 \geq \cdots \geq \lambda_k \geq 1 \,.$$

It is easy to see that there are $\binom{n}{k}$ of such partitions. Further let $\mathcal{B}_k$ be the collection of all subsets of $\{1, 2, \cdots, n-k\}$, and $\mathcal{C}_k$ the collection of all subsets of $\{1, 2, \cdots, k\}$. Now we are ready to give the definition of a basis of operators located on the interval $[1, n]$.

DEFINITION 4.28. The following set is a basis of operators located on $[1, n]$ and belonging to $\mathcal{W}_\alpha$,

$$\left\{ \bar{\mathbf{b}}_R \mathbf{c}_S \mathbf{s}_\lambda B_n\big(\alpha - \#(R) + \#(S) - 2\#(\lambda)\big) \mid (R, S, \lambda) \in \bigcup_{k=0}^{n} \mathcal{B}_k \times \mathcal{C}_k \times \Lambda_k \right\}. \tag{4.81}$$

The parameter α in B_n is shifted in order to have tail α after applying $\bar{\mathbf{b}}_R \mathbf{c}_S \mathbf{s}_\lambda$; spin of the operators is easy to compute.

Obviously, this definition gives 4^n operators localised on $[1, n]$, and we need only to make sure that they are linearly independent.

LEMMA 4.29. *The operators defined in Definition* 4.28 *are linearly independent.*

PROOF. For a partition $\lambda \in \Lambda_k$, define an ordered subset $J(\lambda)$ of the set $\{n, n-1, , \cdots, 1\}$ by

$$J(\lambda) = \{\lambda_1 + (k-1), \lambda_2 + (k-2) \cdots, \lambda_{k-1} + 1, \lambda_k\} \,.$$

We order subsets and partitions lexicographically.

We have

$$\begin{aligned}
\mathbf{s}_1 &= \mathbf{c}_1^* \mathbf{b}_1 + \mathbf{c}_2^* \mathbf{b}_2 + \mathbf{c}_3^* \mathbf{b}_3 + \cdots + \mathbf{c}_k^* \mathbf{b}_k + \cdots \,, \\
\mathbf{s}_2 &= \mathbf{c}_1^* \mathbf{b}_2 + \mathbf{c}_2^* \mathbf{b}_3 + \mathbf{c}_3^* \mathbf{b}_4 + \cdots + \mathbf{c}_k^* \mathbf{b}_{k+1} + \cdots \,, \\
&\cdots \\
\mathbf{s}_{n-k} &= \mathbf{c}_1^* \mathbf{b}_{n-k} + \mathbf{c}_2^* \mathbf{b}_{n-k+1} + \cdots + \mathbf{c}_k^* \mathbf{b}_{n-1} + \cdots \,, \\
\mathbf{s}_{n-k+1} &= \mathbf{c}_1^* \mathbf{b}_{n-k+1} + \mathbf{c}_2^* \mathbf{b}_{n-k+2} + \cdots + \mathbf{c}_k^* \mathbf{b}_n + \cdots \,,
\end{aligned} \tag{4.82}$$

where $\cdots$ stand for terms which do not contain $\mathbf{c}_1^*, \cdots, \mathbf{c}_k^*$. From these formulas one readily concludes that for $\lambda \in \Lambda_k$

$$\mathbf{s}_\lambda = c_\lambda \cdot \mathbf{c}_{\{1,2,\cdots,k\}}^* \mathbf{b}_{J(\lambda)} + \cdots \,, \tag{4.83}$$

where $\cdots$ stands for terms proportional to $\mathbf{c}_I^* \mathbf{b}_K$ (members of I are ordered from left to right, while those of K are ordered oppositely) with either $I \neq \{1, 2, , \cdots, k\}$ or $I = \{1, 2, , \cdots, k\}$, $K \prec J(\lambda)$. The constant c_λ does not vanish being just the product of factorials of cardinalities of all distinct entries of λ. This proves linear independence of $\mathbf{s}_\lambda$.

The relation between the partitions λ and the sets $J(\lambda)$ is one-to-one, so, the triangular structure in (4.83) implies the possibility of introducing linear combinations

$$\tilde{\mathbf{s}}_\lambda = \frac{1}{c_\lambda}\mathbf{s}_\lambda + \sum_{\lambda' \prec \lambda} d_{\lambda,\lambda'}\mathbf{s}_{\lambda'}$$

such that

$$\tilde{\mathbf{s}}_\lambda = c_\lambda \cdot \mathbf{c}^*_{\{1,2,\cdots,k\}}\mathbf{b}_{J(\lambda)} + \cdots , \tag{4.84}$$

where this time we have only one term with $\mathbf{c}^*_{\{1,2,\cdots,k\}}$. It is possible to give explicit formulas for $d_{\lambda,\lambda'}$ in terms of the Littlewood-Richardson coefficients [**BJMS**], but we shall not need them.

The set of operators $\bar{\mathbf{b}}_R\mathbf{c}_S\mathbf{s}_\lambda B_n(\alpha+\#(S)-2\#(\lambda)-\#(R))$ is linearly equivalent to the set

$$B_{R,S,\lambda}(\alpha) = \bar{\mathbf{b}}_R\mathbf{c}_S\tilde{\mathbf{s}}_\lambda B_n(\alpha+\#(S)-2\#(\lambda)-\#(R)) . \tag{4.85}$$

Let us show that the set (4.85) is linearly independent. Suppose there is a linear relation

$$\sum_{R,S,\lambda} A_{R,S,\lambda}B_{R,S,\lambda}(\alpha) = 0$$

with some scalars $A_{R,S,\lambda}$. The sum is taken over λ, $R \subset [1, n-\#(\lambda)]$, $S \subset [1, \#(\lambda)]$.

Recalling that

$$[\bar{\mathbf{b}}_i, \mathbf{b}^*_j]_+ = -\mathbf{t}^*_{j-i+1} ,$$

we observe that the left hand side is a linear combination of monomials containing $n-\#(\lambda)$ number of $\mathbf{b}^*$ and $\mathbf{t}^*$. By Corollary 4.26 the sum is separately zero for each fixed $l = \#(\lambda)$.

Fixing $R_0 \subset [1, n-l]$, $S_0 \subset [1, l]$, we apply $\bar{\mathbf{b}}_{\{1,\ldots,n-l\}\setminus R_0}\mathbf{c}_{\{1,\ldots,l\}\setminus S_0}$ to both sides, obtaining

$$\begin{aligned} 0 &= \sum_{\#(\lambda)=l} A_{R_0,S_0,\lambda}\bar{\mathbf{b}}_{\{1,\ldots,n-l\}}\mathbf{c}_{\{1,\ldots,l\}}\tilde{\mathbf{s}}_\lambda \\ &= \sum_{\#(\lambda)=l} A_{R_0,S_0,\lambda}\det\left(\mathbf{t}^*_{\lambda_a+a-b}\right)_{1\le a,b\le n-l}\left(q^{2(\alpha+n-2l)S(0)}\right) , \end{aligned} \tag{4.86}$$

where we used (4.84). By Corollary 4.26, Schur functions in $\mathbf{t}^*_p$'s applied to a primary field are linearly independent. Hence we conclude that $A_{R_0,S_0,\lambda} = 0$ for all λ. This completes the proof. □

Let us give a simple example. Dropping the primary fields from the notation, we have for $n = 2$

$$B_\emptyset = \mathbf{b}^*_2\mathbf{b}^*_1,\ B_{\{2\}} = \mathbf{c}^*_1\mathbf{b}^*_1,\ B_{\{1\}} = \mathbf{b}^*_2\mathbf{c}^*_1 + \mathbf{c}^*_2\mathbf{b}^*_1,\ B_{\{1,2\}} = \mathbf{c}^*_2\mathbf{c}^*_1 ,$$

from which we derive 16 operators supported on the interval $[1, 2]$

$$\begin{aligned} &\mathbf{b}^*_2\mathbf{b}^*_1 , \\ &\mathbf{b}^*_2,\ \mathbf{b}^*_1,\ \mathbf{t}^*_1\mathbf{b}^*_2 - \mathbf{t}^*_2\mathbf{b}^*_1,\ \mathbf{t}^*_1\mathbf{b}^*_1 , \\ &1,\ \mathbf{t}^*_1,\ \mathbf{t}^*_2,\ (\mathbf{t}^*_1)^2,\ \mathbf{c}^*_1\mathbf{b}^*_1,\ \mathbf{b}^*_2\mathbf{c}^*_1 + \mathbf{c}^*_2\mathbf{b}^*_1 , \\ &\mathbf{c}^*_2,\ \mathbf{b}^*_1,\ \mathbf{t}^*_1\mathbf{c}^*_2 - \mathbf{t}^*_2\mathbf{c}^*_1,\ \mathbf{t}^*_1\mathbf{c}^*_1 , \\ &\mathbf{c}^*_2\mathbf{c}^*_1 . \end{aligned} \tag{4.87}$$

We finish this section with a remark which will be useful later.

LEMMA 4.30. *Consider the operators*

$$\mathbf{y}_m = \sum_{j=1}^{m-1} \mathbf{c}_j \mathbf{b}_{m-j} \,. \tag{4.88}$$

The quasi-local operators localised on the interval $[1,n]$ *belong to the common kernel of* $\mathbf{y}_m$ *with* $m > n$.

PROOF. There are several possibilities to see that this is true. One of them goes back to a property of annihilation operators in the inhomogeneous case. Consider an inhomogeneous chain with inhomogeneities $\xi_1, \cdots, \xi_n$. Denote

$$\mathbf{c}^{(j)} = \mathrm{res}_{\zeta^2=\xi_j^2} \mathbf{c}(\zeta) \frac{d\zeta^2}{\zeta^2} \,, \quad \mathbf{b}^{(j)} = \mathrm{res}_{\zeta^2=\xi_j^2} \mathbf{b}(\zeta) \frac{d\zeta^2}{\zeta^2} \,.$$

The property in question is (3.49), i.e.,

$$\mathbf{b}^{(j)} \mathbf{c}^{(j)} = 0 \,. \tag{4.89}$$

Then for any polynomial $P(\zeta^2)$

$$\oint_\Gamma P(\zeta^2) \prod_{j=1}^{n} \left(1 - \frac{\zeta^2}{\xi_j^2}\right) \mathbf{b}(\zeta) \mathbf{c}(\zeta) \frac{d\zeta^2}{\zeta^2} = 0 \,,$$

where Γ goes around $\xi_1^2, \cdots, \xi_n^2$. Clearly in the homogeneous limit we obtain the statement of Lemma 4.30.

There is another way of doing. We have

$$[\mathbf{y}_m, \mathbf{s}_a] = \sum_{j=1}^{m-a} \mathbf{b}_{m-j} \mathbf{b}_{a+j-1} + \sum_{j=1}^{a-1} \mathbf{b}_{a-j} \mathbf{b}_{m+j-1} \,.$$

The first sum in the right hand side vanishes by anticommutativity, while the second one annihilates B_n for $m > n$.

We gave two proofs of one simple statement in order to show a deep self-consistency of the construction. □

4.5. Summary

The operators $\mathbf{b}^*(\zeta)$, $\mathbf{c}^*(\zeta)$, $\mathbf{t}^*(\zeta)$ are generating functions of $\mathbf{b}^*_p$, $\mathbf{c}^*_p$, $\mathbf{t}^*_p$ which create quasi-local operators:

$$\mathbf{b}^*(\zeta) = \zeta^{2+\alpha+\mathbb{S}} \sum_{p=1}^{\infty} (\zeta^2-1)^{p-1} \mathbf{b}^*_p \,, \quad \mathbf{c}^*(\zeta) = \zeta^{2-\alpha-\mathbb{S}} \sum_{p=1}^{\infty} (\zeta^2-1)^{p-1} \mathbf{c}^*_p \,,$$

$$\mathbf{t}^*(\zeta) = \sum_{p=1}^{\infty} (\zeta^2-1)^{p-1} \mathbf{t}^*_p \,.$$

The operator $\frac{1}{2}\mathbf{t}^*_1 = \boldsymbol{\tau}$ is the shift. Acting on the primary field $q^{2\alpha S(0)}$ by some number of $\mathbf{t}^*$'s and an equal number of $\mathbf{b}^*$'s and $\mathbf{c}^*$'s we create the entire space $\mathcal{W}_{\alpha,0}$.

Moreover, there is an efficient procedure to control the length of the operators as explained in Section 4.4.3. The main theorem states that

$$\begin{aligned}&\mathcal{Z}^{\kappa}_{\mathbf{n}}\{\mathbf{t}^*(\eta_1)\cdots\mathbf{t}^*(\eta_m)\mathbf{b}^*(\zeta_1)\cdots\mathbf{b}^*(\zeta_n)\mathbf{c}^*(\xi_1)\cdots\mathbf{c}^*(\xi_n)(q^{2\alpha S(0)})\}\\&\quad=\prod_{j=1}^{m}2\rho(\eta_j|\alpha)\cdot\det\big(\omega(\zeta_i,\xi_j|\alpha)\big)_{i,j=1,\dots,n},\end{aligned}$$

with functions $\rho(\zeta|\alpha)$, $\omega(\zeta,\xi|\alpha)$ defined in (4.16), (4.35). The former has a simple definition. As for the latter, we would not like to repeat it here because in practice an alternative definition given in the next chapter may be more useful.

CHAPTER 5

Applications and Generalisations

5.1. Function $\omega(\zeta,\xi|\alpha)$ via Integral Equation

The expression (4.35) for the function $\omega(\zeta,\xi|\alpha)$ which we gave in the previous chapter possesses many attractive features. However, it is not very convenient for considering the limit $\mathbf{n}\to\infty$ which is important for physical applications. In this section, we shall provide an alternative description for $\omega(\zeta,\xi|\alpha)$. We shall consider only the case where the *Matsubara* spins equal $1/2$, but we shall keep inhomogeneities $\tau_{\mathbf{m}}$ which play an important role in considering various $\mathbf{n}\to\infty$ limits.

We recall that $\rho(\zeta|\alpha)$ and $\omega(\zeta,\xi|\alpha)$ depend also on κ but this dependence is not shown explicitly; see Remark 4.4. Throughout this section, we write $Q(\zeta,\kappa)=Q^-(\zeta,\kappa)$, $\mathfrak{a}(\zeta)=\mathfrak{a}(\zeta|\kappa)$, $\rho(\zeta)=\rho(\zeta|\alpha)$ to lighten the notation.

We had the equation

$$\log\mathfrak{a}(\zeta)=-2\pi i\nu\kappa+\log\frac{a(\zeta)}{d(\zeta)}+\int_\gamma K_0(\zeta/\eta)\log(1+\mathfrak{a}(\eta))\frac{d\eta^2}{\eta^2}\,, \tag{5.1}$$

for the function

$$\mathfrak{a}(\zeta)=\frac{a(\zeta)Q(\zeta q,\kappa)}{d(\zeta)Q(\zeta q^{-1},\kappa)}\,.$$

The contour γ goes counterclockwise including the zeroes of $Q(\zeta,\kappa)$ inside, and keeping all other singularities outside. The kernel $K_0(\zeta)$ is a particular case of the generalised kernel

$$K_\alpha(\zeta)=\frac{1}{2\pi i}\Delta_\zeta\psi(\zeta,\alpha)\,. \tag{5.2}$$

This finally explains the notation of Section 2.6. We define the measure is

$$dm(\zeta)=\frac{d\zeta^2}{\zeta^2\rho(\zeta)(1+\mathfrak{a}(\zeta))}\,. \tag{5.3}$$

We denote by $\star$ the convolution with respect to this measure: for two functions F,G in two variables we set

$$(F\star G)(\zeta,\xi)=\int_\gamma F(\zeta,\eta)G(\eta,\xi)dm(\eta)\,. \tag{5.4}$$

In the case when $F(\zeta,\eta)$ (resp. $G(\eta,\xi)$) has a pole along the diagonal, the point $\eta^2=\zeta^2$ (resp. $\eta^2=\xi^2$) should be inside γ as well as the zeroes of $Q(\eta,\kappa)$. All other singularities should stay outside γ.

Introduce the dressed resolvent R_{dress} as the solution to the equation

$$R_{\mathrm{dress}}+K_\alpha\star R_{\mathrm{dress}}=K_\alpha\,. \tag{5.5}$$

Define two functions of two variables

$$f_{\text{left}}(\zeta,\xi)=\frac{1}{2\pi i}\delta^-_\zeta\psi(\zeta/\xi,\alpha)\,,\quad f_{\text{right}}(\zeta,\xi)=\delta^-_\xi\psi(\zeta/\xi,\alpha)\,, \tag{5.6}$$

where δ^- is defined in (4.17). Following [**BG**] we prove an important statement.

LEMMA 5.1. *The following alternative definition of $\omega(\zeta,\xi|\alpha)$ holds:*

$$\begin{aligned}\frac{1}{4}\omega(\zeta,\xi|\alpha)=&-\bigl(f_{\text{left}}\star f_{\text{right}}-f_{\text{left}}\star R_{\text{dress}}\star f_{\text{right}}\bigr)(\zeta,\xi)\\&+\delta^-_\zeta\delta^-_\xi\Delta^{-1}_\zeta\psi(\zeta/\xi,\alpha)\,,\end{aligned} \tag{5.7}$$

where ζ^2 and ξ^2 are inside the integration contours according to the above prescription.

PROOF. Let us introduce an auxiliary function $G_{\text{right}}(\zeta,\xi)$ which satisfies the equation

$$G_{\text{right}}(\zeta,\xi)=f_{\text{right}}(\zeta,\xi)-\bigl(K_\alpha\star G_{\text{right}}\bigr)(\zeta,\xi)\,, \tag{5.8}$$

where the pole ξ^2 is inside the contour of integration. This allows us to rewrite (5.7) as

$$\frac{1}{4}\omega(\zeta,\xi|\alpha)=-\bigl(f_{\text{left}}\star G_{\text{right}}\bigr)(\zeta,\xi)+\delta^-_\zeta\delta^-_\xi\Delta^{-1}_\zeta\psi(\zeta/\xi,\alpha)\,. \tag{5.9}$$

Our nearest goal is to show that this expression has the correct analytic structure.

Denoting the zeroes of $Q(\zeta,\kappa)$ (the Bethe roots) by λ_j, we rewrite the equation (5.8) as

(5.10)

$$\begin{aligned}G_{\text{right}}(\eta,\xi)=&-\rho(\xi)\psi(\eta/\xi,\alpha)+\psi(q\eta/\xi,\alpha)\frac{1}{1+\mathfrak{a}(\xi)}+\psi(q^{-1}\eta/\xi,\alpha)\frac{1}{1+\overline{\mathfrak{a}}(\xi)}\\&-2\pi i\sum_j K_\alpha(\eta/\lambda_j)G_{\text{right}}(\lambda_j,\xi)\operatorname*{res}_{\sigma^2=\lambda_j^2}dm(\sigma)\,,\end{aligned}$$

where we define

$$\overline{\mathfrak{a}}(\xi)=\frac{1}{\mathfrak{a}(\xi)}\,.$$

The equation (5.10) is nothing but the correct way of understanding the equation (5.8): setting $\eta=\lambda_j$ in (5.10) it becomes a matrix equation for $G_{\text{right}}(\lambda_j,\xi)$. At the same time, (5.10) shows also that, as a function of η^2, $G_{\text{right}}(\eta,\xi)$ has poles only at $\eta^2=\xi^2q^{\pm2},\xi^2$ and $\lambda_j^2q^{\pm2}$.

Let us concentrate on

$$\begin{aligned}U(\zeta,\xi)&=\bigl(f_{\text{left}}\star G_{\text{right}}\bigr)(\zeta,\xi)\\&=\frac{1}{2\pi i}\int_\gamma\Bigl(\psi(\zeta q/\eta,\alpha)-\rho(\zeta)\psi(\zeta/\eta,\alpha)\Bigr)G_{\text{right}}(\eta,\xi)dm(\eta)\,.\end{aligned}$$

Denote by μ_j the zeroes of $T(\zeta|\kappa+\alpha)$. Then the singularities of the integrand are as depicted in Fig. 9.

It is important that the poles at $\eta^2=\lambda_j^2q^2$ present in $G_{\text{right}}(\eta,\xi)$ are cancelled by $Q(\eta q^{-1},\kappa)$ which occur in

$$dm(\eta)=\frac{d(\eta)Q(\eta q^{-1},\kappa)}{T(\eta,\kappa+\alpha)Q(\eta,\kappa)}\frac{d\eta^2}{\eta^2}\,. \tag{5.11}$$

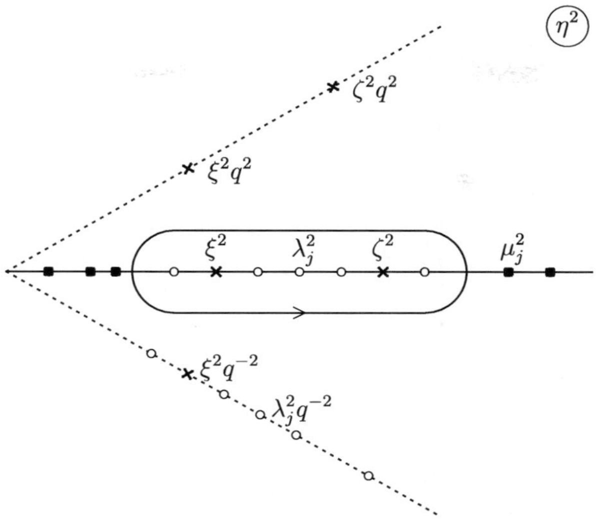

FIG. 9.

So, in principle we may have poles at $\zeta^2=\mu_j^2$, $\zeta^2=\lambda_jq^{-2}$, $\zeta^2=\xi^2q^{\pm2}$ which appear when singularities pinch the contour of integration. In addition, we have poles at zeroes of $T(\zeta,\kappa)$ which are explicitly present in $\rho(\zeta)$. Let us compute their residues.

- The poles at $\zeta^2=\mu_j^2$ come only from the pole at $\eta^2=\zeta^2$ inside the contour, but the latter is accompanied with the multiplier $\rho(\zeta)$ which vanishes at $\zeta^2=\mu_j^2$.
- The poles at $\zeta^2=\lambda_j^2q^{-2}$ come from two places: the pole at $\eta^2=\zeta^2$ and $\eta^2=\lambda_j^2$. We compute

$$\begin{aligned}\operatorname*{res}_{\zeta^2=\lambda_j^2q^{-2}}U(\zeta,\xi)&=-\frac{1}{1+\mathfrak{a}(\lambda_jq^{-1})}G_{\mathrm{right}}(\lambda_j,\xi)\operatorname*{res}_{\eta^2=\lambda_j^2}dm(\eta)\\&\quad+G_{\mathrm{right}}(\lambda_j,\xi)\operatorname*{res}_{\eta^2=\lambda_j^2}dm(\eta)=0\,,\end{aligned}$$

  where we use $\mathfrak{a}(\lambda_jq^{-1})=0$.
- The pole at $\zeta^2=\xi^2q^2$ comes from the pole at $\eta^2=\zeta^2$ only. We have

$$\operatorname*{res}_{\zeta^2=\xi^2q^2}U(\zeta,\xi)=\frac{1}{(1+\mathfrak{a}(\xi q))(1+\overline{\mathfrak{a}}(\xi))}=\frac{a(\xi)d(\xi q)}{T(\xi,\kappa)T(\xi q,\kappa)}\,.$$

- The pole at $\zeta^2=\xi^2q^{-2}$ comes from two poles at $\eta^2=\zeta^2$ and at $\eta^2=\xi^2$. We have

$$\begin{aligned}\operatorname*{res}_{\zeta^2=\xi^2q^2}U(\zeta,\xi)&=\frac{1}{(1+\mathfrak{a}(\xi q^{-1}))(1+\mathfrak{a}(\xi))}-\frac{1}{1+\mathfrak{a}(\xi)}\\&=-\frac{d(\xi)a(\xi q^{-1})}{T(\xi,\kappa)T(\xi q^{-1},\kappa)}\,.\end{aligned}$$

A similar computation can be performed for the singularities in ξ. In addition it is clear that $(\xi/\zeta)^\alpha U(\zeta,\xi)$, which is a rational function of ζ^2, ξ^2, behaves as $O(1)$

for $\zeta,\xi\to 0,\infty$. In summary, we obtain that

$$U(\zeta,\xi)=\frac{(-1)}{T(\zeta,\kappa)T(\xi,\kappa)}\Big(P(\zeta^2,\xi^2)+\psi(q\zeta/\xi,\alpha)a(\zeta)d(\xi)-\psi(q^{-1}\zeta/\xi,\alpha)d(\zeta)a(\xi)\Big)\,,$$

where $P(\zeta^2,\xi^2)$ is a polynomial of degree $\mathbf{n}$ in both variables. Comparing with the definition (4.35) of $\omega(\zeta,\xi|\alpha)$ we see that (5.7) gives the correct analytic structure.

We have to think about the polynomial $P(\zeta^2,\xi^2)$. The main property of $\omega(\zeta,\xi|\alpha)$ is that it has vanishing $\mathbf{n}+1$ periods (4.32). In the case of spin 1/2, $\mathbf{n}$ of these equations can be rewritten as follows. Introduce the operation

$$\delta^+_\zeta f(\zeta)=f(\zeta)+\rho(\zeta)f(\zeta q^{-1})\,. \tag{5.12}$$

It is easy to see that the normalisation conditions (4.32) for $\mathbf{m}=\mathbf{1},\dots,\mathbf{n}$ can be rewritten as follows:

$$\delta^+_\zeta\omega(\zeta,\xi|\alpha)\Big|_{\zeta=\tau_{\mathbf{m}}q^{1/2}}=0,\qquad \mathbf{m}=\mathbf{1},\dots,\mathbf{n}\,. \tag{5.13}$$

First of all, we want to explain how this equation is understood for the transcendental part of $\omega(\zeta,\xi|\alpha)$. Certainly, this repeats what has been explained in Section 4.2, but let us be careful. We have

$$\delta^+_\zeta\delta^-_\zeta f(\zeta)=f(\zeta q)-\rho(\zeta)\rho(\zeta q^{-1})f(\zeta q^{-1})\,.$$

Since

$$\rho(\tau_{\mathbf{m}}q^{1/2})\rho(\tau_{\mathbf{m}}q^{-1/2})=1\,, \tag{5.14}$$

we obtain

$$\big(\delta^+_\zeta\delta^-_\zeta f\big)(\tau_{\mathbf{m}}q^{1/2})=\big(\Delta_\zeta f\big)(\tau_{\mathbf{m}}q^{1/2})\,.$$

Hence

$$\delta^+_\zeta\delta^-_\zeta\delta^-_\xi\Delta^{-1}_\zeta\psi(\zeta/\xi,\alpha)\Big|_{\zeta=\tau_{\mathbf{m}}q^{1/2}}=\delta^-_\xi\psi(\tau_{\mathbf{m}}q^{1/2}/\xi,\alpha)=f_{\mathrm{right}}(\tau_{\mathbf{m}}q^{1/2},\xi)\,.$$

So, the transcendentality disappears. Thus we have to prove that

$$\delta^+_\zeta U(\zeta,\xi)\Big|_{\zeta=\tau_{\mathbf{m}}q^{1/2}}=f_{\mathrm{right}}(\tau_{\mathbf{m}}q^{1/2},\xi)\,. \tag{5.15}$$

Carefully tracing analytic continuation and using the equation (5.8), one derives

$$\begin{aligned}\delta^+_\zeta U(\zeta,\xi)&=f_{\mathrm{right}}(\zeta,\xi)-G_{\mathrm{right}}(\zeta,\xi)\frac{1}{1+\bar{\mathfrak{a}}(\zeta)}+G_{\mathrm{right}}(\zeta q^{-1},\xi)\frac{\rho(\zeta)}{1+\mathfrak{a}(\zeta q^{-1})}\\&\quad+\Big(1-\rho(\zeta)\rho(\zeta q^{-1})\Big)\frac{1}{2\pi i}\int_\gamma\psi(q^{-1}\zeta/\eta,\alpha)G_{\mathrm{right}}(\eta,\xi)dm(\eta)\,.\end{aligned}$$

Now, noticing that $\bar{\mathfrak{a}}(\tau_{\mathbf{m}}q^{1/2})=\mathfrak{a}(\tau_{\mathbf{m}}q^{-1/2})=\infty$ and using (5.14) we arrive at (5.15).

It remains to prove the normalisation condition for $\mathbf{m}=\mathbf{0}$. From Remark 4.14 we find

$$\frac{1}{2\pi i}\int_{\Gamma_{\mathbf{0}}}T(\zeta,\kappa)\delta^-_\zeta\delta^-_\xi\Delta^{-1}_\zeta\psi(\zeta/\xi,\alpha)d\mu^+(\zeta)=-\frac{1}{2}q^{-\kappa+s}\xi^{-\alpha}(q^{-\alpha}-\rho(\xi))\,.$$

So, we have to prove that

$$U(\zeta,\xi)=-\frac{q^{-\alpha}-\rho(\xi)}{2(1+q^{2(\kappa-s)})}(\zeta/\xi)^\alpha(1+O(\zeta^2))\,,\quad \zeta^2\to 0. \tag{5.16}$$

In order to prove this, it is convenient to use an alternative expression for $U(\zeta,\xi)$:

$$U(\zeta,\xi)=\big(G_{\mathrm{left}}\star f_{\mathrm{right}}\big)(\zeta,\xi)\,,$$

where G_{left} solves the equation

$$G_{\mathrm{left}}(\zeta,\xi)=f_{\mathrm{left}}(\zeta,\xi)-\big(G_{\mathrm{left}}\star K_\alpha\big)(\zeta,\xi)\,. \tag{5.17}$$

Define $g_{\mathrm{left}}(\xi)$ by the leading term in the expansion:

$$G_{\mathrm{left}}(\zeta,\xi)=\zeta^\alpha g_{\mathrm{left}}(\xi)(1+O(\zeta^2))\,.$$

Deriving the equation for $g_{\mathrm{left}}(\xi)$ one should be careful: recall that the point $\eta^2=\zeta^2$ is assumed to be inside γ while the point $\eta^2=0$ is outside. So, something should be done about the pinching of the contour for $\zeta^2\to 0$. For any ζ^2 let us denote by $\tilde\gamma$ the contour with $\eta^2=\zeta^2$ (and $\eta^2=0$ as usual) being outside the contour. For the integral over $\tilde\gamma$ the limit $\zeta^2\to 0$ is straightforward. For any $F(\eta)=\eta^\alpha R(\zeta^2)$ regular at $\eta^2=\zeta^2$ we have

$$\int_\gamma G(\zeta,\eta)F(\eta)dm(\eta)=\int_{\tilde\gamma} G(\zeta,\eta)F(\eta)dm(\eta)-\rho(\zeta)\zeta^{-\alpha}F(\zeta)\,. \tag{5.18}$$

Using the latter identity and the definition of $f_{\mathrm{left}}(\zeta,\xi)$ one finds the equation

$$g_{\mathrm{left}}(\xi)=-\frac{1}{4\pi i}(q^\alpha-q^{-\alpha})\frac{1-q^{2(\kappa-s)}}{1+q^{2(\kappa-s)}}\xi^{-\alpha}-\int_{\tilde\gamma} g_{\mathrm{left}}(\eta)K_\alpha(\eta/\xi)dm(\eta)\,. \tag{5.19}$$

A key observation of [**BG**] is that it is possible to solve this equation 'explicitly'. Consider the function

$$H(\eta)=h(\eta q^{-1})-h(\eta q)\,,\quad h(\eta)=\frac{Q(\eta,\kappa+\alpha)}{Q(\eta,\kappa)}\,.$$

Using Baxter equation one proves the identity

$$H(\eta)dm(\eta)=\Big(h(\eta)-\frac{1}{\rho(\eta)}h(\eta q)\Big)\frac{d\eta^2}{\eta^2}\,.$$

Using this and the formula (5.11) one computes the integral going around the Bethe roots. If σ^2 is outside γ, then

(5.20)

$$\frac{1}{2\pi i}\int_{\tilde\gamma} H(\eta)\psi(\eta/\sigma,\alpha)dm(\eta)=\frac{1}{2\pi i}\int_{\tilde\gamma}\psi(\eta/\sigma,\alpha)h(\eta)\frac{d\eta^2}{\eta^2}=-h(\sigma)-C\sigma^{-\alpha}\,,$$

where

$$C=-\frac12\lim_{\eta\to 0}\eta^\alpha h(\eta)+\frac12\lim_{\eta\to\infty}\eta^\alpha h(\eta)\,.$$

Similarly if σ^2 is inside γ, then

$$\frac{1}{2\pi i}\int_{\tilde\gamma} H(\eta)\psi(\eta/\sigma,\alpha)dm(\eta)=-\frac{1}{\rho(\sigma)}h(\sigma q)-C\sigma^{-\alpha}\,. \tag{5.21}$$

From (5.20), we find that

$$g_{\mathrm{left}}(\xi)=-\frac{1}{4\pi i}\frac{1}{C}\frac{1-q^{2(\kappa-s)}}{1+q^{2(\kappa-s)}}H(\xi)\,.$$

It remains to compute

$$U(\zeta,\xi)=\zeta^\alpha u(\xi)(1+O(\zeta^2))\,, \tag{5.22}$$

$$u(\xi)=\int_{\tilde\gamma}g_{\mathrm{left}}(\eta)f_{\mathrm{right}}(\eta,\xi)dm(\eta)-\frac{1}{2}\frac{q^{-\alpha}-\rho(\xi)}{1+q^{2(\kappa-s)}}q^{2(\kappa-s)}\xi^{-\alpha}$$

$$=-\frac{1}{2}\frac{q^{-\alpha}-\rho(\xi)}{1+q^{2(\kappa-s)}}\xi^{-\alpha}\,.$$

Here we use (5.20), (5.21), choosing ξ^2 inside and ξ^2q^2 outside γ.

This finishes the proof. □

For physical applications we shall consider the limit $\mathbf{n}\to\infty$. In that case for $|q|=1$ the Bethe roots occupy the entire half-axis $\mathbb{R}_+$, and it would be much more convenient to have kernels which decrease for $\zeta^2\to 0,\infty$. This is possible.

COROLLARY 5.2. *The formula* (5.7) *remains valid if we replace everywhere in* (5.2), (5.5), (5.6) *the function* $\psi(\zeta,\alpha)$ *by*

$$\psi_+(\zeta,\alpha)=\psi(\zeta,\alpha)-\frac{1}{2}\zeta^\alpha=\frac{\zeta^\alpha}{\zeta^2-1}\,. \tag{5.23}$$

PROOF. From the proof of the above Lemma we conclude that

$$\int\eta^{-\alpha}G_{\mathrm{right}}(\eta,\xi)dm(\eta)=\frac{1}{q^\alpha-q^{-\alpha}}\xi^{-\alpha}\left(q^{-\alpha}-\rho(\xi)\right).$$

It is easy to see that this identity allows us to change ψ to ψ_+ in (5.8) and in (5.9). □

Notice that for $0<\alpha<2$ the function $\psi_+(\zeta,\alpha)$ decreases for $\zeta^2\to 0,\infty$, and we have a nicer formula for the Mellin transform:

$$\Delta_\zeta^{-1}\psi_+(\zeta,\alpha)=VP\int_{-\infty}^{\infty}\zeta^{2ik}\coth\pi(k+\frac{i\alpha}{2})\frac{i}{4\sinh 2\pi\nu k}dk\,, \tag{5.24}$$

where the principal value concerns only the pole at $k=0$.

For considering the limit $\mathbf{n}\to\infty$, it is convenient to proceed further assuming, as in Section 2.6, that the Bethe roots λ_j^2 are real and that $|\mathfrak{a}(\zeta)|\ll 1$ for $\zeta^2\in\mathbb{C}^+$. The function $\rho(\zeta)$ stays finite. Consider the equation for the dressed resolvent (5.5)

$$R_{\mathrm{dress}}+K_\alpha\star R_{\mathrm{dress}}=K_\alpha\,.$$

Suppose the Bethe roots are as shown in Fig. 9. Under the present assumptions, the measure on the lower part of the contour is small while the measure on the upper part of the contour is close to

$$dm_0(\zeta)=\frac{d\zeta^2}{\zeta^2\rho(\zeta)}\,. \tag{5.25}$$

In the same notation as in (5.4), introduce the convolution

$$(F\circ G)(\zeta,\xi)=\int_a^b F(\zeta,\eta)G(\eta,\xi)dm_0(\eta)\,. \tag{5.26}$$

Similarly to what has been done in Section 2.6, we extract the integral with dm_0:

$$R_{\mathrm{dress}}-K_\alpha\circ R_{\mathrm{dress}}+K_\alpha * R_{\mathrm{dress}}=K_\alpha\,,$$

where $*$ stands for the convolution

$$(F * G)(\zeta,\xi) = \int_a^b F(\zeta,\eta)G(\eta,\xi)d\tilde{m}(\eta)\,, \tag{5.27}$$

with the measure

$$d\tilde{m}(\zeta) = \frac{1}{\rho(\zeta)}\Bigl(\frac{1}{1+\overline{\mathfrak{a}}(\zeta+i0)} + \frac{1}{1+\mathfrak{a}(\zeta-i0)}\Bigr)\,.$$

The advantage of this measure is that it is small for $\zeta^{\pm 1}\to\infty$. Now defining the resolvent R by

$$R - K_\alpha \circ R = K_\alpha\,,$$

we obtain

$$R_{\rm dress} + R * R_{\rm dress} = R\,. \tag{5.28}$$

This certainly implies

$$R_{\rm dress} + R_{\rm dress} * R = R\,. \tag{5.29}$$

We want to rewrite the equation for $\omega(\zeta,\xi|\alpha)$ (5.7) in the same fashion. Introduce the functions

$$F_{\rm left} = f_{\rm left} + f_{\rm left}\circ R\,,\quad F_{\rm right} = f_{\rm right} + R\circ f_{\rm right}\,. \tag{5.30}$$

According to the general prescription, we understand that the first (resp. second) argument of $F_{\rm left}$ (resp. $F_{\rm right}$) belongs to $\mathbb{C}^-$; certainly analytic continuation is possible. After a simple calculation, we obtain

$$\begin{aligned} &\omega(\zeta,\xi|\alpha) = \omega^{(1)}(\zeta,\xi|\alpha) + \omega^{(2)}(\zeta,\xi|\alpha)\,,\\ &\frac{1}{4}\omega^{(1)}(\zeta,\xi|\alpha) = \bigl(-F_{\rm left} * F_{\rm right} + F_{\rm left} * R_{\rm dress} * F_{\rm right}\bigr)(\zeta,\xi)\,,\\ &\frac{1}{4}\omega^{(2)}(\zeta,\xi|\alpha) = \bigl(f_{\rm left}\circ F_{\rm right}\bigr)(\zeta,\xi) + \delta^-_\zeta\delta^-_\xi\Delta^{-1}_\zeta\psi(\zeta/\xi,\alpha)\,. \end{aligned} \tag{5.31}$$

These formulas will be very useful later.

5.2. Main Theorem and Inverse Problem

5.2.1. General Idea. We have found the fermionic basis of quasi-local operators. The main theorem states that the expectation values on a cylinder with arbitrary *Matsubara* data are simple for elements of the fermionic basis. But in physical applications we are interested in expectation values of given, physically relevant, operators. Hence an important problem is to express a given quasi-local operator in the fermionic basis. A straightforward idea would be to follow the definitions of Section 3.4, namely, to compute the creation operators in terms of spin operators, to apply them to the primary field obtaining a linear combination of quasi-local operators and then to invert the corresponding formulas. The problem is that the definitions in Section 3.4 are rather complicated, and the above programme is hard to implement even on four lattice sites. In the present section we propose an alternative, more witty, method.

Consider a quasi-local operator $q^{2\alpha S(0)}\mathcal{O}$, with $\mathcal{O}$ being a spinless operator located on the interval $[1,n]$. In Section 4.4 we constructed the basis of quasi-local operators

$$v_{R,S,\lambda} = \bar{\mathbf{b}}_R\mathbf{c}_S\mathbf{s}_\lambda B_n(\alpha-n)\,.$$

REMARK 5.3. For computational goals, it is more practical to use the operators $\mathbf{b}^*_{\mathrm{rat}}(\zeta), \mathbf{c}^*_{\mathrm{rat}}(\zeta)$ (3.109) instead of $\mathbf{b}^*(\zeta), \mathbf{c}^*(\zeta)$. By so doing, obviously the construction of Section 4.4 does not change, but the transcendental functions which are rather spurious here disappear.

The Main Theorem has a simple corollary which makes the operators $\mathbf{b}^*_{\mathrm{rat}}(\zeta)$, $\mathbf{c}^*_{\mathrm{rat}}(\zeta)$ all the more convenient:

$$\mathrm{Tr}\big(\mathbf{b}^*_{\mathrm{rat}}(\zeta)\ \bullet\big)=0\,,\quad \mathrm{Tr}\big(\mathbf{c}^*_{\mathrm{rat}}(\zeta)\ \bullet\big)=0\,. \tag{5.32}$$

In other words, the functional Tr plays the role of left vacuum for the fermion creation-annihilation operators. Recall that the role of the right vacuum is played by the primary fields.

In order to avoid heavy notation, we shall not write the suffix 'rat' below; instead we shall give comments on necessary changes when needed. For spinless operators, there is a restriction on lengths of λ, R, S following from the properties: B_n has spin n, $\bar{\mathbf{b}}_j$ ($\mathbf{c}_j$) lowers (raises) spin by 1, $\mathbf{s}_j$ lowers spin by 2. It is easy to compute the transformation

$$v_{R,S,\lambda}=\sum_{I^+,I^0,I^-}C^{I^+,I^0,I^-}_{R,S,\lambda}\ \ \mathbf{b}^*_{I^+}\mathbf{t}^*_{I^0}\mathbf{c}^*_{I^-}q^{2\alpha S(0)}\,.$$

The exact number of terms in the right hand side is not easy to compute in the general case, for $n=2,3,4,5,6$ it is respectively $7,33,166,898,5089$ while the number of spinless operators is $\binom{2n}{n}$ which is in our cases $6,20,70,252,924$. There are two limitations on I^+,I^0,I^- which are clear from general construction: Russian doll implies $\#(I^+)+\#(I^0)+\#(I^-)\le n$, from the estimate (3.86) it follows that $\max(I^+)\le n$, and similarly for I^0, I^-.

Completeness ensures that any quasi-local operator can be written as

$$q^{2\alpha S(0)}\mathcal{O}=\sum_{R,S,\lambda}C_{R,S,\lambda}\ v_{R,S,\lambda}\,, \tag{5.33}$$

with some coefficients $C_{R,S,\lambda}$ depending on $\mathcal{O}$. This is the inverse problem in the framework of our method. For some *Matsubara* data we compute the functions $\rho(\zeta|\alpha)$ and $\omega(\zeta,\xi|\alpha)$ and the Taylor coefficients

$$\rho(\zeta|\alpha)=\sum_{p=1}^{\infty}(\zeta^2-1)^{p-1}\rho_p(\alpha)\,,$$

$$\omega(\zeta,\xi|\alpha)=\zeta^{2+\alpha}\xi^{2-\alpha}\sum_{p=1,r=1}^{\infty}(\zeta^2-1)^{p-1}(\xi^2-1)^{r-1}\omega_{p,r}(\alpha)\,.$$

The main theorem tells that

$$\begin{aligned}&\frac{\langle\kappa+\alpha|\mathrm{Tr}_{[1,n]}\left(T_{[1,n],\mathbf{M}}q^{2\kappa S_{[1,n]}}\mathcal{O}\right)|\kappa\rangle}{T(1,\kappa)^n\langle\kappa+\alpha|\kappa\rangle}\\ &\qquad=\sum_{R,S,\lambda}C_{R,S,\lambda}\prod_{i\in I^0}\rho_i(\alpha)\ \det\left(\omega_{i,j}(\alpha)\right)_{i\in I^+,j\in I^-}\,.\end{aligned} \tag{5.34}$$

The idea is that if the *Matsubara* chain is not very long one should be able to find sufficiently efficient way of direct computation of the left hand side of (5.34). Then (5.34) is an equation for the coefficients $C_{R,S,\lambda}$. One can hope to find in this way a sufficient number of independent equations which would allow us to fix the coefficients. Moreover, the number of different *Matsubara* data is infinite, so, we would deal with overdetermined systems, and the existence of a solution would be a powerful check of the entire procedure. Let us be more precise.

5.2.2. Matsubara Data. Consider a *Matsubara* chain of length $\mathbf{n}$ with spins $s_1, \cdots, s_{\mathbf{n}}$, inhomogeneities $\tau_1, \cdots, \tau_{\mathbf{n}}$, and the twist κ. We use the polynomials

$$a(\zeta) = \prod_{\mathbf{m}=1}^{\mathbf{n}} a_{s_{\mathbf{m}}}(\zeta/\tau_{\mathbf{m}}), \quad d(\zeta) = \prod_{\mathbf{m}=1}^{\mathbf{n}} d_{s_{\mathbf{m}}}(\zeta/\tau_{\mathbf{m}}),$$
$$a_s(\zeta) = 1 - q^{-2s}\zeta^2, \quad d_s(\zeta) = 1 - q^{2s}\zeta^2. \tag{5.35}$$

There are two transfer-matrices: one with twist κ, and the other with twist $\kappa + \alpha$. Because of the spin conservation we consider eigenvectors of the same spin, in other words with the same number of Bethe roots m. For technical reasons we shall assume $m \le \mathbf{n}/2$. The eigenvalues of the Q-operators are

$$Q(\zeta, \kappa) = \zeta^{\kappa - m} \prod_{j=1}^{m}\Bigl(1 - \frac{\zeta^2}{\beta_j^2}\Bigr), \quad Q(\zeta, \kappa + \alpha) = \zeta^{\kappa + \alpha - m} \prod_{j=1}^{m}\Bigl(1 - \frac{\zeta^2}{\gamma_j^2}\Bigr),$$

Since they depend only on the squares of β_j, γ_j it is often convenient to use

$$b_j = \beta_j^2, \quad c_j = \gamma_j^2.$$

The Bethe roots are found from the Bethe equations

$$a(\zeta)Q(\zeta q, \kappa) + d(\zeta)Q(\zeta q^{-1}, \kappa)\big|_{\zeta^2 = b_j} = 0,$$
$$a(\zeta)Q(\zeta q, \kappa + \alpha) + d(\zeta)Q(\zeta q^{-1}, \kappa + \alpha)\big|_{\zeta^2 = c_j} = 0,$$
$$j = 1, 2, \cdots, k. \tag{5.36}$$

Historically one uses the Bethe equations for physically interesting situations like, for instance, the finite temperature case considered in Section 1.6. However, in the present context our goals are quite different: we need sufficient stock of *Matsubara* data for an unphysical situation of small $\mathbf{n}$ and rather arbitrary parameters. Still, the problem looks complicated: for given $a(\zeta)$ and $d(\zeta)$ we have to solve a system of algebraic equations in order to find the Bethe roots. In order to make our procedure efficient we have to find a possibility to avoid this problem.

Introduce the coefficients

$$a(\zeta) = \mathsf{a}(\zeta^2), \qquad \mathsf{a}(z) = 1 + \sum_{k=1}^{\mathbf{n}} a_k z^k,$$
$$d(\zeta) = \mathsf{d}(\zeta^2), \qquad \mathsf{d}(z) = 1 + \sum_{k=1}^{\mathbf{n}} d_k z^k.$$

These coefficients are not arbitrary complex numbers because the spins $s_1, \cdots, s_{\mathbf{n}}$ take discrete, half-integer, values. However, the dependence of solutions to (5.36) on the values of spins is analytical, for that reason we shall consider a_k, d_k as arbitrary complex numbers. So, the *Matsubara* data are

$$\mathrm{Md} = \{\{b_1, \cdots, b_m\}, \{c_1, \cdots, c_m\}, \{a_1, \cdots a_{\mathbf{n}}, d_1, \cdots d_{\mathbf{n}}\}\},$$

which satisfy the system of equations (5.36). The traditional way of proceeding requires to start with

$$\text{input} = \{a_1, \cdots a_{\mathbf{n}}, d_1, \cdots d_{\mathbf{n}}\}, \tag{5.37}$$

and to find the Bethe roots solving the algebraic equations. This is hard. For our goals we obtain unphysical solutions differently. Start with the data:

$$\text{input} = \{\{b_1, \cdots, b_m\}, \{c_1, \cdots, c_m\}, \{a_{2m+1}, \cdots a_{\mathbf{n}}, d_1, \cdots d_{\mathbf{n}}\}\}\,. \tag{5.38}$$

The number of continuous parameters is the same as in (5.37). The unknowns now are $a_1, \cdots, a_{2m}$, for which the equations (5.36) *are linear*! So, starting from (5.38) one easily finds the *Matsubara* data.

Consider an example. Denote $x = q^{\kappa+\alpha}$, $y = q^{\kappa}$, and take the input data

$$\text{input} = \{\{1,2\},\{3,4\},\{9,1,2,6\}\}\,.$$

then we find

$$\begin{aligned}
\text{Md} = \Big\{&\{1,2\},\{3,4\},\\
&\Big\{\frac{1}{6}\Big(-60 + \frac{-1205+1148q^2-230q^4}{(2-q^2)(1-2q^2)x^2} + \frac{-4477+6q^2(-692+1467q^2)}{(4-3q^2)(3-4q^2)y^2}\Big),\\
&\frac{1}{2}\Big(70 + \frac{3207+12q^2(-253+49q^2)}{(2-q^2)(1-2q^2)x^2} + \frac{5849+12q^2(1382-1927q^2)}{(4-3q^2)(3-4q^2)y^2}\Big),\\
&-50 - \frac{19(599-560q^2+101q^4)}{3(2-q^2)(1-2q^2)x^2} + \frac{-10651-60096q^2+73251q^4}{3(4-3q^2)(3-4q^2)y^2},\\
&24 + \frac{4(593-536q^2+77q^4)}{(2-q^2)(1-2q^2)x^2} + \frac{14(98+3q^2(296-341q^2))}{(4-3q^2)(3-4q^2)y^2}, 9, 1, 2, 6\Big\}\Big\}\,.
\end{aligned}$$

5.2.3. Making Equations. Having the *Matsubara* data we construct

$$T(\zeta,\kappa) = \frac{a(\zeta)Q(\zeta q,\kappa) + d(\zeta)Q(\zeta q^{-1},\kappa)}{Q(\zeta,\kappa)}\,,$$

$$T(\zeta,\kappa+\alpha) = \frac{a(\zeta)Q(\zeta q,\kappa+\alpha) + d(\zeta)Q(\zeta q^{-1},\kappa+\alpha)}{Q(\zeta,\kappa+\alpha)}\,,$$

which are polynomials of ζ^2 due to (5.36). The function $\rho(\zeta|\alpha)$ is given by (4.16).

For the computation of $\omega(\zeta,\xi|\alpha)$, we use the method of Section 5.1 whose notation we follow. We have the 'integral equations' (5.8):

$$\begin{aligned}
G_{\text{right}}(\zeta,\xi) &= f_{\text{right}}(\zeta,\xi)\\
&- 2\pi i\Big(\sum_{j=1}^{m} K_\alpha(\zeta/\beta_j,\alpha)G_{\text{right}}(\beta_j,\xi)\operatorname*{res}_{\eta^2=\beta_j^2} dm(\eta) - \frac{K_\alpha(\zeta/\xi,\alpha)}{\rho(\xi)(1+\mathfrak{a}(\xi,\kappa))}\Big)\,.
\end{aligned} \tag{5.39}$$

First, setting $\zeta = \beta_j$ we use (5.39) as a system of m equations for m unknowns $G_{\text{right}}(\beta_j,\xi)$. Second, knowing $G_{\text{right}}(\beta_j,\xi)$ we use (5.39) to find $G_{\text{right}}(\zeta,\xi)$ as a function of ζ,ξ. Now we slightly rewrite (5.9):

$$\begin{aligned}
\omega(\zeta,\xi|\alpha) &= \omega_{\text{rat}}(\zeta,\xi|\alpha) + \overline{D}_\zeta\overline{D}_\xi\Delta_\zeta^{-1}\psi(\zeta/\xi,\alpha)\,,\\
\omega_{\text{rat}}(\zeta,\xi|\alpha) &= -4(f_{\text{left}} \star G_{\text{right}})(\zeta,\xi) + \big(2\rho(\zeta|\alpha) - 2\rho(\xi|\alpha) - \Delta_\zeta\big)\psi(\zeta/\xi,\alpha)\,,
\end{aligned} \tag{5.40}$$

where

$$\overline{D}_\zeta f(\zeta) = f(\zeta q) + f(\zeta q^{-1}) - 2\rho(\zeta|\alpha)f(\zeta)\,. \tag{5.41}$$

The Bogolubov transformation (3.109) implies that working with $\mathbf{b}^*_{\text{rat}}(\zeta), \mathbf{c}^*_{\text{rat}}(\zeta)$ one should use $\omega_{\text{rat}}(\zeta,\xi|\alpha)$ instead of $\omega(\zeta,\xi|\alpha)$.

Now that we have found the Taylor series, we are able to write down the right hand side of (5.34).

If the *Matsubara* chain is not too long, the left hand side of (5.34) allows a rather efficient direct computation. A local operator $\mathcal{O}$ can be presented as a linear combination of

$$E_{\epsilon_1',\epsilon_1}\otimes E_{\epsilon_2',\epsilon_2}\otimes\cdots\otimes E_{\epsilon_n',\epsilon_n}\,,$$

where $E_{\pm,\pm}=\tau^{\pm}$, $E_{\pm,\mp}=\sigma^{\pm}$. We have

$$\begin{aligned}&\mathrm{Tr}_{[1,n]}\left(T_{[1,n],\mathbf{M}}q^{2\kappa S_{[1,n]}}E_{\epsilon_1',\epsilon_1}\otimes E_{\epsilon_2',\epsilon_2}\otimes\cdots\otimes E_{\epsilon_n',\epsilon_n}\,,\right)\\ &=q^{\kappa\sum_{j=1}^n\epsilon_j'}T_{\epsilon_1,\epsilon_1'}T_{\epsilon_2,\epsilon_2'}\cdots T_{\epsilon_n,\epsilon_n'}\,,\end{aligned}\tag{5.42}$$

where $T_{\epsilon,\epsilon'}$ are blocks in the block decomposition of $T_{j,\mathbf{M}}$. They are related as

$$T_{++}=A(1)\,,\ T_{+-}=B(1)\,,\ T_{-+}=C(1)\,,\ T_{--}=D(1)\,,$$

to the Algebraic Bethe Ansatz notation (Section 2.1):

$$T_{j,\mathbf{M}}(\xi)=\begin{pmatrix}A(\xi)&B(\xi)\\C(\xi)&D(\xi)\end{pmatrix}_j\,.$$

In what follows we shall write $A=A(1)$, etc.

Our goal is to compute

$$\langle\kappa+\alpha|T_{\epsilon_1,\epsilon_1'}T_{\epsilon_2,\epsilon_2'}\cdots T_{\epsilon_n,\epsilon_n'}|\kappa\rangle\,.\tag{5.43}$$

We shall use the notation

$$|\xi_1,\cdots,\xi_m\rangle=\prod_{j=1}^m\xi_j\cdot C(\xi_1)\cdots C(\xi_m)|{\downarrow}\rangle\,,$$

$$\langle\xi_1,\cdots,\xi_m|=\prod_{j=1}^m\xi_j^{-1}\cdot\langle{\downarrow}|B(\xi_1)\cdots B(\xi_m)\,.$$

We multiplied by $\prod_{j=1}^m\xi_j^{\pm1}$ in order to make these vectors polynomials of ξ_j^2. Rewrite (5.43) as

$$\langle\gamma_1,\cdots,\gamma_m|T_{\epsilon_1,\epsilon_1'}T_{\epsilon_2,\epsilon_2'}\cdots T_{\epsilon_n,\epsilon_n'}|\beta_1,\cdots,\beta_m\rangle\,.$$

We shall proceed inductively computing

$$\langle\gamma_1,\cdots,\gamma_m|T_{\epsilon_1,\epsilon_1'}T_{\epsilon_2,\epsilon_2'}\cdots T_{\epsilon_k,\epsilon_k'}|\xi_1,\cdots,\xi_{l(k)}\rangle\,,\tag{5.44}$$

where ξ_j are taken to be arbitrary numbers up to the very last step when we identify them with the Bethe roots β_j, and

$$l(k)=m+\frac{1}{2}\sum_{j=1}^k(\epsilon_j-\epsilon_j')$$

takes care of the spin. Using arbitrary ξ_j is a well-known trick often called off-shell Bethe Ansatz [**KIB**, **Sl**].

In what follows we use the commutation relations (2.4), (2.5), (2.6). It is convenient to use variables $x_i=\xi_i^2$, so we use instead of u, v the functions

$$\mathsf{u}(x)=\frac{xq-q^{-1}}{x-1}\,,\quad\mathsf{v}(x)=\frac{q-q^{-1}}{x-1}\,.$$

For any vector $\langle\Phi|$ independent of $\xi_1,\cdots\xi_l$, both $\langle\Phi|\xi_1,\cdots,\xi_l\rangle$ and $\langle\Phi|A|\xi_1,\cdots,\xi_l\rangle$ are symmetric polynomials of $x_1,\cdots,x_l$. We apply the same technique as the one used in the derivation of (2.9) to compute $A|\xi_1,\cdots,\xi_l\rangle$. For future convenience we would like to formalise the procedure. Denote by $\mathfrak{S}_l$ the linear space of symmetric polynomials of l variables. Introduce the operator $\mathcal{A}:\mathfrak{S}_l\to\mathfrak{S}_l$:

$$(\mathcal{A}f)(x_1,\cdots,x_l)=\mathsf{a}(1)\prod_{j=1}^{l}\mathsf{u}(1/x_j)f(x_1,\cdots,x_l)$$
$$-\sum_{j=1}^{l}\mathsf{a}(x_j)\mathsf{v}(1/x_j)\prod_{r\neq j}^{l}\mathsf{u}(x_j/x_r)f(x_1,\cdots,\widehat{x_j},\cdots,x_l,1)\,. \tag{5.45}$$

It is easy to see that this definition is consistent: the left hand side is a symmetric polynomial provided $f(x_1,\cdots,x_l)$ is such. The matrix element $\langle\Phi|A|\xi_1,\cdots,\xi_l\rangle$ is obtained by applying $\mathcal{A}$ to $\langle\Phi|\xi_1,\cdots,\xi_l\rangle$.

Similarly

$$(\mathcal{D}f)(x_1,\cdots,x_l)=\mathsf{d}(1)\prod_{j=1}^{l}\mathsf{u}(x_j)f(x_1,\cdots,x_l)$$
$$+\sum_{j=1}^{l}\mathsf{d}(x_j)\mathsf{v}(1/x_j)\prod_{r\neq j}^{l}\mathsf{u}(x_r/x_j)f(x_1,\cdots,\widehat{x_j},\cdots,x_l,1)\,. \tag{5.46}$$

The operator $\mathcal{B}$ acts from $\mathfrak{S}_{l-1}$ to $\mathfrak{S}_l$:

$$(\mathcal{B}f)(x_1,\cdots,x_l)=\sum_{j=1}^{l}\mathsf{v}(1/x_j)\Big(\mathsf{a}(1)\mathsf{d}(x_j)\prod_{r\neq j}^{l}\mathsf{u}(1/x_r)\mathsf{u}(x_r/x_j)$$
$$-\,\mathsf{d}(1)\mathsf{a}(x_j)\prod_{r\neq j}^{l}\mathsf{u}(x_r)\mathsf{u}(x_j/x_r)\Big)f(x_1,\cdots,\widehat{x_j},\cdots,x_l)$$
$$-\sum_{j>i}\mathsf{v}(1/x_i)\mathsf{v}(1/x_j)\Big(\mathsf{d}(x_i)\mathsf{a}(x_j)\mathsf{u}(x_j/x_i)\prod_{r\neq i,j}^{l}\mathsf{u}(x_r/x_i)\mathsf{u}(x_j/x_r)$$
$$+\,\mathsf{a}(x_i)\mathsf{d}(x_j)\mathsf{u}(x_i/x_j)\prod_{r\neq i,j}^{l}\mathsf{u}(x_i/x_r)\mathsf{u}(x_r/x_j)\Big)$$
$$\times\, f(x_1,\cdots,\widehat{x_i},\cdots,\widehat{x_j},\cdots,x_l,1)\,. \tag{5.47}$$

The operator $\mathcal{C}$ is simple, it acts from $\mathfrak{S}_{l+1}$ to $\mathfrak{S}_l$:

$$(\mathcal{C}f)(x_1,\cdots,x_l)=f(x_1,\cdots,x_l,1) \tag{5.48}$$

These relations allow us to compute inductively (5.44), provided the base case

$$N_{c_1,\cdots,c_m}(x_1,\cdots,x_m)=\langle\gamma_1,\cdots,\gamma_m|\xi_1,\cdots,\xi_m\rangle$$

is known, with γ_j being Bethe roots and ξ_j being arbitrary (recall that $c_i=\gamma_i^2$, $x_i=\xi_i^2$). We had different formulas for the scalar products in Section 2.4 and Section 4.2, but the first kind of formulas does not allow analytical continuation with respect to spins, while the second one gives expressions which are not very convenient for our goals here. Fortunately, there is a third way which is due to

Slavnov [**Sl**]. Consider the polynomials

$$(5.49)\quad P_j(x) = \frac{1}{x-c_j} \times \left(q^{\kappa+\alpha-m+1}\mathsf{a}(x)\prod_{k\neq j}\left(1-\frac{x^2}{c_k^2}q^2\right) - q^{-(\kappa+\alpha-m+1)}\mathsf{d}(x)\prod_{k\neq j}\left(1-\frac{x^2}{c_k^2}q^{-2}\right)\right).$$

The formula in question is

(5.50)

$$N_{c_1,\cdots,c_m}(x_1,\cdots,x_m) = C(c_1,\cdots,c_m)\cdot\frac{1}{\prod_{i<j}(x_i-x_j)}\det\left(P_j(x_i)\right)_{i,j=1,\cdots,m},$$

where the multiplier $C(c_1,\cdots,c_m)$ is independent of x_j and cancels in our final formulas. This is a symmetric polynomial of x_j. Here is an example of final computation with these definitions:

$$\langle\gamma_1,\cdots,\gamma_m|ABCD|\xi_1,\cdots,\xi_m\rangle = (\mathcal{DCBA}N_{c_1,\cdots,c_m})(x_1,\cdots,x_m).$$

Notice that in the right hand side the action is in the opposite direction.

Applying the formulas above to computations, we find that they have some drawbacks. The variables x_j have to be symbolic. Though it is easy to prove that the right hand sides of the formulas (5.45), (5.46), (5.47) are symmetric polynomials, it is not simple to use them in practice: factorisation is needed which is a very time consuming procedure. However, analysing the formulas one concludes that the procedure can be considerably simplified if we use Schur polynomials and Young diagrams. Let us be more precise at this point.

Consider Young diagrams Y_λ where $\lambda = (\lambda_1,\cdots,\lambda_l)$, $\lambda_1\geq\cdots\geq\lambda_l>0$, is a partition. We set $\#(\lambda)=l$. It is called the length of Y_λ. For a given l consider a Young diagram of length at most l. If necessary we complement it with zeroes up to length l. Then we identify Y_λ with the Schur polynomial

$$s_\lambda(x_1,\cdots,x_l) = \frac{\det\left(x_j^{\lambda_i+l-i}\right)}{\det\left(x_j^{l-i}\right)}.$$

This establishes an isomorphism between $\mathfrak{B}_l$ (space of bosons) whose elements are

$$Y = \sum_{\#(\lambda)\leq l} c_\lambda Y_\lambda$$

and the space $\mathfrak{S}_l$. For fixed $x_1,\cdots,x_l$ we have the linear functional defined by

$$\mathrm{Schur}_{x_1,\cdots,x_l}(Y_\lambda) = s_\lambda(x_1,\cdots,x_l),$$

and then extended by linearity.

Consider the Grassmann algebra $\mathfrak{F}_l$ (space of fermions) with basis $\psi^*_{k_1}\cdots\psi^*_{k_l}$ $(k_1>\cdots>k_l\geq 0)$. We have the usual isomorphism between the spaces $\mathfrak{B}_l$ and $\mathfrak{F}_l$.

$$(5.51)\quad \begin{aligned} \psi^*_{k_1}\psi^*_{k_2}\cdots\psi^*_{k_l} &\mapsto (k_1-(l-1),k_2-(l-2),\cdots,k_l)_0\,,\\ (\lambda_1,\cdots,\lambda_n) &\mapsto \psi^*_{\lambda_1+l-1}\cdots\psi^*_{\lambda_n+l-n}\psi^*_{l-n-1}\cdots\psi^*_0\,,\ \text{where } n\leq l. \end{aligned}$$

In the above, $()_0$ means removing all entries equal to 0.

For a given polynomial of one variable $P(x)=\sum_{j=0}^d p_jx^j$ we define the operator $P\wedge\,\mathfrak{F}_{l-1}\subset\mathfrak{F}_l$ multiplying by $\sum_{j=0}^d p_j\psi^*_j$, this operator is defined as $P\wedge\mathfrak{B}_{l-1}\subset\mathfrak{B}_l$

by the isomorphism (5.51). This definition can be generalised to polynomials of several variables in an obvious way. Certainly the polynomials anti-symmetrise themselves automatically.

We shall also need the simplest Littlewood-Richardson formula for multiplication of a Schur polynomial by elementary symmetric functions σ_j, which translates as action on $\mathfrak{B}_l$

$$\sigma_j \circ (\lambda_1, \cdots, \lambda_n) = \sum_{I^-}^{\binom{n+\min(j,l-n)}{j}} \Big((\lambda_1, \cdots, \lambda_n, \underbrace{0, \cdots, 0}_{\min(j,l-n)}) + e_{I^-} \Big)_{\text{order}},$$

where e_{I^-} are all vectors of dimension $n + \min(j, l-n)$ with j elements equal to 1 other elements being 0, 'order' means that we have to drop all the tables in which elements happen to be not ordered, and we also drop all zeroes in the final table.

In what follows we shall also need an operation $\mathrm{cut}_l(Y)$; given a linear combination Y of Young diagrams, cut_l erases all diagrams appearing in Y whose length is greater than l.

For a partition λ, define the coefficients $e_{\lambda,\lambda'}$ via

$$s_\lambda(x_1, \cdots, x_l, 1) = \sum_{\lambda'} e_{\lambda,\lambda'}\ s_{\lambda'}(x_1, \cdots, x_l).$$

This gives rise to an operator

$$EY_\lambda = \sum_{\lambda'} e_{\lambda,\lambda'} Y_{\lambda'}.$$

The easiest algorithm for finding $e_{\lambda,\lambda'}$ is the following: express $s_\lambda(x_1, \cdots, x_l, 1)$ through Jacobi-Trudi formula using the transposed λ. This formula is given in terms of the elementary symmetric functions $\sigma_k(x_1, \cdots, x_l, 1)$. Using

$$\sigma_k(x_1, \cdots, x_l, 1) = \sigma_k(x_1, \cdots, x_l) + \sigma_{k-1}(x_1, \cdots, x_l),$$

expand the Jacobi-Trudi determinant getting a sum of Schur polynomials.

Now Slavnov formula looks as

$$N_{c_1,\cdots,c_m}(x_1, \cdots, x_m) = C(c_1, \cdots, c_m) \cdot \Big(P_1 \wedge \cdots \wedge P_m \wedge \emptyset \Big), \tag{5.52}$$

and we proceed to the definition of the operators $\mathcal{T}_{++}$, $\mathcal{T}_{+-}$, $\mathcal{T}_{-+}$, $\mathcal{T}_{--}$ which translate the action of $\mathcal{A}$, $\mathcal{B}$, $\mathcal{C}$, $\mathcal{D}$ to the language of Young diagrams.

We begin with the operators $\mathcal{T}_{++}$, $\mathcal{T}_{--}$ which act from $\mathfrak{B}_l$ to itself as follows

$$\mathcal{T}_{++}Y = \sum_{k=0}^{l} \mathrm{cut}_l\Big(\sigma_k \circ EA_k \wedge Y\Big), \qquad A_k(x) = (-1)^k q^{l-2k} x^{l-k} \mathsf{a}(x), \tag{5.53}$$

$$\mathcal{T}_{--}Y = \sum_{k=0}^{l} \mathrm{cut}_l\Big(\sigma_k \circ ED_k \wedge Y\Big), \qquad D_k(x) = (-1)^k q^{-l+2k} x^{l-k} \mathsf{d}(x). \tag{5.54}$$

More complicated is $\mathcal{T}_{+-} : \mathfrak{B}_{l-1} \to \mathfrak{B}_l$:

$$\mathcal{T}_{+-}Y = \sum_{j,k=0}^{l} \mathrm{cut}_l\Big(\sigma_j \circ \sigma_k \circ EB_{j,k} \wedge Y\Big) + \sum_{k=0}^{l} \mathrm{cut}_l\Big(\sigma_k \circ B_k \wedge Y\Big), \tag{5.55}$$

where $B_{j,k}$, B_k are polynomials of respectively two and one variables:

$$B_{j,k}(x,y)=(-1)^{j+k}q\mathsf{d}(y)\mathsf{a}(x)(q^{-2}y)^{l-k}\frac{(xq^2)^{l-j}-y^{l-j}}{xq^2-y}\,,$$

$$B_k(x)=(-1)^{k+1}q^{-1}\mathsf{d}(1)\mathsf{a}(x)\frac{q^{-2(l-k)}-x^{l-k}}{xq^{-2}-x}\,.$$

Finally, we have the simplest operator $\mathcal{T}_{-+}:\mathfrak{B}_{l+1}\to\mathfrak{B}_l$:

$$\mathcal{T}_{-+}Y=EY\,. \tag{5.56}$$

Now we are ready to write down an expression for the left hand side of (5.34) in the form which allows very fast computation for reasonably small *Matsubara* data. Recall that

$$|\kappa\rangle=|\beta_1,\cdots,\beta_m\rangle\,,\quad \langle\kappa+\alpha|=\langle\gamma_1,\cdots,\gamma_m|\,.$$

We have

$$\frac{\langle\kappa+\alpha|T_{\epsilon_1,\epsilon_1'}T_{\epsilon_2,\epsilon_2'}\cdots T_{\epsilon_n,\epsilon_n'}|\kappa\rangle}{\langle\kappa+\alpha|\kappa\rangle}=\frac{\mathrm{Schur}_{b_1,\cdots,b_m}\left(\mathcal{T}_{\epsilon_n,\epsilon_n'}\cdots\mathcal{T}_{\epsilon_2,\epsilon_2'}\mathcal{T}_{\epsilon_1,\epsilon_1'}\mathcal{N}_{c_1,\cdots,c_m}\right)}{\mathrm{Schur}_{b_1,\cdots,b_m}\left(\mathcal{N}_{c_1,\cdots,c_m}\right)}\,. \tag{5.57}$$

5.2.4. Examples. As an example we consider the operator

$$\mathcal{O}=q^{2\theta S_{[1,n]}}\,.$$

Here θ is a parameter (not to be confused with the charge conjugation (1.23)). Denote

$$x=q^\alpha,\quad y=q^\theta\,.$$

Our goal is to find the coefficients in the decomposition

$$q^{2\alpha S(0)+2\theta S_{[1,n]}}=\sum_{R,S,\lambda}X_{R,S,\lambda}(x,y)\;v_{R,S,\lambda}\,. \tag{5.58}$$

Using the procedure above, we were able to compute them up to $n=6$. The dependence on y being very simple we keep it generic, but specialise all other parameters to numerical values and invoke interpolation. It is clear from the previous consideration that we may have an overdetermined system since the number of possible *Matsubara* data is unlimited. The problem is rather to find independent equations. The main characteristics of our data are the length $\mathbf{n}$ of the *Matsubara* chain and the number m of Bethe roots. We combine them into $(\mathbf{n},m)$. The simplest possibility is to take $m=0$ (i.e., the ferromagnetic vacuum in the *Matsubara* direction). This is, however, not sufficient even for $\mathbf{n}=2$; no matter how large, the data with $m=0$ do not fix all the coefficients in (5.58). Experiments show that for $n=2,\cdots,6$ the following sets of the *Matsubara* data are sufficient (and are actually overdetermined)

(5.59)

$n=2$ 10 (1,0), 5 (2,1)

$n=3$ 10 (1,0), 15 (2,1)

$n=4$ 30 (1,0), 10 (2,0), 50 (2,1), 10 (3,1), 10 (4,2)

$n=5$ 30 (1,0), 10 (2,0), 140 (2,1), 50 (3,1), 70 (4,2)

$n=6$ 50 (1,0), 50 (2,0), 10 (3,0), 300 (2,1), 200 (3,1), 50 (4,1), 350 (4,2), 100 (5,2), 25 (6,3)

As has been said, we collect sufficient amount of numerical x, q and find the coefficients in (5.58) taking the *Matsubara* data in the number and type prescribed in (5.59). Then we want to interpolate in x, q but the coefficients are rational functions, so, we need to know the denominators. Some experiments convince us that the common denominator of the coefficients $X_{R,S,\lambda}(x,y)$ is

$$D_n(x) = q^{2\left([\frac{n-1}{2}]^2+[\frac{n-2}{2}]^2\right)}(x^2-1)^{[\frac{n+1}{2}]}\prod_{j=1}^{[\frac{n}{2}]}\prod_{p=1}^{n-(2j-1)}(x^2-q^{2p})(1-x^2q^{2p})\,. \tag{5.60}$$

Up to $n = 4$ we computed the entire density matrices, i.e., the decomposition in the fermionic basis for all the quasi-local operators. This will be useful in the next section which deals with the case $\alpha = 0$. It is physically interesting and easier for computation because it contains one parameter less.

5.3. The Case $\alpha = 0$

5.3.1. General Remarks. Let us discuss what happens to our construction in the limit $\alpha \to 0$. The operator $\mathbf{t}^*(\zeta)$ is obviously a function regular at $\alpha = 0$. For the fermionic operators this is less clear. Consider the example of two fermionic operators $\mathbf{b}^*(\zeta_2)\mathbf{c}^*(\zeta_1)$ acting on the primary field. We continue working with $\mathbf{b}^*_{\mathrm{rat}}(\zeta)$, $\mathbf{c}^*_{\mathrm{rat}}(\zeta)$. By the Russian doll construction, this is equivalent to considering

$$\begin{aligned} g^+_{\mathrm{rat},c_2}(\zeta_2)g^-_{\mathrm{rat},c_1}(\zeta_1)(I) &= -\frac{1}{2}\frac{q^\alpha+q^{-\alpha}}{q^\alpha-q^{-\alpha}}\psi(\zeta,\alpha)\left(q^{\alpha\sigma^3_{c_1}}-\mathbb{R}_{c_2,c_1}(\zeta)\big(q^{\alpha\sigma^3_{c_2}}\big)\right)\\ &+\psi(q\zeta,\alpha)\tau^+_{c_1}\tau^-_{c_2}-\psi(q^{-1}\zeta,\alpha)\tau^-_{c_1}\tau^+_{c_2}-\Big(\frac{(q\zeta)^\alpha}{q\zeta-q^{-1}\zeta^{-1}}-\frac{(q^{-1}\zeta)^\alpha}{q^{-1}\zeta-q\zeta^{-1}}\Big)\sigma^-_{c_1}\sigma^+_{c_2}\,. \end{aligned} \tag{5.61}$$

This explicit formula is not very hard to derive from the general definitions of Section 3.5. The right hand side is regular at $\alpha = 0$. Indeed, the only dangerous denominator $q^\alpha - q^{-\alpha}$ is accompanied by the multiplier $q^{\alpha\sigma^3_{c_1}} - \mathbb{R}_{c_2,c_1}(\zeta)\big(q^{\alpha\sigma^3_{c_2}}\big)$ which vanishes at $\alpha = 0$. It is not obvious how to prove directly that the fermionic operators are regular at $\alpha = 0$ in general. Nevertheless the regularity of the function $\omega_{\mathrm{rat}}(\zeta,\xi|\alpha)$ implies that they are regular in the weak sense. This weak statement is sufficient for our goals.

So, the operators $\mathbf{t}^*(\zeta)$, $\mathbf{b}^*(\zeta)$, $\mathbf{c}^*(\zeta)$ are regular at the point $\alpha = 0$. However, if we decompose a given quasi-local operator in the fermionic basis, the coefficients may be singular. The simplest example is

$$\sigma^3_1 = \frac{1}{q^\alpha - q^{-\alpha}}\left(\mathbf{t}^*_1 - q^\alpha - q^{-\alpha}\right). \tag{5.62}$$

More sophisticated computations show that the degree of the pole at $\alpha = 0$ grows with the length of operators, the example of this is given by the denominator (5.60) (recall that $x = q^\alpha$).

The main content of the rest of this section consists in treatment of these singularities.

5.3.2. Reduction to the Quotient Space. We have

$$\rho(\zeta|0) = 1\,, \tag{5.63}$$

which means that in the weak sense $\mathbf{t}^*(\zeta) = 2$ for $\alpha = 0$. However, the example (5.62) shows that in order to deal with all local operators we have to evaluate indeterminate forms. For (5.62) this would force us to introduce a new operator at $\alpha = 0$: the derivative of $\mathbf{t}^*(\zeta)$. Then we may need higher derivatives and the situation soon becomes uncontrollable. So, up to now the general impression is that the easiest way is to compute everything for $\alpha \neq 0$ and take the limit. In the presence of descendants created by $\mathbf{t}^*_p$ this is probably the only possibility. Let us restrict our consideration to operators created by the fermions $\mathbf{b}^*_p$, $\mathbf{c}^*_p$ before taking the limit. In other words we consider the quotient space

$$\mathcal{W}^{\rm quo} = \mathcal{W}/ \ \Bigl\{\bigoplus_{p=1}^{\infty}(\mathbf{t}^*_p - 2\delta_{p,1})\mathcal{W}\Bigr\},$$

its subspaces are denoted by $\mathcal{W}^{\rm quo}_{\alpha,s}$.

Using (5.32) we define a functional which takes the same values as $\mathcal{Z}^{\kappa}_{\mathbf{n}}$ on $\mathcal{W}^{\rm quo}_{\alpha,0}$:

$$\mathcal{Z}^{\kappa}_{\mathbf{n}}\{X\} = \mathrm{Tr}\left(e^{\Omega}(X)\right), \quad X \in \mathcal{W}^{\rm quo}_{\alpha,0}, \tag{5.64}$$

where

$$\Omega = \frac{1}{(2\pi i)^2}\oint_\Gamma\oint_\Gamma \omega_{\rm rat}(\zeta_2,\zeta_1|\alpha)\mathbf{c}(\zeta_1,\alpha)\mathbf{b}(\zeta_2,\alpha+1)\frac{d\zeta_1^2}{\zeta_1^2}\frac{d\zeta_2^2}{\zeta_2^2}, \tag{5.65}$$

and $\omega_{\rm rat}(\zeta,\xi|\alpha)$ is given in (5.40).

For the operator Ω we have the following statement.

LEMMA 5.4. *The operator Ω allows three alternative representations in terms of traces over two two-dimensional representations and one over a Verma module:*

$$\begin{aligned}\Omega = &\frac{1}{(2\pi i)^2}\oint_\Gamma\oint_\Gamma \omega_{\rm rat}(\zeta_2,\zeta_1|\alpha)\\ &\times \mathrm{Tr}_{a,b}\{B^{\varepsilon}_{a,b}(\zeta)\mathbb{T}_a(\zeta_1)\mathbb{T}_b(\zeta_2)\}\mathrm{Tr}^{\Lambda}\{q^{(1-2\varepsilon)H}\mathbb{T}_\star(\overline{\zeta},\alpha)\}\frac{d\zeta_1^2}{\zeta_1^2}\frac{d\zeta_2^2}{\zeta_2^2}.\end{aligned} \tag{5.66}$$

Here $\varepsilon = 0, 1/2, 1$, and we use the notation of Section 2.3: $\zeta = \zeta_1/\zeta_2$, $\overline{\zeta} = \sqrt{\zeta_1\zeta_2}$, $\mathbb{T}_\star(\overline{\zeta},\alpha)$ is the twisted adjoint monodromy matrix in the Verma module with lowest weight $-\Lambda$ such that $q^{-\Lambda} = q\zeta$. The matrix $B^{\varepsilon}(\zeta)$ takes the form in the usual tensor notation

$$B^{\varepsilon}(\zeta) = \frac{\zeta - \zeta^{-1}}{(\zeta q - \zeta^{-1}q^{-1})(\zeta q^{-1} - \zeta^{-1}q)}\begin{pmatrix}0 & 0 & 0 & 0\\ 0 & \multicolumn{2}{c}{\multirow{2}{*}{\boxed{B^{\varepsilon}_{\rm mid}(\zeta)}}} & 0\\ 0 & & & 0\\ 0 & 0 & 0 & 0\end{pmatrix}. \tag{5.67}$$

The non-trivial middle blocks are

$$B^{1}_{\rm mid}(\zeta) = \begin{pmatrix} q^{-1} & -\zeta\\ -\zeta^{-1} & q\end{pmatrix}, \quad B^{0}_{\rm mid}(\zeta) = \begin{pmatrix} q & -\zeta^{-1}\\ -\zeta & q^{-1}\end{pmatrix},$$

$$B^{1/2}_{\rm mid}(\zeta) = \frac{1}{2}\begin{pmatrix} \zeta+\zeta^{-1} & -(q+q^{-1})\\ -(q+q^{-1}) & \zeta+\zeta^{-1}\end{pmatrix}.$$

PROOF. On the quotient space, we can replace $\mathbf{b}(\zeta)$, $\mathbf{c}(\zeta)$ in the definition of Ω by $\bar{\mathbf{b}}(\zeta)$, $\bar{\mathbf{c}}(\zeta)$ due to Theorem 3.23. The operators $\bar{\mathbf{b}}(\zeta)$, $\bar{\mathbf{c}}(\zeta)$ are essentially the singular parts at $\zeta^2 = 1$ of the operators $\mathbf{k}(\zeta)$, $\phi(\mathbf{k})(\zeta)$. Doing some redundant integrations, and recalling the definitions of $\mathbf{k}(\zeta)$, $\phi(\mathbf{k})(\zeta)$ (3.16), (3.24), we obtain after some minor computation

$$\Omega = -\frac{1}{(2\pi i)^2}\oint_\Gamma\oint_\Gamma \omega_{\rm rat}(\zeta_2,\zeta_1|\alpha)\mathbf{c}(\zeta_1,\alpha)\mathbf{b}(\zeta_2,\alpha+1)\frac{d\zeta_1^2}{\zeta_1^2}\frac{d\zeta_2^2}{\zeta_2^2}$$
$$= \frac{q^{\alpha-1}-q^{-\alpha-1}}{(2\pi i)^2}\oint_\Gamma\oint_\Gamma \omega_{\rm rat}(\zeta_2,\zeta_1|\alpha)\zeta^{\alpha+1}\mathrm{Tr}_{A,B}\{q^{-2D_B}\mathbb{C}_A(\zeta_1,\alpha)\mathbb{B}_B(\zeta_2,\alpha)\}\frac{d\zeta_1^2}{\zeta_1^2}\frac{d\zeta_2^2}{\zeta_2^2},$$

where $\mathbb{C}_A(\zeta,\alpha)$ is defined in (3.15). The definition of $\mathbb{B}_B(\zeta,\alpha)$ is similar, but uses the matrix $\mathbb{T}^-_{b,B}(\zeta,\alpha)$, see (3.5) and (3.24).

Let us transform the right hand side of (5.66). Changing the order of multipliers, using fusion and the cyclicity of trace, we compute

$$\text{(5.68)}\quad \mathrm{Tr}_{a,b}\{B^\varepsilon_{a,b}(\zeta)\mathbb{T}_a(\zeta_1)\mathbb{T}_b(\zeta_2)\}\mathrm{Tr}^\Lambda\{q^{(1-2\varepsilon)H}\mathbb{T}_\star(\overline{\zeta},\alpha)\}$$
$$= \zeta^{\alpha+1}(q^{\alpha+1}-q^{-\alpha-1})\mathrm{Tr}_{a,b,A,B}\Big\{q^{2(D_A-D_B)}(q^{-1}Y_{A,B}(\zeta))^{-2\varepsilon}B^\varepsilon_{a,b}(\zeta)$$
$$\times\,\mathbb{T}_a(\zeta_1,\alpha)\mathbb{T}_b(\zeta_2,\alpha)\mathbb{T}^+_A(\zeta_1,\alpha)\mathbb{T}^-_B(\zeta_2,\alpha)\Big\}$$
$$= \zeta^{\alpha+1}\mathrm{Tr}_{a,b,A,B}\Big\{S^\varepsilon_{a,b,A,B}(\zeta)\mathbb{T}_{a,A}(\zeta_1,\alpha)\mathbb{T}^-_{b,B}(\zeta_2,\alpha)\Big\},$$

where

$$S^\varepsilon_{a,b,A,B}(\zeta)$$
$$= (F^+_{a,A})^{-1}(F^-_{b,B})^{-1}R^+_{A,b}(\zeta)^{-1}B^\varepsilon_{a,b}(\zeta)(q^{-1}Y_{A,B}(\zeta))^{1-2\epsilon}R^+_{A,b}(\zeta)F^-_{b,B}F^+_{a,A}$$

and $Y_{A,B}(\zeta)$ is defined in (2.65). Recall that under the trace $Y_{A,B}(\zeta)$ equals q^{H+1}. In what follows we denote $Y=Y_{A,B}(\zeta)$. Deriving (5.68) we were able to change $\mathbb{T}_a(\zeta_1)\mathbb{T}_b(\zeta_2)$ to $\mathbb{T}_a(\zeta_1,\alpha)\mathbb{T}_b(\zeta_2,\alpha)$ because $B^\varepsilon_{a,b}(\zeta)$ acts in the space of weight 0.

The diagonal entries of $\mathbb{T}_{a,A}(\zeta_1,\alpha)$ are $\mathbb{T}_A(\zeta_1 q,\alpha)q^{-\mathbb{S}}$ and $\mathbb{T}_A(\zeta_1 q^{-1},\alpha)q^{\mathbb{S}}$, so, they are *regular* at $\zeta^2=1$. Hence under the integral we can replace $\mathbb{T}_{a,A}(\zeta_1,\alpha)$ by off-diagonal piece: $\mathbb{C}_A(\zeta_1,\alpha)\sigma^-_a$. Similarly we can replace $\mathbb{T}_{b,B}(\zeta_2,\alpha)$ by $\mathbb{B}_B(\zeta_2,\alpha)\sigma^+_b$. Then the right hand side of (5.68) becomes [1]

$$\zeta^{\alpha+1}\mathrm{Tr}_{A,B}\Big\{\mathrm{Tr}_{a,b}\{\sigma^-_a\sigma^+_b S^\varepsilon_{a,b,A,B}(\zeta)\}\mathbb{C}_A(\zeta_1,\alpha)\mathbb{B}_B(\zeta_2,\alpha)\Big\}.$$

A simple computation shows that

$$\mathrm{Tr}_{a,b}\{\sigma^-_a\sigma^+_b S^\varepsilon_{a,b,A,B}(\zeta)\} = \frac{\zeta}{(1-\zeta^2q^2)(1-\zeta^2q^{-2})}$$
$$\times\, q^{-D_A}\big(1-q^{-2}\zeta Y,\ \zeta q^{-1}-q^{-1}Y\big)B^\varepsilon_{\rm mid}(\zeta)(q^{-1}Y)^{1-2\varepsilon}\begin{pmatrix}\zeta q-q^{-1}Y\\ 1-\zeta Y\end{pmatrix}q^{-D_A}$$
$$= -q^{-2D_A}(q^{-1}Y).$$

Now we use our favourite trick (2.77) replacing under the trace $q^{-1}Y_{AB}(\zeta)$ by $(q^{\alpha+1}-q^{-\alpha-1})q^{2(D_A-D_B)}$. This finishes the proof. □

REMARK 5.5. Later we shall need some freedom in the definition of $B^{1/2}(\zeta)$. Namely, let us add to $B^{1/2}(\zeta)$ a scalar with a coefficient, regular at $\zeta=0$. Then $S^{1/2}_{a,b,A,B}(\zeta)$ acquires the same addition, and $\mathrm{Tr}_{a,b}\{\sigma^-_a\sigma^+_b S^{1/2}_{a,b,A,B}(\zeta)\}$ remains, obviously, unchanged.

---

[1] For this trick to be possible one should require that $S^\varepsilon_{a,b,A,B}(\zeta)$ be regular at $\zeta=1$.

The use of the label $\varepsilon = 0, 1/2, 1$ may look bizarre at first glance, but the reason will be made clear in Section 5.3.6.

Consider the trace $\mathrm{Tr}^{\Lambda}\{q^{(1-2\varepsilon)H}\mathbb{T}_{\star}(\overline{\zeta},\alpha)\}$ in (5.66). The procedure of computing this trace was explained in Section 2.2.2. By using the commutation relations, cyclicity and the Casimir, which in our case equals

$$C = \frac{\zeta+\zeta^{-1}}{(q-q^{-1})^2}\,,$$

we reduce any element of $U_q(\mathfrak{sl}_2)$ to a polynomial of $q^{\pm H}$. Then we use the formula

$$\mathrm{Tr}^{\Lambda}\{q^{(m+\alpha)H}\} = \frac{(\zeta q)^{\alpha+m}}{1-q^{2(\alpha+m)}}\,. \tag{5.69}$$

So, the only possible singularity at $\alpha = 0$ is the simple pole coming from terms with $m = 0$. Let us separate the pole from the rest:

$$\begin{aligned}\mathrm{Tr}_{a,b}\{B^{\varepsilon}_{a,b}(\zeta)\mathbb{T}_a(\zeta_1)\mathbb{T}_b(\zeta_2)\}\mathrm{Tr}^{\Lambda}\{q^{(1-2\varepsilon)H}\mathbb{T}_{\star}(\overline{\zeta},\alpha)\}\\ = \frac{1}{\alpha}X^{\varepsilon}_1(\zeta_1,\zeta_2) + X^{\varepsilon}_0(\zeta_1,\zeta_2|\alpha)\,,\end{aligned} \tag{5.70}$$

the second term in the right hand side is regular at $\alpha = 0$.

5.3.3. The Case $\kappa = 0$. Consider the symmetric case $\epsilon = 1/2$ (5.67). Due to the Yang-Baxter equation, $\mathrm{Tr}_{a,b}\{B^{1/2}_{a,b}(\zeta)\mathbb{T}_a(\zeta_1)\mathbb{T}_b(\zeta_2)\}$ is antisymmetric with respect to ζ_1, ζ_2. The singular part of $\mathrm{Tr}^{\Lambda}\{\mathbb{T}_{\star}(\overline{\zeta},\alpha)\}$ comes only from terms with $m = 0$ in (5.69). So it is a polynomial of C, and as such is symmetric with respect to ζ_1, ζ_2. Altogether the $X^{1/2}_1(\zeta_1,\zeta_2)$ is antisymmetric.

Later we shall consider $\omega(\zeta_1,\zeta_2|\alpha)$ at $\kappa = 0$ in more detail, but we shall need immediately some rather simple properties. We have generally

$$\omega_{\mathrm{rat}}(\zeta_2,\zeta_1|\alpha) = \omega_0(\zeta_2,\zeta_1) + \alpha\,\omega_1(\zeta_2,\zeta_1) + O(\alpha^2)\,, \tag{5.71}$$

where

$$\omega_0(\zeta_2,\zeta_1) = \omega_{\mathrm{rat}}(\zeta_2,\zeta_1|0)\,.$$

In the case $\kappa = 0$, the function $\omega_0(\zeta_2,\zeta_1)$ is symmetric while the function $\omega_1(\zeta_2,\zeta_1)$ is antisymmetric. Denote

$$X^{1/2}_0(\zeta_1,\zeta_2) = X^{1/2}_0(\zeta_1,\zeta_2|0)\,.$$

The symmetry of $\omega_0(\zeta_2,\zeta_1)$ implies that

$$\begin{aligned}\Omega = \frac{1}{(2\pi i)^2}\oint_{\Gamma}\oint_{\Gamma}\Big\{X^{1/2}_0(\zeta_1,\zeta_2)\omega_0(\zeta_2,\zeta_1) + X^{1/2}_1(\zeta_1,\zeta_2)\omega_1(\zeta_2,\zeta_1)\Big\}\frac{d\zeta_1^2}{\zeta_1^2}\frac{d\zeta_2^2}{\zeta_2^2}\\ + O(\alpha)\,.\end{aligned} \tag{5.72}$$

Let us return to fermions.

In what follows we consider α as an infinitesimal parameter which will be set to 0 eventually. We have seen that the expectation values of operators from $\mathcal{W}^{\mathrm{quo}}_{\alpha,0}$ are expressed in terms of $\omega(\zeta_1,\zeta_2|\alpha), \omega_1(\zeta_1,\zeta_2)$, the coefficients may contain singularities in α. Consider $\mathcal{O}_\alpha \in \mathcal{W}^{\mathrm{quo}}_{\alpha,0}$. By definition

$$\mathcal{O}_\alpha = \sum_{\#(I^+)=\#(I^-)} C_{I^+,I^-}(\alpha)\,\mathbf{b}^*_{I^+}\mathbf{c}^*_{I^-}q^{2\alpha S(0)}\,. \tag{5.73}$$

As in Section 5.2 we denote the modes of $\mathbf{b}^*_{\mathrm{rat}}(\zeta)$, $\mathbf{c}^*_{\mathrm{rat}}(\zeta)$ by $\mathbf{b}^*_j$, $\mathbf{c}^*_j$ skipping the index 'rat'. The coefficients $C_{I^+,I^-}(\alpha)$ may be singular at $\alpha=0$. Let us discuss the possible structure of their singularities for operators $\mathcal{O}_\alpha$ which have finite limit as $\alpha\to 0$. The considerations of the previous subsection suggest that for such operators

$$\lim_{\alpha\to 0} Z^\kappa_{\mathbf{n}}\{\mathcal{O}_\alpha\} = \lim_{\alpha\to 0}\mathrm{Tr}\bigl(e^{\Omega_0+\alpha\Omega_1}\mathcal{O}_\alpha\bigr)\,, \tag{5.74}$$

where

$$\Omega_0=\frac{1}{(2\pi i)^2}\oint\oint \omega_0(\xi,\zeta)\mathbf{c}(\zeta)\mathbf{b}(\xi)\frac{d\zeta^2}{\zeta^2}\frac{d\xi^2}{\xi^2}\,, \tag{5.75}$$

$$\Omega_1=\frac{1}{(2\pi i)^2}\oint\oint \omega_1(\xi,\zeta)\mathbf{c}(\zeta)\mathbf{b}(\xi)\frac{d\zeta^2}{\zeta^2}\frac{d\xi^2}{\xi^2}\,. \tag{5.76}$$

Recall (5.32):

$$\mathrm{Tr}(\mathbf{b}^*_j\bullet)=\mathrm{Tr}(\mathbf{c}^*_j\bullet)=0\,.$$

Let us discuss the restrictions which our construction imposes on the decomposition (5.73). Recall the operators defined in (4.88)

$$\mathbf{y}_m=\sum_{j=1}^{m-1}\mathbf{c}_j\mathbf{b}_{m-j}\,.$$

The following requirement follows from the general construction:

$$\mathbf{y}_m\mathcal{O}_\alpha=0\,,\quad m>\mathrm{length}(\mathcal{O}_\alpha) \tag{5.77}$$

One more operator which we had before (4.82) will be crucial:

$$\mathbf{s}_1=\sum_{j=1}^{\infty}\mathbf{c}^*_j\mathbf{b}_j\,. \tag{5.78}$$

It is easy to see that for all m

$$[\mathbf{s}_1,\mathbf{y}_m]=0\,.$$

More importantly

$$\mathrm{Tr}\,(\mathbf{s}_1\ \bullet)=0\,,\quad [\Omega_0,\mathbf{s}_1]=0\,. \tag{5.79}$$

Define the charge as

$$\mathcal{Q}=\frac12\sum_{j=1}^{\infty}(\mathbf{b}^*_j\mathbf{b}_j-\mathbf{c}^*_j\mathbf{c}_j)\,. \tag{5.80}$$

Fix the length n of local operators, and consider the space $\mathcal{H}_{n,p}$ of charge p which is the linear span of $\mathbf{b}^*_{I^+}\mathbf{c}^*_{I^-}q^{2(\alpha+2p)S(0)}$ such that $\#(I^+)-\#(I^-)=2p$, $\max(I^+)\le n$, $\max(I^-)\le n$, with coefficients *independent of* α. Denote by $\mathcal{N}_{n,p}$ the common kernel of operators $\mathbf{y}_m$, $m>n$ acting on $\mathcal{H}_{n,p}$. We are interested in the space $\mathcal{N}_{n,0}$, but it is not sufficient for our goals. Introduce the dependence on α defining the space

$$\mathcal{N}_{n,0}(\alpha)=\bigoplus_{p=0}^{[\frac n2]}\alpha^{-p}\ \mathbf{s}_1^p\bigl(\mathcal{N}_{n,p}/\mathbf{s}_1\mathcal{H}_{n,p+1}\bigr)\,. \tag{5.81}$$

We convince ourselves that this is the only possibility to avoid infinities in the limit $\alpha\to 0$: due to (5.79) the operator Ω_0 does not interact with $\mathbf{s}_1$, so we need at least p operators Ω_1 in order to kill $\mathbf{s}_1^p$. Actually we need exactly p operators Ω_1, otherwise a positive degree of α remains which vanishes when $\alpha \to 0$. Notice that

$$[\Omega_1, \mathbf{s}_1] = \widetilde{\Omega}_1, \quad \widetilde{\Omega}_1 = \frac{1}{(2\pi i)^2} \oint \oint \omega_1(\zeta, \xi) \mathbf{b}(\zeta) \mathbf{b}(\xi) \frac{d\zeta^2}{\zeta^2} \frac{d\xi^2}{\xi^2}.$$

So the final answer is that

$$\text{for} \quad \mathcal{O}_\alpha = \sum_{p=0}^{[\frac{n}{2}]} \alpha^{-p} \mathbf{s}_1^p X_p\,, \quad X_p \in \mathcal{N}_{n,p}/\mathbf{s}_1 \mathcal{H}_{n,p+1}\,, \tag{5.82}$$

$$\lim_{\alpha \to 0} \mathcal{Z}_\mathbf{n}^\kappa \{\mathcal{O}_\alpha\} = \sum_{p=0}^{[\frac{n}{2}]} \mathrm{Tr}\left(\widetilde{\Omega}_1^p \, e^{\Omega_0}(X_p)\right).$$

This answer is cooked out of the minors of the infinite matrix $\left(\omega_{0,i,j}\right)$ and Pfaffians of square submatrices of the infinite skew-symmetric matrix $\left(\omega_{1,i,j}\right)$.

It is not easy to find an *a priori* description of operators which satisfy the requirements above. At this point we proceed experimentally. Consider the simplest case $n = 2$ and generic α. Following the procedure described in Section 5.2 we find the basis of spinless quasi-local operators (4.87):

$$1,\ \mathbf{t}_1^*,\ \mathbf{t}_2^*,\ (\mathbf{t}_1^*)^2,\ \mathbf{c}_1^*\mathbf{b}_1^*,\ \mathbf{b}_2^*\mathbf{c}_1^* + \mathbf{c}_2^*\mathbf{b}_1^*\,. \tag{5.83}$$

In the limit $\alpha \to 0$ the operators $\mathbf{t}_j^*$ have the following weak limits:

$$\mathbf{t}_1^* = 2 + \alpha\tau_1^*\,; \quad \mathbf{t}_j^* = \alpha\tau_j^* \quad j > 1\,.$$

Due to the computation performed in Section 5.2 in the basis (5.83) the operator $\mathcal{O}_1 = \sigma_1^+\sigma_2^-$ has the components:

$$\mathcal{O}_1 = \frac{1}{2(x^2 - q^2)(x^2q^2 - 1)} \Big\{ \frac{(1+q^2)(1+x^2)^2}{16},\ -(1+q^2)(1+x^2),\ \frac{(q^2-1)(-2q^2+x^2+q^4x^2)}{q^2(x^2-1)},\ (1+q^2)x^2,\ 2(q^2-1)(1+x^2),\ -\frac{2(q^2-1)^2(1+q^2)x^2}{q^2(x^2-1)} \Big\},$$

where $x = q^\alpha$. Dangerous terms here are those containing the pole at $\alpha = 0$, there are two of them: the third and the last. In fact the last one fits (5.82) since

$$\mathbf{b}_2^*\mathbf{c}_1^* + \mathbf{c}_2^*\mathbf{b}_1^* = \mathbf{s}_1\mathbf{b}_2^*\mathbf{b}_1^*\,.$$

But the third term in the decomposition of $\mathcal{O}_1$ is really bad for out goals: it provides in the limit τ_2^*, so, it cannot be decomposed in the pure fermionic basis.

Similarly for $\mathcal{O}_2 = \sigma_1^-\sigma_2^+$:

$$\mathcal{O}_2 = \frac{1}{2(x^2 - q^2)(x^2q^2 - 1)} \Big\{ \frac{(1+q^2)(1+x^2)^2}{16},\ -(1+q^2)(1+x^2),\ -\frac{(q^2-1)x^2(1+q^4-2q^2x^2)}{q^2(x^2-1)},\ (1+q^2)x^2,\ 2(q^2-1)(1+x^2),\ -\frac{2(q^2-1)^2(1+q^2)x^2}{q^2(x^2-1)} \Big\}.$$

Again the third term is bad.

Let us take the sum:

$$\mathcal{O}_1+\mathcal{O}_2=\frac{1}{2(x^2-q^2)(x^2q^2-1)}\Big\{\frac{(1+q^2)(1+x^2)^2}{8},\ -2(1+q^2)(1+x^2),\ 2(q^2-1)(1+x^2),\ 2(1+q^2)x^2,\ 4(q^2-1)(1+x^2),\ -\frac{4(q^2-1)^2(1+q^2)x^2}{q^2(x^2-1)}\Big\}.$$

We see that the singularity in the third term canceled!

We have the density matrices for $\alpha\neq 0$ up to $n=4$, they provide a considerable amount of data. Analysing these data we come to the following assumption.

ASSUMPTION 5.6. *Consider three transformations:*

- Spin reversal $\mathbb{J}$.
- Charge conjugation θ defined as usual as the tensor product of (1.23).
- *Space* reflection $\mathbb{P}$ which reflects the interval $[1,n]$.

The operators invariant under these three symmetries satisfy (5.82). Moreover, they obey an additional restriction: in the term $\mathbf{s}_1^p\mathbf{b}^*_{I^+}\mathbf{c}^*_{I^-}$ one can impose $|I^+|+|I^-|\equiv p$ (mod 2).

Now we start to compute and, simultaneously, to check our assumption. The computation is just a simplified version of the one performed in Section 5.2. We do not go into details providing only the most important data.

The number of operators possessing all three symmetries is

n	2	3	4	5	6	7	8
#	3	5	14	30	84	210	594

The following amount of *Matsubara* data suffices for different n:

$$\begin{aligned}
n=2\ \ & 5\ (1,0)\,,\\
n=4\ \ & 5\ (1,0),\ \ 5\ (2,0),\ \ 1\ (2,1)\,,\\
n=4\ \ & 10\ (1,0),\ \ 10\ (2,0),\ \ 5\ (2,1)\,,\\
n=5\ \ & 10\ (1,0),\ \ 10\ (2,0),\ \ 10\ (2,1),\ \ 10\ (3,1)\,,\\
n=6\ \ & 20\ (1,0),\ \ 20\ (2,0),\ \ 40\ (2,1),\ \ 20\ (3,1),\ \ 5\ (4,2)\,,\\
n=7\ \ & 35\ (1,0),\ \ 40\ (2,0),\ \ 10\ (3,0),\ \ 10\ (4,0),\ \ 80\ (2,1),\\
& 60\ (3,1),\ \ 10\ (4,1),\ \ 25\ (4,2)\,,\\
n=8\ \ & 30\ (1,0),\ \ 80\ (2,0),\ \ 50\ (3,0),\ \ 20\ (4,0),\ \ 150\ (2,1),\\
& 200\ (3,1),\ \ 50\ (4,1),\ \ 150\ (4,2),\ \ 5\ (5,2)\,.
\end{aligned}$$

It is important to know the greatest common denominators of the coefficients. Below we present the denominators omitting the integer multipliers and also powers of q and q^2-1, denoted by $D(n)$. Our goal is to understand the singularities at the roots of unity, but $q=\pm1$ and especially $q=0$ are exceptional, on one of them we shall comment later.

$$D(2) = 1\,,\quad D(3) = [2]\,,\quad D(4) = [2][3]\,,\quad D(5) = [2][3][4]\,,\quad D(6) = [2][3]^2[4][5]\,,\quad D(7) = [3][4]^2[5][6]\,,\quad D(8) = [2][3]^2[4]^2[5]^2[6][7]\,. \tag{5.84}$$

where

$$[k] = \frac{q^k - q^{-k}}{q - q^{-1}}\,.$$

We are interested in the case when the expectation values of the operators satisfying all of our symmetries are sufficient to obtain the entire density matrix introduced in Section 1.7. As an example of such a situation we consider the XXZ spin chain at finite temperature (Section 1.6). Recall that in addition we set

$$\kappa = 0\,. \tag{5.85}$$

The symmetries considered above are the symmetries of the Hamiltonian, but the spin reversal changes the sign of κ hence the requirement (5.85). We consider the anti-ferromagnetic case assuming the uniqueness of the maximal Matsrubara eigenvector, which is confirmed by numerical experience. Under this assumption the spontaneous symmetry breaking is impossible.

5.3.4. Computation of the Function ω. In the present volume we use multiplicative variables like ζ. Up to now it has been rather convenient. However, when we do numerical computations it is often more adequate to use the additive variable

$$x = \frac{1}{\pi\nu}\log\zeta\,.$$

For example, in the absence of twist κ, the Bethe roots are symmetric around zero in the additive variable. In the present section we shall try to pass between Scylla and Charybdis using both kinds of variables.

In the following sections we shall need only the case of real twist κ, we shall mostly restrict the exposition to this case. For real κ the Bethe roots for the ground state are real. In the additive variable they are located on some interval, symmetric with respect to 0 for $\kappa = 0$. The zeroes accumulate to the point $x = 0$ (or $\zeta = 1$). At this point the function $\mathfrak{a}(\zeta)$ has an essential singularity.

Introduce the map

$$Z(x) = e^{\pi\nu x}\,. \tag{5.86}$$

Suppose all the Bethe roots in the additive variable are located inside the interval $[-r, r]$. We denote by $I(r)$ the image of $[-r, r]$ under the map (5.86). We start from the resolvent

$$R(\zeta,\xi) - \int_{I(r)} K(\zeta/\eta)R(\eta,\xi)\frac{d\eta^2}{\eta^2} = K(\zeta/\xi)\,, \tag{5.87}$$

taking

$$r = 1, 2, 3\,,$$

more explanation on this point is given below. The equation (5.87) can be solved numerically for $\zeta,\xi\in Z(I(r))$, we shall give details of the numerical method soon. When needed $R(\zeta,\xi)$ is continued analytically via

$$R(\zeta,\xi)=K(\zeta/\xi)+\int_{I(r)}\int_{I(r)}K(\zeta/\eta)R(\eta,\eta')K(\eta'/\xi)\frac{d\eta^2}{\eta^2}\frac{d\eta'^2}{\eta'^2}\,.$$

To formulate the DdV equation, we need to deform the contour of integration in the upper or lower half-planes. Define two half-ellipses

(5.88) $$x=r\sin\pi t\pm ih\cos\pi t,\quad -1/2\le t\le 1/2\,.$$

For $\nu\le 1/2$ the height h can be chosen as $h=1/2$, for $\nu>1/2$ we set $h=1/2(\nu^{-1}-1)$, $C_\pm(r)$ are their images under (5.86). Both choices are done in order that there are no poles of $K(\zeta)$ below C_+ on the one hand and the contour is sufficiently far from the essential singularity at $\zeta=1$ on the other hand. The DdV equation is

(5.89)

$$\log\mathfrak{a}(\zeta)=f(\zeta)-\int_{C_+(r)}R(\zeta,\eta)\log(1+\mathfrak{a}(\eta))\frac{d\eta^2}{\eta^2}+\int_{C_-(r)}R(\zeta,\eta)\log\Bigl(1+\frac{1}{\mathfrak{a}(\eta)}\Bigr)\frac{d\eta^2}{\eta^2}\,,$$

where

$$f(\zeta)=((I+R)\phi)(\zeta)\,,\quad \phi(\zeta)=\frac{i}{T}\left(\psi(\zeta q^{-1},0)-\psi(\zeta,0)\right)-2\pi i\nu\kappa\,.$$

For real κ the function $\log\mathfrak{a}(\zeta)$ satisfies the the Schwarz principle (2.114) and it is sufficient to consider $\zeta\in C_+(r)$ in (5.89). For imaginary κ, which corresponds to real magnetic field, the locus of the Bethe roots is slightly deformed into the upper or lower half planes [**BGKS**]. In that case $C_\pm(r)$ should be deformed, the Schwarz principle cannot be used and one is forced to solve a system of equations for $\mathfrak{a}(\zeta)$, $\zeta\in C_\pm(r)$. In what follows we consider real κ only.

Now we come to an important point: r and T must be consistent. This means that there are no Bethe roots outside the interval $[-r,r]$. So, we solve the equation and then check that

(5.90) $$\frac{1}{i}\log\mathfrak{a}(Z(-r))<\pi,\quad \frac{1}{i}\log\mathfrak{a}(Z(r))>-\pi\,.$$

Before going further let us discuss the problem of precision for numerical solutions of the equations (5.87), (5.89). In the latter equation we make the change of variables (5.88) in the integral. The main problem is that we have to integrate over a finite interval, so, one has to take care of the ends of the interval. In order to avoid this problem we use the double exponential method [**MS**]. For numerical integration of a smooth function $f(t)$ from $-a$ to a we introduce

(5.91) $$g(u)=-1+\frac{4}{\pi}\arctan(\exp(c\sinh(u)))\,,$$

and use the approximation

$$\int_{-a}^{a}f(t)dt\simeq ha\sum_{k=-N}^{N}f(ag(hk))g'(hk)\,.$$

Actually, the function (5.91) is different from the ones used traditionally, it was introduced rather recently [**Oo**]. It makes the numerical integration procedure rather fast and marvellously precise. We always take the parameter c equal to

1/10. In the integral in (5.87) we change the variable to $t = \frac{1}{\pi\nu}\log\eta$, and apply the double exponential method with $a = r$ and the parameters

$$h = 1/20,\quad N = 200,\quad \text{for}\ \ r = 1\,,$$
$$h = 1/25,\quad N = 250,\quad \text{for}\ \ r = 2\,,$$
$$h = 1/40,\quad N = 400,\quad \text{for}\ \ r = 3\,.$$

For the integrals over $C_\pm(r)$ we use the parametrisation (5.88) and set $a = 1/2$. To ensure a good precision we have to take more points than previously:

$$h = 1/30,\quad N = 300,\quad \text{for}\ \ r = 1\,,$$
$$h = 1/40,\quad N = 400,\quad \text{for}\ \ r = 2\,,$$
$$h = 1/60,\quad N = 600,\quad \text{for}\ \ r = 3\,.$$

For numerical computations, one has to take r satisfying (5.90) as small as possible: there is no need to integrate over long Bethe roots-free parts of the interval.

Now let us turn to the function $\omega_{\mathrm{rat}}(\zeta,\xi|\alpha)$. We have already considered the case $\kappa = 0$. In what follows we shall also consider $\kappa = -1$. In both cases we shall need $\omega_{\mathrm{rat}}(\zeta,\xi|0)$. Also only in the first case we need the first Taylor coefficient $\omega_1(\zeta,\xi)$ (5.71). For $\kappa = 0$

$$\rho(\zeta|\alpha) = \frac{T(\zeta|\alpha)}{T(\zeta|0)} = 1 + O(\alpha^2)\,. \tag{5.92}$$

Indeed, due to

$$JT_{\mathbf{M}}(\zeta|\alpha)J = T_{\mathbf{M}}(\zeta|-\alpha)\,,$$

the spectra of $T_{\mathbf{M}}(\zeta|\alpha)$ and $T_{\mathbf{M}}(\zeta|-\alpha)$ coincide. In particular, the maximal eigenvalues are the same. The maximal eigenvalue is analytical in the vicinity of $\alpha = 0$. All that ensures (5.92). So, considering the zeroth order in α generally and the first order in α for $\kappa = 0$, we can drop ρ (set it equal to 1) throughout equations defining ω, thereby simplifying them considerably.

In fact, working on the interval of length n, we need not the functions $\omega(\zeta,\xi|\alpha)$ itself but rather its first Taylor coefficients

$$\omega(\zeta,\xi|\alpha) = (\zeta/\xi)^\alpha \sum_{j,k=0}^{n} \omega_{j,k}(\alpha)(\zeta^2-1)^j(\xi^2-1)^k + \cdots\,.$$

To have direct access to the coefficients we proceed as follows. We have the Taylor series

$$f_{\mathrm{right}}(\eta,\xi) = \xi^{-\alpha}\sum_{j=0}^{\infty}(\xi^2-1)^j f_{\mathrm{right},j}(\eta)\,,$$
$$f_{\mathrm{left}}(\zeta,\eta) = \zeta^{\alpha}\sum_{j=0}^{\infty}(\zeta^2-1)^j f_{\mathrm{left},j}(\eta)\,,$$

and the corresponding functions $F_{\mathrm{right},j}(\eta)$, $F_{\mathrm{left},j}(\eta)$ defined from (5.30) in an obvious way. The functions $f_{\mathrm{right},j}(\eta)$, $f_{\mathrm{left},j}(\eta)$ $F_{\mathrm{right},j}(\eta)$, $F_{\mathrm{left},j}(\eta)$ have poles of order j at the points $\eta^2 = 1, q^{\pm 2}$. Integrating this kind of functions numerically, one loses the precision very fast. This is especially bad for the integrals coming from the convolution in the formula for $\omega^{(2)}$ (5.31), where the singularities of $f_{\mathrm{left},j}(\eta)$ and $F_{\mathrm{right},j}(\eta)$ double. This problem can be cured, at least partly, using an exact solution available in our case ($\rho(\zeta|\alpha) = 1$) for $r = \infty$, where the equation for $F_{\text{right}}(\eta, \zeta)$ (5.30) has the solution:

$$F^{\infty}_{\text{right}}(\eta, \zeta) = \frac{\nu(\zeta\eta)^{\frac{1}{\nu}}}{\zeta^{\frac{2}{\nu}} - \eta^{\frac{2}{\nu}}},$$

as usual $F^{\infty}_{\text{right},j}(\eta)$ are the Taylor coefficients in $\zeta^2 - 1$. Let us introduce the difference

$$\Delta F_{\text{right},j}(\eta) = F^{\infty}_{\text{right},j}(\eta) - F_{\text{right},j}(\eta),$$

which can be computed as

$$\Delta F_{\text{right},j}(\eta) = d_{\text{right},j}(\eta) + \int_{I(r)} R_\alpha(\eta, \xi) d_{\text{right},j}(\xi) \frac{d\xi^2}{\xi^2}, \tag{5.93}$$

where R_α is the resolvent of K_α on the interval $I(r)$ computed up to the first order in α, and the driving terms $d_{\text{right},j}(\eta)$ are

$$d_{\text{right},j}(\eta) = \int_{\overline{I}(r)} K_\alpha(\eta/\xi) F^{\infty}_{\text{right},j}(\xi) \frac{d\xi^2}{\xi^2}, \tag{5.94}$$

where $\overline{I}(r)$ is the complement of $I(r)$ in the positive real half-line. In the integrals (5.93), (5.94) the singularities are far from the contours of integration.

Now we are ready to present the formulas for $\omega^{(1,2)}$ (5.31) convenient for computations in the case under consideration (first order in α for $\kappa = 0$, zeroth order in α for $\kappa \neq 0$) for which $\rho(\zeta|\alpha) = 1$. We start with $\omega^{(2)}$ for which it is straightforward to compute

$$\omega^{(2)}_{i,j} = -4\Big(\int_{\overline{I}(r)} f_{\text{left},i}(\eta) F^{\infty}_{\text{right},j}(\eta) \frac{d\eta^2}{\eta^2} + \int_{I(r)} f_{\text{left},i}(\eta) \Delta F_{\text{right},j}(\eta) \frac{d\eta^2}{\eta^2} \Big). \tag{5.95}$$

The integrand of the first integral in the right hand side does not have poles close to the contour of integration, while the integrand of the second integral does contain the poles of $f_{\text{left},i}(\eta)$. So, we did not completely eliminate the singularities, but we did eliminate the overlap of singularities form $f_{\text{left},i}(\eta)$ and $F_{\text{right},j}(\eta)$ which we had at the beginning. The singularities from $f_{\text{left},i}(\eta)$ only are manageable for the numerical study.

Now let us turn to $\omega^{(1)}$. It is convenient to proceed as in Section 5.1, introducing the functions $G_{\text{right},j}(\eta)$ as solutions to the equations:

$$\begin{aligned} G_{\text{right},j}(\eta) = F_{\text{right},j}(\eta) &- \int_{C_+(r)} R_\alpha(\eta/\xi) G_{\text{right},j}(\xi) \frac{\mathfrak{a}(\xi) d\xi^2}{\xi^2(1+\mathfrak{a}(\xi))} \\ &+ \int_{C_-(r)} R_\alpha(\eta/\xi) G_{\text{right},j}(\xi) \frac{d\xi^2}{\xi^2(1+\mathfrak{a}(\xi))}. \end{aligned} \tag{5.96}$$

For the numerical work this equation should be understood as two equations for $G_{\text{right},j}(\eta)$ with $\eta^2 \in C_\pm$. In the driving term we have the singularities close to the integration contour, however, this time one should not care too much because the function $\mathfrak{a}(\eta)$ has an essential singularity at $\eta = 1$, and the measure of integration damps very much the integrands near potentially dangerous points. Our final formula is

$$\omega^{(1)}_{i,j} = 4\Big(\int_{C_+(r)} F_{\mathrm{left},i}(\xi)G_{\mathrm{right},j}(\xi)\frac{\mathfrak{a}(\xi)d\xi^2}{\xi^2(1+\mathfrak{a}(\xi))} - \int_{C_-(r)} F_{\mathrm{left},i}(\xi)G_{\mathrm{right},j}(\xi)\frac{d\xi^2}{\xi^2(1+\mathfrak{a}(\xi))}\Big). \tag{5.97}$$

We apply the numerical procedure explained above to the equations (5.96), (5.97), (5.95), the parameters of the double-exponential method are the same. This allows one to compute the matrix $\omega_{i,j}$ with very good precision (up to 50 digits). Such precision is needed because the coefficients in the decomposition of the local operators in the fermionic basis grow quite fast with the length of operators. This is similar to the OPE in the QFT.

Up to now we considered the function $\omega(\zeta,\xi|\alpha)$, but it was convenient to find the coefficients of the density matrix in the basis created by $\mathbf{b}^*_{\mathrm{rat}}, \mathbf{c}^*_{\mathrm{rat}}$, which requires the use of ω_{rat}. For $\rho(\zeta)=1$ the relation between ω and ω_{rat} is especially simple:

$$\omega_{\mathrm{rat},i,j}(\alpha) = \omega_{i,j}(\alpha) + \Delta\omega_{i,j}(\alpha)\,, \tag{5.98}$$

where $\Delta\omega_{i,j}$ are constructed by a simplified version of our usual procedure. Consider the function

$$\sigma(z|\alpha) = \int_{-\infty}^{\infty} z^{ik-\alpha/2}\frac{\sinh\pi\big((1-\nu)k+\frac{i\alpha}{2}\big)}{2i\sinh\pi\big(k+\frac{i\alpha}{2}\big)\cosh(\pi\nu k)}dk - \frac{1}{4}\Big(\frac{e^{i\pi\nu\alpha}}{ze^{2i\pi\nu}-1} - \frac{e^{-i\pi\nu\alpha}}{ze^{-2i\pi\nu}-1}\Big). \tag{5.99}$$

This function is regular at $z=1$, so we can define the Taylor series in two arguments

$$\sigma\big((1+u)(1+v)^{-1}|\alpha\big) = \sum_{i,j=0}^{\infty} u^i v^j \Delta\omega_{i,j}(\alpha)\,.$$

Notice also that for $r=\infty$, which corresponds to $T=0$, all the matrix elements of $\omega_{i,j}$ vanish. This circumstance will be important later when we shall describe the scaling limit in terms of CFT. We have the corollary:

$$\omega_{\mathrm{rat},i,j}(\alpha) = \Delta\omega_{i,j}(\alpha)\,, \quad T=0\,. \tag{5.100}$$

5.3.5. Entanglement Entropy. Consider the case $\alpha=\kappa=0$ for which we have access to the entire density matrix. It is interesting and instructive to start the numerical study by considering the case $\nu=1/3$, i.e., the ice model. The denominators (5.84) have for $n=2,3,4,5,6,7,8$ zeroes at $q^3=-1$ of orders $d(n)=0,0,1,1,2,2,3$, respectively. In general, for q being a root of unity, the presence of denominators makes the computational procedure complicated. The case $q^3=-1$ is the most difficult one, that is why we consider it here. First, we present the elements of the density matrix in the form:

$$\sum_{p=0}^{d(n)}\frac{P_{m,p}(q)}{(1-q+q^2)^p}\,,$$

where the polynomial $P_{m,p}(q)$ is of degree 1.

Let us consider an example. The following vector belongs to the space (5.81), more exactly to its subspace with $p=0$:

$$\begin{aligned} v = &\mathbf{b}^*_{\{7,5,3\}}\mathbf{c}^*_{\{7,5,3\}} - \mathbf{b}^*_{\{7,5,2\}}\mathbf{c}^*_{\{7,5,4\}} - \mathbf{b}^*_{\{7,4,3\}}\mathbf{c}^*_{\{7,6,3\}} - \mathbf{b}^*_{\{7,4,3\}}\mathbf{c}^*_{\{7,5,4\}} \\ &+ \mathbf{b}^*_{\{7,4,2\}}\mathbf{c}^*_{\{7,6,4\}} - \mathbf{b}^*_{\{7,3,2\}}\mathbf{c}^*_{\{7,6,5\}} - \mathbf{b}^*_{\{6,5,3\}}\mathbf{c}^*_{\{8,5,3\}} - \mathbf{b}^*_{\{6,5,3\}}\mathbf{c}^*_{\{7,6,3\}} \\ &+ \mathbf{b}^*_{\{6,5,2\}}\mathbf{c}^*_{\{8,5,4\}} + \mathbf{b}^*_{\{6,5,2\}}\mathbf{c}^*_{\{7,6,4\}} + \mathbf{b}^*_{\{6,4,3\}}\mathbf{c}^*_{\{8,6,3\}} + \mathbf{b}^*_{\{6,4,3\}}\mathbf{c}^*_{\{8,5,4\}} \\ &+ \mathbf{b}^*_{\{6,4,3\}}\mathbf{c}^*_{\{7,6,4\}} - \mathbf{b}^*_{\{6,4,2\}}\mathbf{c}^*_{\{8,6,4\}} - \mathbf{b}^*_{\{6,4,2\}}\mathbf{c}^*_{\{7,6,5\}} + \mathbf{b}^*_{\{6,3,2\}}\mathbf{c}^*_{\{8,6,5\}} \\ &- \mathbf{b}^*_{\{5,4,3\}}\mathbf{c}^*_{\{8,7,3\}} - \mathbf{b}^*_{\{5,4,3\}}\mathbf{c}^*_{\{8,6,4\}} - \mathbf{b}^*_{\{5,4,3\}}\mathbf{c}^*_{\{7,6,5\}} + \mathbf{b}^*_{\{5,4,2\}}\mathbf{c}^*_{\{8,7,4\}} \\ &- \mathbf{b}^*_{\{5,3,2\}}\mathbf{c}^*_{\{8,7,5\}} + \mathbf{b}^*_{\{4,3,2\}}\mathbf{c}^*_{\{8,7,6\}}\,. \end{aligned}$$

For example, this vector enters the decomposition of the operator $\tau_1^+\tau_2^+\cdots\tau_8^+$ with the coefficient:

$$\begin{aligned} &\frac{243(88547q-39865)}{67108864(1-q+q^2)^3}+\frac{405(45461q-103198)}{33554432(1-q+q^2)^2} \\ &\quad+\frac{27(80051722q-27362669)}{536870912(1-q+q^2)}-\frac{27(3039933q-29802854)}{268435456}\,. \end{aligned} \tag{5.101}$$

Let us consider the case $T=0$. In that case the matrix $\omega_{\mathrm{rat},i,j}(\alpha)$ is computed from $\sigma(z|\alpha)$ (5.100). We need only the first two Taylor coefficients near $\alpha=0$:

$$\sigma(z|\alpha)=\sigma_0(z)+\sigma_1(z)\alpha+O(\alpha^2)\,.$$

The singularities force us to proceed computing for $i=0,1$

$$\sigma_i(z)=\sum_{p=0}^{3}(1-q+q^2)^p\sigma_i^{(p)}(z)+O((\nu-1/3)^4)\,. \tag{5.102}$$

We couple the formulas for ω's obtained from here with the formulas of the type (5.101), to obtain for any operator and for $T=0$

$$\langle\mathcal{O}\rangle=\sum_{p=0}^{3}\frac{1}{(1-q+q^2)^p}X_p\,.$$

First of all we have to make sure that

$$X_p=0\,,\quad p=1,2,3\,, \tag{5.103}$$

then the final result for the expectation value equals X_0.

The ideal laboratory is provided by the operator $\tau_1^+\tau_2^+\cdots\tau_8^+$ which we mentioned already. The expectation value of this operator is called the Emptiness Formation Probability, for the case $\nu=1/3$ an explicit formula is known [**KMST**]:

$$v(n)=2^{-n^2}\prod_{k=0}^{n-1}\frac{(3k+1)!}{(n+k)!}\,.$$

In particular for $n=8$ we have a rather long exact decimal number:

$$v(8)=5.8819138795684899889693042496219277381896972656\,25\cdot10^{-13}\,. \tag{5.104}$$

We check numerically the consistency (5.103) and compare X_0 with the exact value of $v(8)$ working with 60 digits. The agreement holds with all 49 digits in (5.104), so we are satisfied with our numerics.

Now we come to the more serious business which is the computation of the entanglement entropy. We continue with $\nu=1/3$, $T=0$. In the process of computation we check (5.103) for all the operators. According to Section 1.7, we expect that the entropy behaves for large n as

$$s(n) \simeq \frac{1}{3}\log n + A\,,$$

A being a constant, the central charge equals 1. The result of our computation is given in the following table.

n	$s(n) - \frac{1}{3}\log n$
2	0.7196998669
3	0.7266679480
4	0.7286649228
5	0.7294027960
6	0.7297573460
7	0.7299590479
8	0.7300875384

We see that the difference $s(n) - \frac{1}{3}\log n$ approaches constant rather fast. It may even look surprising that the agreement with the exponential formula is so good for rather small n, but as has been said in Section 1.7, the entropy is a very robust measure of the number of degrees of freedom.

Now we would like to consider the maximal available length of interval and non-zero temperature. The numerical computations which follow the procedure of Section 5.3.4 are rather involved. In principle it is possible to consider the case with cancelling of singularities as the one treated above. However, this would be hard, so, we prefer to consider q being a root of unity but for which the denominator $d(8)$ does not vanish. The simplest example is

$$q = e^{\frac{2j-1}{8}\pi i}\,.$$

More precisely we shall consider the case $\nu = 5/8$, for which one has to be careful with the choice of the heights h (see the discussion after (5.88)). We take $h = 1/5$, and follow the procedure of Section 5.3.4. Asymptotically for $n \gg 1$ and nT being finite, we are supposed to have

$$s(n,T) - s(n,0) = \frac{1}{3}\log\left(\frac{\sinh(nT)}{nT}\right)\,.$$

On Fig. 10 the points are our numerical results, and the grey line is $\frac{1}{3}\log\left(\frac{\sinh(8T)}{8T}\right)$. We observe quite good agreement.

We performed similar computations for $\nu = 1/8, 3/8$. The agreement is a little worse for $\nu = 1/8$, it is generally worse closer to the isotropic case $\nu = 0$. The latter, XXX, case will be considered later.

5.3.6. Invariant Operators. In Lemma 5.4 we had three representations labeled by $\varepsilon = 0, 1/2, 1$ for the expectation values for the XXZ model. In this section we clarify the importance of the cases $\varepsilon = 0, 1$ in the case $\alpha = 0$.

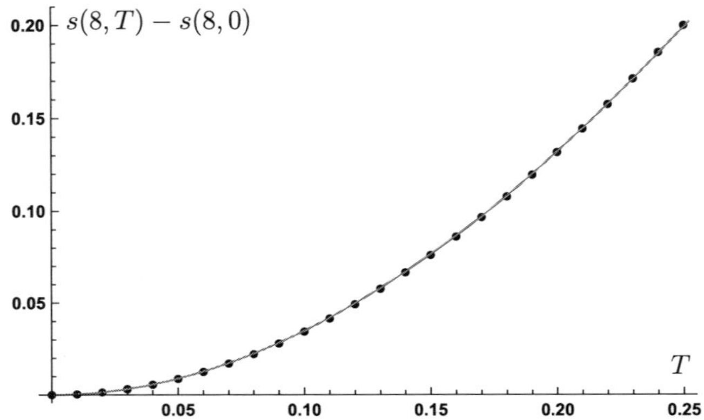

FIG. 10.

The quantum algebra $\mathcal{U} = U_q(\widehat{\mathfrak{sl}}_2)$ contains two subalgebras $\mathcal{U}_i$ isomorphic to $U_q(\mathfrak{sl}_2)$. They are generated respectively by e_i, f_i, t_i, $i = 0, 1$. We shall see that the formulas (5.66) with $\alpha = 0$ and $\varepsilon = 0, 1$ are useful to describe the action on the operators invariant under one of these two subalgebras. We shall concentrate on the case $\varepsilon = 1$; the other one is similar.

It is convenient to make a gauge transformation with $\zeta^{-H/2}$ in the evaluation homomorphism (2.43):

$$\tilde{ev}_\zeta(e_0) = \zeta^2 F\,, \quad \tilde{ev}_\zeta(f_0) = \zeta^{-2} E\,, \quad \tilde{ev}_\zeta(t_0) = K^{-1}\,,$$
$$\tilde{ev}_\zeta(e_1) = E\,, \quad \tilde{ev}_\zeta(f_1) = F\,, \quad \tilde{ev}_\zeta(t_1) = K\,. \tag{5.105}$$

Using $\tilde{ev}_\zeta$, we identify the subalgebra $\mathcal{U}_1 = \langle e_1, f_1, t_1^{\pm 1}\rangle$ with $U_q(\mathfrak{sl}_2) = \langle E, F, K^{\pm 1}\rangle$. The square of antipode is an inner automorphism:

$$S^2(x) = K^{-1} \cdot x \cdot K \quad (x \in U_q(\mathfrak{sl}_2))\,. \tag{5.106}$$

This is a general property of quantum groups called 'quantum double'.

Quite generally, for a Hopf algebra $(\mathcal{A}, \epsilon, \Delta, S)$ one has the notion of adjoint action of $\mathcal{A}$ onto itself. Namely we have a homomorphism ad : $\mathcal{A} \to \mathrm{End}(\mathcal{A})$, $x \mapsto \mathrm{ad}_x$, given by

$$\mathrm{ad}_x(y) = \sum x'_j \cdot y \cdot S(x''_j) \qquad (x, y \in \mathcal{A},\ \Delta(x) = \sum x'_j \otimes x''_j). \tag{5.107}$$

In the notation above, we have

$$\mathrm{ad}_x(y_1 y_2) = \sum \mathrm{ad}_{x'_j}(y_1)\mathrm{ad}_{x''_j}(y_2)\,. \tag{5.108}$$

If $\mathcal{A}$ is equipped with a linear map $\mathrm{Tr} : \mathcal{A} \to \mathbb{C}$ satisfying $\mathrm{Tr}(uv) = \mathrm{Tr}(vu)$, and an element K satisfying (5.106), then with respect to the pairing

$$\langle u, v\rangle = \mathrm{Tr}\big(K^{-1} uv\big) \tag{5.109}$$

we have $\langle \mathrm{ad}_x u, v\rangle = \langle u, \mathrm{ad}_{S(x)} v\rangle$. Note that the pairing (5.109) is not symmetric.

Consider (5.70) for $\varepsilon = 1$. For the sake of generality, consider an interval $[1, l]$ in the space direction with the inhomogeneity parameters $\xi_1, \ldots, \xi_l$. Applying the gauge transformation with $\prod \xi_j^{-\sigma_j^3/2}$ we come to

$$\operatorname{Tr}_{a,b}\{\tilde{B}_{a,b}(\zeta)\tilde{\mathbb{T}}_a(\zeta_1)\tilde{\mathbb{T}}_b(\zeta_2)\}\operatorname{Tr}^{\Lambda}\{q^{-H}\tilde{\mathbb{T}}(\overline{\zeta},\alpha)\} = \frac{1}{\alpha}\cdot \tilde{X}_1(\zeta_1,\zeta_2) + \tilde{X}_0(\zeta_1,\zeta_2|\alpha)\,, \quad q^{-\Lambda} = q\zeta\,. \tag{5.110}$$

In this formula, $\tilde{\mathbb{T}}$'s are constructed via

$$\tilde{R}_{\mathrm{aux},j}(\zeta) = \zeta^{\sigma_j^3/2} R_{\mathrm{aux},j}(\zeta)\zeta^{-\sigma_j^3/2}\,.$$

The operator

$$\tilde{B}_{a,b}(\zeta) = \frac{(\zeta-\zeta^{-1})(q+q^{-1})}{(\zeta q-\zeta^{-1}q^{-1})(\zeta q^{-1}-\zeta^{-1}q)}\mathcal{P}^{q,-}_{a,b} \tag{5.111}$$

is proportional to the q-antisymmetriser $\mathcal{P}^{q,-}_{a,b}$ given by

$$\mathcal{P}^{q,-} = \frac{1}{q+q^{-1}}\begin{pmatrix} 0 & 0 & 0 & 0\\ 0 & q^{-1} & -1 & 0\\ 0 & -1 & q & 0\\ 0 & 0 & 0 & 0\end{pmatrix},$$

which has the properties

$$\mathcal{P}^{q,-}(\pi^{(1)}\otimes\pi^{(1)})(x) = (\pi^{(1)}\otimes\pi^{(1)})(x)\,\mathcal{P}^{q,-} = \epsilon(x)\,\mathcal{P}^{q,-}, \quad x\in U_q(\mathfrak{sl}_2)\,. \tag{5.112}$$

Finally $\tilde{X}_1(\zeta_1,\zeta_2)$ and $\tilde{X}_0(\zeta_1,\zeta_2|\alpha)$ are rational functions of ζ_1^2, ζ_2^2.

The operator (5.110) acts on local operators which are gauge transformed with respect to the original case. The algebra $\mathcal{U}_1$ acts on local operators by adjoint. Let us consider the subspace $\mathcal{W}^{\mathrm{inv}}$ of local operators which are invariant under this action. Due to (5.108), $\mathcal{W}^{\mathrm{inv}}$ is a subalgebra of the algebra of local operators.

We shall need a general lemma.

LEMMA 5.7. *Let V_1, V_2 be $U_q(\mathfrak{sl}_2)$ modules with V_1 finite dimensional. Let $T\in \operatorname{End}(V_1\otimes V_2)$ be an intertwiner in the sense that*

$$T\Delta(x) = \Delta'(x)T\,, \qquad \forall x\in U_q(\mathfrak{sl}_2). \tag{5.113}$$

Let further $A\in\operatorname{End}(V_1)$ be a $U_q(\mathfrak{sl}_2)$-invariant operator. Then the operator

$$\operatorname{Tr}_{V_1}\{(AK^{-1}\otimes I)T\}\,: V_2\to V_2 \tag{5.114}$$

preserves the submodule $V_2^{\mathrm{inv}}\subset V_2$ of $U_q(\mathfrak{sl}_2)$-invariants.

PROOF. The intertwining relation (5.113) can equivalently be rewritten as

$$(1\otimes x)T = \sum (S^{-1}(x_j''')\otimes 1)T(x_j'\otimes x_j'')\,, \quad \forall x\in U_q(\mathfrak{sl}_2),$$

where $(\Delta\otimes\mathrm{id})\Delta(x) = \sum x_j'\otimes x_j''\otimes x_j'''$. Multiplying by $AK^{-1}\otimes I$ from the left and taking trace, we have

$$\begin{aligned} x\operatorname{Tr}_{V_1}(AK^{-1}\otimes I)T &= \sum \operatorname{Tr}_{V_1}\{(AK^{-1}S^{-1}(x_j''')\otimes I)T(x_j'\otimes I)\}\cdot x_j''\\ &= \sum \operatorname{Tr}_{V_1}\{(x_j' AS(x_j''')K^{-1}\otimes I)T\}\cdot x_j''\,. \end{aligned}$$

On the space of invariants V_2^{inv}, x''_j acts by $\epsilon(x''_j)$. Along with the assumption $\mathrm{ad}_x(A) = \epsilon(x)A$, we obtain

$$x\,\mathrm{Tr}_1\{(AK^{-1}\otimes I)T\}\big|_{V_2^{\mathrm{inv}}} = \mathrm{Tr}_{V_1}\{(\mathrm{ad}_x(A)K^{-1}\otimes I)T\}\big|_{V_2^{\mathrm{inv}}}$$
$$= \epsilon(x)\mathrm{Tr}_{V_1}\{(AK^{-1}\otimes I)T\}\big|_{V_2^{\mathrm{inv}}},$$

which shows that $\mathrm{Tr}_{V_1}\{(AK^{-1}\otimes I)T\}(V_2^{\mathrm{inv}})\subset V_2^{\mathrm{inv}}$. □

Notice that the statement of this Lemma is much weaker than the one in the case of symmetry under an ordinary Lie algebra; in that case, for an invariant operator A, $\mathrm{Tr}_{V_1}\{(A\otimes I)T\}$ acts from every irreducible (not only one-dimensional) representation to itself.

The action of the quantum group on $(\mathbb{C}^2)^{\otimes 2l}$ is equivalent to its adjoint action on $\mathrm{End}((\mathbb{C}^2)^{\otimes l})$, to which the local operators belong. On two separate occasions we shall need the exact form of the equivalence. First, notice that for $x\in U_q(\mathfrak{sl}_2)$

$$\pi^{(1)}\big(S(x)\big) = c^{-1}\cdot\pi^{(1)}(x)^t\cdot c,\quad c=\begin{pmatrix}0 & -q^{-1}\\ 1 & 0\end{pmatrix}. \tag{5.115}$$

With any vector $Y\in(\mathbb{C}^2)^{\otimes 2l}$ of the form $Y=f\otimes g$ with $f,g\in(\mathbb{C}^2)^{\otimes l}$, we associate

$$Y^{\vee} = f\otimes\{(g^*)^t c^{\otimes l}\}\quad\in\mathrm{End}((\mathbb{C}^2)^{\otimes l}), \tag{5.116}$$

where we set $g^* = v_l\otimes\cdots\otimes v_1$ for $g=v_1\otimes\cdots\otimes v_l$. We extend this definition to the entire space $(\mathbb{C}^2)^{\otimes 2l}$ by linearity. Due to (5.115), the action of $\mathcal{U}_1$ (*and only this subalgebra*) on the vector Y indeed translates to the adjoint action on the operator $Y^{\vee}$ via the coproduct.

Now we come to an important property of operators restricted to $\mathcal{W}^{\mathrm{inv}}$.

LEMMA 5.8. *On the space $\mathcal{W}^{\mathrm{inv}}$ the operator $\tilde{X}_1(\zeta_1,\zeta_2)$ in (5.110) vanishes:*

$$\tilde{X}_1(\zeta_1,\zeta_2)\big|_{\mathcal{W}^{\mathrm{inv}}} = 0. \tag{5.117}$$

PROOF. We have to show that $\mathrm{Tr}^{\Lambda}\{q^{-H}\tilde{\mathbb{T}}_{\star}(\overline{\zeta}|\alpha)\}$ acting on invariant operators does not contain terms singular in α. This amounts to proving that the action of $\tilde{\mathbb{T}}(\overline{\zeta}|\alpha)$ on $\mathcal{W}^{\mathrm{inv}}$ does not produce q^H. Let us show that.

It is convenient to consider the inhomogeneous case. Usually we numerate the sites (tensor components) by $j=1,\dots,l$. Now that we have twice more of them, the new spaces will be enumerated by $\overline{j}=\overline{1},\dots,\overline{l}$.

For $Y\in\mathbb{C}^{\otimes 2l}$ we compute:

$$\big(\tilde{\mathbb{T}}_{\star}(\overline{\zeta})(Y^{\vee})\big) = f(\overline{\zeta})\cdot\big[\tilde{R}_{\star,l}(\overline{\zeta}/\xi_l)\cdots\tilde{R}_{\star,1}(\overline{\zeta}/\xi_1)\tilde{R}_{\star,\overline{1}}(q^{-1}\overline{\zeta}/\xi_1)\cdots\tilde{R}_{\star,\overline{l}}(q^{-1}\overline{\zeta}/\xi_l)Y\big]^{\vee},$$

where $f(\overline{\zeta})$ is a certain function irrelevant for us, which comes from the crossing symmetry relation. The monodromy matrix in the right hand side provides an equivalent 'vector' formulation. It is convenient to put this into a more general and symmetric context by considering

$$\tilde{\mathcal{T}}_{\star}(\overline{\zeta}) = \tilde{R}_{\star,2l}(\overline{\zeta}/\xi_{2l})\tilde{R}_{\star,2l-1}(\overline{\zeta}/\xi_{2l-1})\cdots\tilde{R}_{\star,1}(\overline{\zeta}/\xi_1).$$

Let us see what will happen when acting by $\tilde{\mathcal{T}}_{\star}(\overline{\zeta})$ on invariant vectors. One of these vectors is immediately clear:

$$S = s_{1,2}s_{3,4}\cdots s_{2l-1,2l},$$

where

$$s = v_+ \otimes v_- - q v_- \otimes v_+ \tag{5.118}$$

is the singlet for which

$$(\pi^{(1)} \otimes \pi^{(1)})(x)s = \epsilon(x)s, \quad x \in U_q(\mathfrak{sl}_2)\,.$$

We compute by noting some important cancellations:

$$\begin{aligned} &R_{\star,2}(\eta_2)R_{\star,1}(\eta_1)s_{1,2} \\ &= \begin{pmatrix} (q-q^{-1})g(\eta_1,\eta_2)F \\ -q^2\eta_1^2(q-q^{-1})^2C - qK^{-1}g(\eta_1,\eta_2) + h(\eta_1,\eta_2) \\ q\Big(\eta_2^2(q-q^{-1})^2C + q^{-1}K^{-1}g(\eta_1,\eta_2) - h(\eta_1,\eta_2)\Big) \\ -(q-q^{-1})g(\eta_1,\eta_2)EK^{-1} \end{pmatrix}, \end{aligned} \tag{5.119}$$

where C is the Casimir, $g(\eta_1,\eta_2) = \eta_1^2 q^2 - \eta_2^2$, $h(\eta_1,\eta_2) = (\eta_1\eta_2)^2 q^2 + 1$.

The important property of (5.119) is that it contains only F, EK^{-1}, K^{-1}. The weight zero subalgebra of the algebra generated by these three generators is generated by C and K^{-1}. So, it does not contain positive degrees of K, which implies that $\tilde{\mathcal{T}}_\star(\overline{\zeta})S$ has no terms proportional to the dimension. Applying the intertwiners $\check{R}_{i,i+1}(\xi_i/\xi_{i+1}) = P_{i,i+1}\tilde{R}_{i,i+1}(\xi_i/\xi_{i+1})$ one generates the entire space of invariant vectors from S. This finishes the proof. □

Combining Lemma 5.7 and Lemma 5.8 we draw the corollary:

COROLLARY 5.9. *On $\mathcal{W}^{\mathrm{inv}}$ the operator*

(5.120)
$$\tilde{X}(\zeta_1,\zeta_2|0) = \lim_{\alpha\to 0} \mathrm{Tr}_{a,b}\{\tilde{B}_{a,b}(\zeta)\tilde{\mathbb{T}}_a(\zeta_1)\tilde{\mathbb{T}}_b(\zeta_2)\}\mathrm{Tr}^{\Lambda}\{q^{-H}\tilde{\mathbb{T}}_\star(\overline{\zeta}|\alpha)\}\,, \quad q^{-\Lambda} = q\zeta\,,$$

exists and preserves the space $\mathcal{W}^{\mathrm{inv}}$.

It looks quite plausible now that the local operators invariant under the quantum group $\mathcal{U}_1$ are decomposable for $\alpha = 0$ in the usual fermionic basis. The only transcendental function which will show up is $\omega_0(\zeta_1,\zeta_2)$. We still have the operators $\mathbf{y}_m$ and $\mathbf{s}_1$ (Section 5.3.3), and it is natural to assume that the invariant operators belong to the space

$$\mathcal{V}_n = \mathcal{N}_{n,0}/\mathbf{s}_1\mathcal{H}_{n,1}\,.$$

Notice that both C and P symmetries are broken; this is not a very pleasant property of quantum groups.

For invariant vectors we shall use two bases.

One is the *A-basis* constructed as in the proof of Lemma 5.8. We draw non-intersecting arcs joining pairs of points $1,\cdots,2n$ (an example of $n=3$ is given on Fig. 11). To every such picture we associate a tensor product of singlets (5.118) for all joined pairs; for example, Fig. 11 corresponds to $s_{1,4}s_{2,3}s_{5,6}$. We define a Bratteli diagram to be a sequence of non-negative half integers $\{j_p\}_{p=0,1,\dots,n}$, such that $j_0 = 0$ and $|j_p - j_{p+1}| = 1/2$ for $p = 0,\dots,n-1$. For the invariant vectors considered here we have $j_n = 0$, but this is not a part of the definition. The equivalence of our pictures and Bratteli diagrams is achieved by attaching one half of the number of arcs going over every interval between two neighbouring points, see Fig. 11. We shall denote by $s(J)$ the invariant vector corresponding to a Bratteli diagram J. The A-basis of invariant operators consists of

$$\widetilde{O}(J) = s(J)^{\vee} \,. \tag{5.121}$$

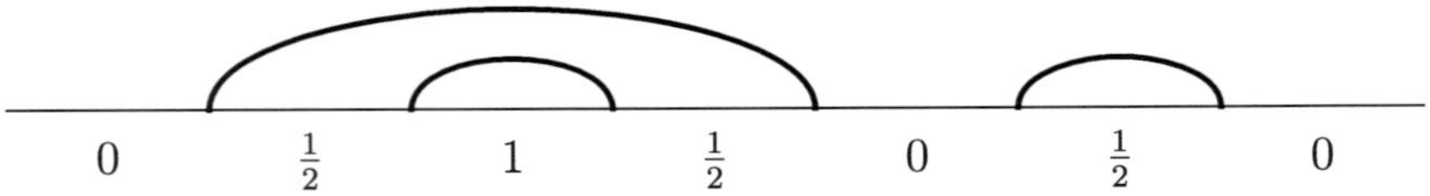

FIG. 11.

Being very simple, the A-basis is useful for computations.

Another basis, the *3j-basis*, is constructed using the $3j$-symbols. We shall need this basis in a more general case than just for invariant vectors. We define the $3j$-symbols as

$$\begin{pmatrix} j & 1/2 & j+1/2 \\ m & -1/2 & m-1/2 \end{pmatrix} = 1\,, \quad \begin{pmatrix} j & 1/2 & j+1/2 \\ m & 1/2 & m+1/2 \end{pmatrix} = \frac{q^{2(m+1)} - q^{-2j}}{q^2-1}\,,$$
$$\begin{pmatrix} j & 1/2 & j-1/2 \\ m & -1/2 & m-1/2 \end{pmatrix} = 1\,, \quad \begin{pmatrix} j & 1/2 & j-1/2 \\ m & 1/2 & m+1/2 \end{pmatrix} = \frac{q^{2(m+1)} - q^{2(j+1)}}{q^2-1}\,.$$

This definition differs from the usual one by normalisation; the point is that we avoid denominators from vanishing when q is a root of unity, except ± 1.

Consider the space$(\mathbb{C}^2)^{\otimes k}$, and fix a Bratteli diagram $J = \{0, j_1, \cdots, j_k\}$. Then the following vectors with $m = -j_k, -j_k+1, \cdots, j_k$ form a basis of the irreducible spin-j_k representation in the tensor product:

$$v_{k,J,m} = \sum_{\substack{\epsilon_1,\cdots,\epsilon_k=\pm \\ \sum_{p=1}^k \epsilon_p = 2m}} \prod_{p=1}^{k} \begin{pmatrix} j_{p-1} & 1/2 & j_p \\ m_{p-1} & \epsilon_p/2 & m_p \end{pmatrix} v_{\epsilon_1} \otimes \cdots \otimes v_{\epsilon_k} \,. \tag{5.122}$$

Here we set $m_p = \sum_{s=1}^{p} \epsilon_s/2$. Denote the totality of Bratteli diagrams $\{0, j_1, \cdots, j_k\}$ with $j_k = j$ by $B_\infty(k,j)$ (the meaning of the index ∞ will be clear soon). It is easy to see that

$$\operatorname{card}(B_\infty(k,j)) = \begin{pmatrix} k \\ \frac{k-2j}{2} \end{pmatrix} - \begin{pmatrix} k \\ \frac{k-2j}{2} - 1 \end{pmatrix} . \tag{5.123}$$

In particular, for $j = 0$ the integer k is even and we have the Catalan number $C_{k/2}$ in the right hand side. For generic q we have the decomposition

$$\left(\mathbb{C}^2\right)^{\otimes k} = \bigoplus_{J \in B_\infty(k,j)} M_J \otimes V_j \,, \tag{5.124}$$

where V_j are the irreducible representations and M_J are the corresponding space of multiplicities. The 3j-basis of invariant operators is obtained by applying $\vee$ to vectors constructed via the above procedure using Bratteli diagrams corresponding to $J = \{0, j_1, \cdots, j_{2n-1}, 0\}$. The 3j-basis of operators consists of

$$O(J) = v_{2n,J,0}^{\vee} \,. \tag{5.125}$$

Bratteli diagrams are lexicographically ordered, and the matrix U defined by

$$O(J) = \sum_{J'} U_{J,J'} \widetilde{O}(J')$$

is triangular. It is not hard to compute U inductively. We define the pairing

$$(O_1, O_2) = \frac{\mathrm{Tr}_{[k,l]}(q^{-2S_{[k,l]}} O_1 O_2)}{(q+q^{-1})^{l-k+1}},$$

where $[k, l]$ is an interval containing supports of both O_1 and O_2. We have

$$(O(J), O(J')) = \delta_{J^*,J'} \prod_{k=1}^{2n} [2j_k + 1], \tag{5.126}$$

where we set $J^* = \{0, j_{2n-1}, \ldots, j_1, 0\}$ for $J = \{0, j_1, \ldots, j_{2n-1}, 0\}$.

Later we shall be interested in the case when q is a primitive $2r$-th root of unity: $q^{2r} = 1$. In this case the rank of the matrix $\big((v_{2n,J_a,0}, v_{2n,J_b,0})\big)_{a,b=1,\cdots,C_n}$ is known to be equal to

$$\mathrm{card}\,(B_r(2n, 0))$$

where $B_r(2n, 0)$ is the totality of Bratteli diagrams with spins restricted as

$$j_k \le \frac{r-2}{2}, \quad \forall\, k\,. \tag{5.127}$$

It is also known that the $v_{2n,J,0}$'s with J satisfying (5.127) are rank$\,(B_r(2n, 0))$ linearly independent vectors which are orthogonal with respect to the pairing. The formula (5.126) for the scalar product holds true for J, J' satisfying (5.127).

We proceed with the same machinery as in Section 5.2 to find the decomposition of invariant operators in the fermionic basis. There is nothing special to comment on here. The denominators are the same as before (5.84). This calculation allows one to compute the expectation values of quantum group invariant operators for any κ and arbitrary *Matsubara* data. Let us show that for a particular value of κ the situation is more interesting than in the generic case.

Lemma 5.7 implies that the *Matsubara* transfer matrix $T_{\mathbf{M}}(\zeta, -1)$ acts from the $\mathcal{U}_1$-invariant subspace to itself. Let us emphasise once again a great difference from the case of invariance under a Lie algebra, for which the transfer-matrix acts from any irreducible representation to itself. For $T_{\mathbf{M}}(\zeta, -1)$ it is not at all the case.

Since $T_{\mathbf{M}}(\zeta, -1)$ acts from the $\mathcal{U}_1$-invariant subspace to itself, we can as usual distinguish two cases: that of unbroken symmetry when the maximal Mastubara eigenvector is also $\mathcal{U}_1$-invariant, and the case of broken symmetry for which it is not invariant. We are mostly interested in the first case. Numerical experiments show that it takes place for $\nu < 1/2$.

More generally, Lemma 5.7 implies that for a $\mathcal{U}_1$-invariant operator $\mathcal{O}_{[1,n]}$ localised on the space interval $[1, n]$, the operator

$$\mathrm{Tr}_{[1,n]}\left(\mathcal{O}_{[1,n]} q^{-2S_{[1,n]}} T_{[1,n],\mathbf{M}}\right)$$

acts from the $\mathcal{U}_1$-invariant subspace of the *Matsubara* space to itself. So, if we restrict our consideration to invariant operators (which by the way constitute a subalgebra), then the *Matsubara* space reduces itself automatically to the invariant subspace. In what follows we consider only the self-consistent situation (completely invariant case) in which the local operators and the *Matsubara* maximal eigenvectors are all invariant. Let us emphasise that similarly to [**RS**] we have to restrict our consideration to invariant operators. This is different from the case of unbroken symmetry under an ordinary Lie algebra in which the restriction to invariant operators is automatic.

In the completely invariant case with generic q, we introduce the density matrix at a finite temperature

$$\rho = \sum_{J \in B(2n,0)} \frac{\mathrm{Trs}\left(q^{-2S} e^{-\beta H_{XXZ}} O(J)\right)}{\mathrm{Trs}\left(q^{-2S} e^{-\beta H_{XXZ}}\right)} \frac{O(J^*)}{\prod_{j \in J} [2j+1]} . \tag{5.128}$$

As usual the density matrix ρ provides a formal way to compute the expectation values

$$\frac{\mathrm{Trs}\left(q^{-2S} e^{-\beta H_{XXZ}} \mathcal{O}\right)}{\mathrm{Trs}\left(q^{-2S} e^{-\beta H_{XXZ}}\right)} = \mathrm{Trs}\left(q^{-2S} \rho\, \mathcal{O}\right) . \tag{5.129}$$

The difference from the case considered previously is the insertion q^{-2S} under the trace. The 'partition function' resembles the Witten index for supersymmetry, with which it actually coincides in special cases.

The density matrix being $\mathcal{U}_1$-invariant, the Schur Lemma applies, and ρ is block diagonal with respect to the decomposition (5.124): it acts non-trivially only in the space of multiplicities with fixed j, and as a unit matrix in the space V_j. For any function $f(\rho)$ we have

$$\mathrm{Tr}\left(q^{-2S} f(\rho)\right) = \sum_{j=\frac{n}{2}-\left[\frac{n}{2}\right]}^{\frac{n}{2}} [2j+1] \mathrm{Tr}\left(f(\rho_j)\right) , \tag{5.130}$$

where ρ_j is the block acting on spin j multiplicity spaces. The formula (5.130) differs from a classical corollary of the Schur Lemma wherein the dimensions of irreducible representations are changed to q-dimensions $[2j+1]$. For $f(x) = x$ the right hand side must be equal to 1. This is always the first thing we check when doing numerics. We take $f(x) = -x \log x$ in order to compute the von Neumann entropy. Certainly the very meaning of this notion is not quite clear because of the insertion of q^{-2S}; we shall comment on this later.

At roots of unity, 'bad' Bratteli diagrams have to be thrown away.

LEMMA 5.10. *Assume that q is a primitive $2r$-th root of unity with $r \geq 3$. If J does not satisfy the restriction* (5.127), *then*

$$\mathrm{Trs}\left(q^{-2S} e^{-\beta H_{XXZ}} O(J)\right) = 0 .$$

PROOF. For $J = \{0, j_1, \dots, j_{2n-1}, 0\}$, we set $J_L = \{0, j_1, \dots, j_{n-1}, j_n\}$, $J_R = \{0, j_{2n-1}, \dots, j_{n+1}, j_n\}$. Then (5.125) can be rewritten as

$$O(J) = \sum_{m=-j_n}^{j_n} (-q)^{m-n/2} v_{n,J_L,m} \otimes v^{*t}_{n,J_R,m} ,$$

where $v^*_{n,J_R,m}$ is defined by (5.122), using

$$\begin{pmatrix} j & 1/2 & j' \\ m & \epsilon/2 & m' \end{pmatrix}^* = \begin{pmatrix} j' & 1/2 & j \\ m' & -\epsilon/2 & m \end{pmatrix}$$

in place of the $3j$-symbols. For generic q we have

$$\begin{pmatrix} j' & 1/2 & j \\ m' & -\epsilon/2 & m \end{pmatrix} = \frac{c_{j',m'}}{c_{j,m}} \begin{pmatrix} j & 1/2 & j' \\ m & \epsilon/2 & m' \end{pmatrix} ,$$

$$c_{j,m} = (-1)^{j-m} q^{j(j+1)-m(m+1)} \frac{[j-m]!}{[j+m]!} ,$$

so that $v^*_{n,J_R,m} = c_{j,m} v_{n,J_R,m}$. Note that $v^*_{n,J_R,m}$ is well defined at roots of unity even though $c_{j,m}$ may not be.

We use the following fact about representations at roots of unity (see e.g. [**JMT**]). Let us say that a sequence $J_1 = \{0, j_1, \dots, j_n\}$ is good if $j_s \le (r-2)/2$ for all s, and bad otherwise.

Let $V = \mathbb{C}^2$, $n \ge 1$. Then we have a direct sum decomposition $V^{\otimes n} = \mathcal{G}_n \oplus \mathcal{B}_n$ into $U_q\mathfrak{sl}_2$ submodules with the following properties.

(1) $\mathcal{G}_n$ is spanned by vectors $v_{n,J_1,m}$ with good J_1. $\mathcal{B}_n$ is spanned by vectors $v_{n,J_1,m}$ with bad J_1.
(2) $\mathcal{B}_n$ is a direct sum of modules W_i such that $\mathrm{Tr}_{W_i} K^{-1} = 0$.
(3) $\mathcal{G}_n$ and $\mathcal{B}_n$ are orthogonal with respect to the bilinear form $(u, v) = u^t v$.

Property (1) holds true also for $v^*_{n,J_1,m}$.

Now let $\mathcal{H} = \sum_{j=1}^{n-1} h_{j,j+1}$. Since $e^{-\beta\mathcal{H}} O(J)$ commutes with the action of $U_q\mathfrak{sl}_2$, it acts on the space of multiplicities. Hence we have

$$\mathrm{Tr}_{\mathcal{B}_n} K^{-1} e^{-\beta\mathcal{H}} O(J) = 0$$

due to property (2).

By assumption, either J_L or J_R is bad. If J_R is bad, then property (3) implies $(v^*_{n,J_R,m}, u) = 0$ for all $u \in \mathcal{G}_n$, so that

$$\mathrm{Tr}_{\mathcal{G}_n} K^{-1} e^{-\beta\mathcal{H}} O(J) = 0\,.$$

If J_L is bad, then $(u, v_{n,J_L,m}) = 0$ for all $u \in \mathcal{G}_n$, so the same holds by the cyclicity of trace. We conclude that $\mathrm{Tr}_{V^{\otimes n}} K^{-1} e^{-\beta\mathcal{H}} O(J) = 0$. □

Hence (5.130) becomes

$$\mathrm{Tr}\left(q^{-2S} f(\rho)\right) = \sum_{j=\frac{n}{2}-\left[\frac{n}{2}\right]}^{\min\left(\frac{n}{2}, \frac{r-2}{2}\right)} [2j+1] \mathrm{Tr}\left(f(\rho_j)\right)\,, \tag{5.131}$$

and the basis of the space of multiplicities is supposed to be counted by $B_r(k, j)$. For $\nu = \frac{1}{r}$ which means

$$q = e^{\frac{\pi i}{r}}\,, \tag{5.132}$$

it is possible to introduce a new positive definite Hermitian structure in the space of invariants.

The case of CFT will be discussed in detail in Volume II ; for the moment we shall use some notion of CFT without much explanation. We assume that the reader is either familiar with some simple facts from CFT, or he will wait for the explanations in Volume II.

Let us begin with $\nu = \frac{1}{r}$. It has been said that in this case we can introduce a scalar product for the reduced model, and

$$s(n) = -\sum_{j=\frac{n}{2}-\left[\frac{n}{2}\right]}^{\min\left(\frac{n}{2}, \frac{r-2}{2}\right)} [2j+1] \mathrm{Tr}\left(\rho_j \log(\rho_j)\right) \tag{5.133}$$

has the meaning of entropy. The natural conjecture is that the scaling limit provides the unitary minimal model of CFT with the central charge

$$c = 1 - \frac{6}{r(r-1)}\,.$$

The simplest example is given by $r = 4$, for this case $c = 1/2$, so, we are in the universality class of the Ising model. The results of our computations at zero temperature are given in the following table.

n	$s(n) - \frac{1}{6}\log n$
2	0.3584221036
3	0.3615208351
4	0.3623594511
5	0.3626503444
6	0.3627815665
7	0.3628533359
8	0.3628976397

We see quite impressive agreement with the central charge $c = 1/2$.

To have one more example, consider $r = 5$. In this case the results are collected in the table:

n	$s(n) - \frac{7}{30}\log n$
2	0.4983693850
3	0.5028709550
4	0.5041174078
5	0.5045588518
6	0.5047630265
7	0.5048763027
8	0.5049471991

The agreement with $c = 7/10$ is good, but the precision is a little worse than in the foregoing cases. The general rule is: the closer one is to the isotropic case $\nu = 0$ the worse is the precision.

Now we come to intuitively less clear cases. Consider an arbitrary rational $\nu = \frac{p}{r}$ (certainly $p < r$). If $p > 1$ it is impossible to introduce a Hermitian structure, in the scaling limit we have the non-unitary minimal model with the central charge

$$c = 1 - \frac{6\nu^2}{1-\nu} = 1 - \frac{6p^2}{r(r-p)}\,.$$

In this situation we shall call (5.133) the pseudo-entropy. It is not intuitively obvious, but the asymptotics of the pseudo-entropy is given by the same formula as before even if the central charge is negative. To illustrate this, consider $\nu = 2/5$ for which the central charge equals $c = -3/5$. The results of our computations are given in the table:

n	$s(n)+\frac{1}{5}\log n$
2	-0.4978380190
3	-0.5017837324
4	-0.5028616065
5	-0.5032510429
6	-0.5034331553
7	-0.5035349983
8	-0.5035986331

In the cases considered up to now, we had to take care of singular determinants in the same way as in Section 5.3.5.

Now we proceed to the case of non-zero temperature. Recall that in the DdV equations (5.89) we had the driving term

$$f(\zeta)=((I+R)\phi)(\zeta)\,,\quad \phi(\zeta)=\frac{i}{T}\left(\psi(\zeta q^{-1},0)-\psi(\zeta,0)\right)-2\pi i\nu\kappa\,.$$

This is the only place where κ occurs. In Section 5.3.5 we set $\kappa=0$, now we set $\kappa=-1$; the rest of computations is quite parallel.

Take the case $n=8$. In order to simplify the procedure we consider the cases $\nu=1/11,2/11,3/11,4/11,5/11$ for which there are no singularities. The results of our numerical computations for these five cases are presented on Fig. 12 (from up to down). The dots are the numerical results, the grey lines are $\frac{1}{3}c(\nu)\log\left(\frac{\sinh(8T)}{8T}\right)$.

All possible situations are represented here: the unitary case $\nu=1/11$, non-unitary cases with positive central charges $\nu=2/11,3/11$, non-unitary cases with negative central charges $\nu=4/11,5/11$. We always have a good agreement with CFT, but it becomes better the further we are from the isotropic point.

5.4. XXX Case

The last example which we will consider is the isotropic case $\nu=0$. This case has its advantages and disadvantages. First of all, let us describe necessary technical modifications of our procedure.

We have to consider the limit $\nu\to 0$. Like in Section 5.3.4 we begin with introducing the additive variable

$$x=\frac{1}{\pi\nu}\log\zeta\,,$$

and consider the limit $\nu\to 0,\zeta\to 1$ such that x is finite. Now we explain briefly the modifications which should be done in the XXZ formulas.

The Q-functions become

$$Q(x,k)=e^{kx}\prod_{j=1}^{m}(x-b_j)\,,$$

with b_j being the Bethe roots. The Baxter equation is

$$T(x,k)Q(x,k)=a(x)Q(x+i,k)+d(x)Q(x-i,k)\,,$$

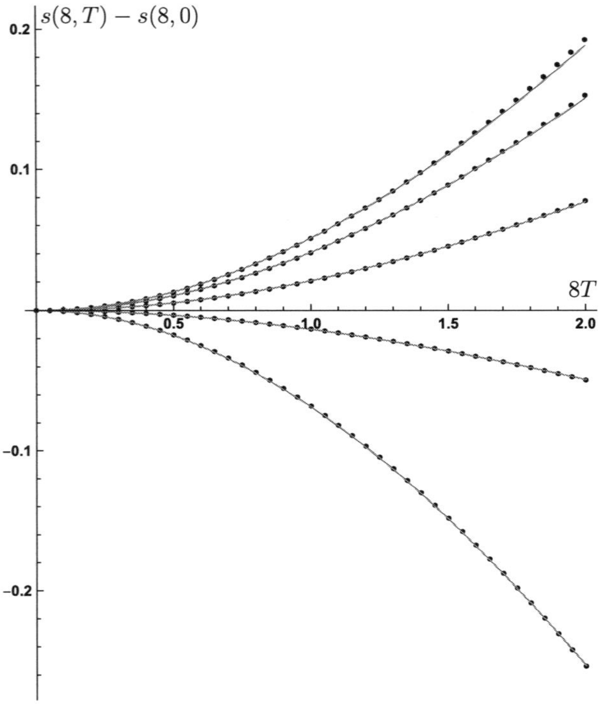

FIG. 12.

where $T(x,k)$ is a polynomial of degree $\mathbf{n}$, $a(x)$ and $d(x)$ are monic polynomials of degree $\mathbf{n}$. The formula (5.49) becomes

$$P_j(x) = \frac{x}{x-b_j}\left(a(x)\frac{Q(x+i)}{x-b_j+i} - d(x)\frac{Q(x-i)}{x-b_j-i}\right).$$

In what follows we work with symmetric polynomials using Young diagrams as explained in Section 5.2.3. The formula (5.52) stays unchanged. The formulas (5.53), (5.55), (5.56), (5.54) become

$$\mathcal{T}_{++}Y = \mathrm{cut}_l\Big(\sum_{k=1}^{l+1}\sigma_{k-1}\circ A_k\wedge Y\Big), \quad A_k(x) = (-1)^{k-1}(x+i)^{l+1-k}a(x),$$

$$\mathcal{T}_{--}Y = \mathrm{cut}_l\Big(\sum_{k=1}^{l+1}\sigma_{k-1}\circ D_k\wedge Y\Big), \quad D_k(x) = (-1)^{k-1}(x-i)^{l+1-k}d(x),$$

$$\begin{aligned}\mathcal{T}_{+-}Y = \mathrm{cut}_l\Big(\sigma_l^{-1}\circ\Big[\sum_{p=0}^{l}\sum_{r=0}^{l-1}\sigma_r\circ\sigma_p\circ\Big(&P_{p,r}\wedge\mathrm{cut}_{l-1}(Y)\\ &+R_{p,r}\wedge\sigma_{l-2}\circ\mathrm{cut}_{l-2}(Y)\Big)\Big]\Big),\end{aligned}$$

$$P_{p,r}(x) = (-1)^r \sum_{s=r}^{l-1} \left[d(0)a(x)(x+i)^{l-p} x^{s-r} - a(0)d(x)(i-x)^{l-p} x^{s-r} \right] ,$$

$$R_{p,r}(x,y) = d(x)a(y)(-1)^{p+r}(y+i)^{l-p} \sum_{s=r}^{l-1} (x-i)^{l-1-s} y^{s-r} ,$$

$$\mathcal{T}_{-+}Y = \mathrm{cut}_l(Y) .$$

As in the anisotropic situation, we would like to study the density matrix for $\alpha = 0$. This is the time to collect advantages of Section 5.3.3 and Section 5.3.6. The quantum group is replaced in the isotropic case by the usual Lie algebra $\mathfrak{sl}_2$. In this case the requirement of $\mathfrak{sl}_2$-invariance is consistent with the discrete symmetries of the Assumptions 5.6: spin reversal, C-symmetry and P-symmetry. The $\mathfrak{sl}_2$-invariance is the limiting case of the quantum group invariance, so, the $\mathfrak{sl}_2$-invariant operators are created by fermions only, for operators located on the interval $[1, n]$ we actually have smaller space $\mathcal{N}_{n,0}/\mathcal{CH}_1$ (we use notations of Section 5.3.3). We start our usual computations using small *Matsubara* data. The definitions of the Section 5.1 should be very modestly modified. For example $K_0(\zeta/\eta)$ is replaced by

$$K_0(x-y) = -\frac{1}{\pi((x-y)^2+1)} .$$

Now there is no parameter q, and the number of operators which satisfies the $\mathfrak{sl}_2$-invariance and discrete symmetries is smaller than before, so, we hope to proceed further than in the anisotropic case.

Performing experiments we come to the conjecture that a smaller subspace of the fermionic space contains all the local $\mathfrak{sl}_2$-invariant operators which are C-symmetric, P-symmetric, and symmetric under the spin reversal. Namely, take $\mathbf{b}^*_{I^+}\mathbf{c}^*_{I^-}$ with $\#(I^+) = \#(I^-)$ satisfying the requirement $i^+_p \le i^-_p$, $p = 1, \cdots, \#(I^+)$. Require also

$$\sum i^+_p + \sum i^-_p \equiv 0 \pmod 2 . \tag{5.134}$$

In order to narrow down to operators of length n we impose also

$$\max(I^+, I^-) \le n , \qquad \#(I^+) \le \left[\frac{n}{2}\right] . \tag{5.135}$$

Denote the space with such basis by $\mathfrak{H}_n$. The space $\tilde{\mathfrak{H}}_n$ satisfies all the requirement above except (5.134). The charge 2 space $\tilde{\mathfrak{H}}_{n,2}$ is spanned by $\mathbf{b}^*_{I^+}\mathbf{c}^*_{I^-}$ with $\#(I^+) + 1 = \#(I^-) - 1 \le \left[\frac{n}{2}\right]$. Define the subspace $\mathfrak{V}_n$ of $\mathfrak{H}_n$ by

$$\mathfrak{V}_n = \{ v \in \mathfrak{H}_n \mid \mathcal{G}_m v \in \mathcal{C}\tilde{\mathfrak{H}}_{n,2} \text{ for } m = n+1, n+2, \cdots \}, \tag{5.136}$$

where $\mathcal{G}_m$ is defined in (4.88). We conjecture, and check up to $n = 10$, that $\mathfrak{V}_n$ contains all the local operators with the properties described above. The dimensions of these spaces for n up to 11 are given in the table:

n	2	3	4	5	6	7	8	9	10	11	12
$\dim(\mathfrak{V}_n)$	1	2	6	12	31	79	178	434	1141	2946	7888

We manage to solve the equations up to $n = 10$.

The *Matsubara* transfer matrix commutes with the action of $\mathfrak{sl}_2$. More generally, if an operator $\mathcal{O}_{[1,n]}$ is a part of spin j representation of $\mathfrak{sl}_2$ acting in the space direction, then the operator $\mathrm{Tr}_{[1,n]}\left(T_{[1,n],\mathbf{M}}\mathcal{O}_{[1,n]}\right)$ is a part of spin j representation of $\mathfrak{sl}_2$ in the *Matsubara* direction. Suppose that the *Matsubara* maximal eigenvector is invariant (unbroken symmetry). Then the properties above imply that the expectation values vanish for operators which transform under all irreducible representations except the one-dimensional one. So the expectation values of the invariant operators provide the entire density matrix. This situation is different from the one described for the quantum group case; certainly it is much more natural from the physicist point of view.

The maximal eigenvector is $\mathfrak{sl}_2$ invariant for $\kappa = 0$ and any temperature.

In the case of zero temperature we have

$$\begin{aligned} &\omega(x,y) = \omega(x-y)\,, \\ &\omega(x) = -\frac{1}{2(x^2+1)} + i\frac{d}{dx}\log\left(\frac{\Gamma\left(1+\frac{x}{2i}\right)\Gamma\left(\frac{1}{2}-\frac{x}{2i}\right)}{\Gamma\left(1-\frac{x}{2i}\right)\Gamma\left(\frac{1}{2}+\frac{x}{2i}\right)}\right) \\ &\quad = 2\log 2 - \frac{1}{2} + \sum_{k=1}^{\infty}(-1)^k x^{2k}\left(2\zeta(2k+1)(1-2^{-2k}) - \frac{1}{2}\right). \end{aligned} \tag{5.137}$$

This formula explains why the expectation values in this case are expressed in terms of odd ζ-values [**BK**].

For the entanglement entropy at zero temperature we have:

n	$s(n) - \frac{1}{3}\log n$
2	0.7226225664
3	0.7306966875
4	0.7333763535
5	0.7345480889
6	0.7352244493
7	0.7356681827
8	0.7359913623
9	0.7362367768
10	0.7364324939

So, in order to achieve the same precision as in the XXZ case (for large enough ν) we have to go up to 10 sites instead of 8.

The numerical procedure for non zero temperature does not differ much from the XXZ case. The results for $n = 10$ are summarised on the following picture, where as usual the grey line corresponds to the conformal limit: $\frac{1}{3}\log\left(\frac{\sinh(10T)}{10T}\right)$.

We observe that on 10 sites the agreement with the CFT curve is not bad, but it is worse than on 8 sites for XXZ with $n = 5/8$ (Fig. 9).

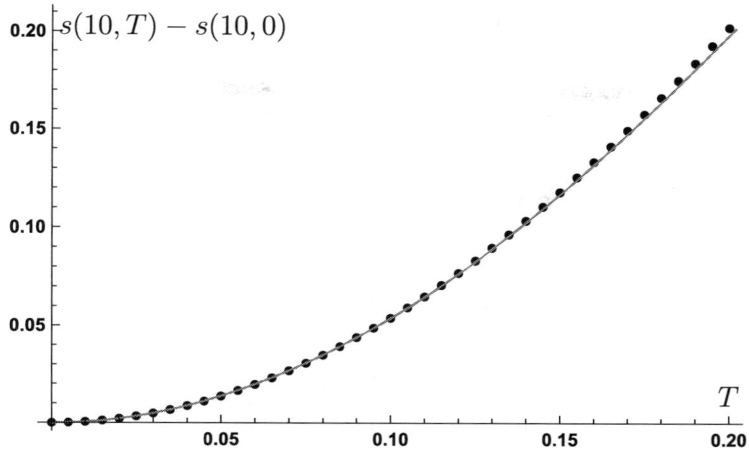

FIG. 13.

Now let us return to zero temperature and discuss the correlation functions

$$G(n) = \langle \sum_a \sigma_1^a \sigma_n^a \rangle = 3\langle \sigma_1^3 \sigma_n^3 \rangle ,$$

the last equality is due to the $\mathfrak{sl}_2$-invariance. The most interesting question concerns the behaviour of the correlation functions at large n. The best known formula for the large n asymptotics was obtained by Lukyanov [**Lk1**, **LT**] in the framework of perturbed CFT. Related questions will be discussed in Volume II. Here we would just like to say that the perturbation in question is irrelevant and it breaks the rotational invariance. All that is not quite usual. In particular, the perturbation series contain an arbitrary parameter c. The result of summation of the series with sufficiently many terms is supposed to be independent of c, but if we take only several terms the value of c may be chosen as an adjusting parameter. The formula is

$$\begin{aligned}\frac{(-1)^{n-1}}{12}G(n) = \sqrt{\frac{1}{8\pi^3}}\frac{1}{(n-1)\sqrt{g}}\Big\{(1+\Big(\frac{3}{8}-\frac{c}{2}\Big)g+\Big(\frac{5}{128}-\frac{c}{16}-\frac{c^2}{8}\Big)g^2\\ +\Big(\frac{21}{1024}+\frac{7c}{256}-\frac{7c^2}{64}-\frac{c^3}{16}+\frac{13}{32}\zeta(3)\Big)g^3+O(g^4)\Big\}\\ -\frac{(-1)^{n-1}}{4\pi^2(n-1)^2}\Big\{1+\frac{g}{2}+\frac{1}{2}\Big(c+\frac{3}{4}\Big)g^2+\frac{c(c+2)}{2}g^3+O(g^4)\Big)\Big\}+\cdots,\end{aligned}$$

where the coupling g is defined from the equation

$$\sqrt{g}e^{\frac{1}{g}} = 2\sqrt{2\pi}(n-1)e^{c+\gamma_E} .$$

Lukyanov proposes to take c from the interval $[-2,-1]$. We take bluntly

$$c = -\frac{3}{2} .$$

The left column of the following table presents 24 digits exact results for the correlation function computed by our method. The right column provides the Lukyanov formula.

N	$(-1)^{n-1}\frac{1}{12}G(n)$	Lukyanov
2	0.1477157268533151031390774	not applicable
3	0.0606797699564353014934941	0.06346368378
4	0.0502486272572352479593931	0.05083191412
5	0.0346527769827281656556596	0.03482741316
6	0.0308903666476093257628751	0.03096613477
7	0.0244467383279589065417695	0.02447862849
8	0.0224982227633722183770986	0.02251356472
9	0.0189734169587321977494075	0.01898064986
10	0.0177751064604679461357958	0.01777789616
11	0.0155478493216075886422179	0.01554899383
12	0.0147330671331807547552987	0.01473265930

We observe quite good agreement of our exact results with Lukyanov's asymptotics. For finite temperature the results are going to be even better. So, it seems that we should not go much further with exact computations.

5.5. Remarks on XYZ Case

The completely anisotropic XYZ case is much more complicated than the previously considered XXZ case. In XYZ case the $U(1)$-symmetry is broken to $\mathbb{Z}_2$, which forbids us to introduce the parameters α, κ which were so useful for us. On the other hand the mathematical structures needed to study the XYZ case are much more sophisticated. As a result of these complications only partial results for the zero temperature case are available. That is why we find it appropriate to discuss the XYZ case here, after similar situation ($\alpha = \kappa = T = 0$) was considered for the XXZ and XXX cases. The goal of our exposition will be rather in pointing out the problems which one has to overcome working with the XYZ case than in giving complete and rigorous proofs.

5.5.1. Another Way of Presenting the XXZ Results. For $\alpha = \kappa = 0$ and zero temperature the results of Section 5.3 can be presented as

$$\frac{\langle \mathrm{vac}|\mathcal{O}|\mathrm{vac}\rangle}{\langle \mathrm{vac}|\mathrm{vac}\rangle} = \mathrm{Tr}\left(e^{\Omega}(\mathcal{O})\right), \tag{5.138}$$

where Ω is given by (5.72):

$$\Omega = \frac{1}{(2\pi i)^2}\oint_\Gamma\oint_\Gamma \Bigl\{X_0(\zeta_1,\zeta_2)\omega_0(\zeta_2/\zeta_1) + X_1(\zeta_1,\zeta_2)\omega_1(\zeta_2/\zeta_1)\Bigr\}\frac{d\zeta_1^2}{\zeta_1^2}\frac{d\zeta_2^2}{\zeta_2^2}. \tag{5.139}$$

The operators $X_\epsilon(\zeta_1,\zeta_2)$ were obtained form the general case by taking the limit $\alpha\to 0$ we have (5.70):

$$\begin{aligned}&\mathrm{Tr}_{a,b}\{B_{a,b}(\zeta)\mathbb{T}_a(\zeta_1)\mathbb{T}_b(\zeta_2)\}\mathrm{Tr}^\Lambda\{q^{(1-2\varepsilon)H}\mathbb{T}_\star(\overline{\zeta},\alpha)\}\\ &=\frac{1}{\alpha}X_1(\zeta_1,\zeta_2)+X_0(\zeta_1,\zeta_2)+O(\alpha)\,,\end{aligned}$$

where $q^{-\Lambda}=q\zeta_1/\zeta_2$, $\bar{\zeta}=(\zeta_1\zeta_2)^{\frac{1}{2}}$. For future convenience we modify $B^{1/2}(\zeta)$ using Remark 5.5

$$\begin{aligned}B(\zeta)&=B^{1/2}(\zeta)-\frac{\zeta^2-\zeta^{-2}}{2(\zeta q-\zeta^{-1}q^{-1})(\zeta q^{-1}-\zeta^{-1}q)}\\ &=\frac{(\zeta-\zeta^{-1})(q+q^{-1})}{2(\zeta q-\zeta^{-1}q^{-1})(\zeta q^{-1}-\zeta^{-1}q)}\Big(\sigma^1\otimes\sigma^1+\sigma^2\otimes\sigma^2+\frac{\zeta+\zeta^{-1}}{q+q^{-1}}\sigma^3\otimes\sigma^3\Big)\,.\end{aligned}$$

The function $\omega_0(\zeta)$ is even while the function $\omega_1(\zeta)$ is odd. Hence the definition (5.139) can be rewritten as

$$\Omega=\frac{1}{(2\pi i)^2}\oint_\Gamma\oint_\Gamma\Big\{Y_0(\zeta_1,\zeta_2)\omega_0(\zeta_2/\zeta_1)+Y_1(\zeta_1,\zeta_2)\omega_1(\zeta_2/\zeta_1)\Big\}\frac{d\zeta_1^2}{\zeta_1^2}\frac{d\zeta_2^2}{\zeta_2^2}\,. \tag{5.140}$$

with

$$Y_0(\zeta_1,\zeta_2)=\frac{1}{2}(X_0(\zeta_1,\zeta_2)+X_0(\zeta_2,\zeta_1))\,,\quad Y_1(\zeta_1,\zeta_2)=\frac{1}{2}(X_1(\zeta_1,\zeta_2)-X_1(\zeta_2,\zeta_1))\,.$$

These operators allow an alternative definition in terms of trace $\mathrm{Tr}^{\{d\}}$ over d-dimensional representation analytically continued with respect to dimension d. Let us give a formal definition.

Like in Section 2.2.2 the linear functional $\mathrm{Tr}^{\{d\}}$ on $U_q(\mathfrak{sl}_2)$ is defined to satisfy the properties

(1) $\mathrm{Tr}^{\{d\}}A\big|_{d=k+1}=\mathrm{tr}_{\mathcal{V}^{(k)}}\pi^{(k)}(A)\quad(k\in\mathbb{Z}_{\geq 0})$,

(2) Cyclicity

$$\mathrm{Tr}^{\{d\}}(xy)=\mathrm{Tr}^{\{d\}}(yx),$$

(3) Linearity over the ring generated by the Casimir element

$$\mathrm{Tr}^{\{d\}}\,(Cx)=\frac{q^d+q^{-d}}{(q-q^{-1})^2}\mathrm{Tr}^{\{d\}}(x)\quad(i=0,2)\,.$$

As has been explained in Section 2.2.2 it is sufficient to compute the trace of K^m which is

$$\mathrm{Tr}^{\{d\}}\,(K^m)=\begin{cases}\frac{q^{dm}-q^{-dm}}{q^m-q^{-m}}\,, & m\neq 0\,,\\ d\,, & m=0\,.\end{cases} \tag{5.141}$$

It is easy to see that Y_0, Y_1 can be defined using $\mathrm{Tr}^{\{d\}}$ (as usual $\zeta=\zeta_1/z_2$)

(5.142)

$$\mathrm{Tr}_{a,b}\{B_{a,b}(\zeta)\mathbb{T}_a(\zeta_1)\mathbb{T}_b(\zeta_2)\}\mathrm{Tr}^{\{\frac{\log\zeta}{\log q}\}}\{\mathbb{T}_\star(\overline{\zeta})\}=Y_0(\zeta_1,\zeta_2)-\frac{\log\zeta}{\log q}\cdot Y_1(\zeta_1,\zeta_2)\,,$$

with the requirement that $Y_0(\zeta_1,\zeta_2)$, $Y_1(\zeta_1,\zeta_2)$ are rational functions of their arguments. So, the only non-analytical dependence on ζ_1,ζ_2 is in the multiplier in front of $Y_1(\zeta_1,\zeta_2)$ which is traced back to the second case in (5.141).

The functions $\omega_i(\zeta)$ are obtained as first two terms of the Taylor series for $\omega(\zeta|\alpha)$ (the explicit formula for the latter is $\omega(\zeta|\alpha) = \sigma(\zeta^2|\alpha)$ given in (5.99)). It is easy to see that they allow an alternative definition avoiding α:

$$\frac{1}{2} d \log \varphi = \omega_0 d \log \zeta + \omega_1 d \log q \,, \tag{5.143}$$

where $\varphi(\zeta, \nu)$ is defined through the partition function $\rho(\zeta, \nu)$:

$$\varphi(\zeta, \nu) = \rho(\zeta, \nu) \left(\frac{\zeta^2 - q^2}{\zeta^2 q^2 - 1} \right)^{\frac{1}{4}} , \tag{5.144}$$

$$\log \rho(\zeta, \nu) = -\int_0^\infty \frac{1}{2} \left(\zeta^{2ik} - \zeta^{-2ik} \right) \frac{\sinh(\pi(1-\nu)k)}{k \sinh(\pi k) \cosh(\pi \nu k)} dk \,.$$

5.5.2. XYZ Model and Sklyanin Algebra. The Hamiltonian of the XYZ-model is

$$H_{XYZ} = \frac{1}{2} \sum_{j=-\infty}^{\infty} \left(I_1 \sigma_j^1 \sigma_{j+1}^1 + I_2 \sigma_j^2 \sigma_{j+1}^2 + I_3 \sigma_j^3 \sigma_{j+1}^3 \right) , \tag{5.145}$$

where[2]

$$I_a = \frac{\theta_{5-a}(\nu, \tau)}{\theta_{5-a}(0, \tau)} \,.$$

Here and later the indices of θ-functions are counted modulo 4.

In the limit $\tau \to +i\infty$ we have

$$I_1 \to 1 \,, \; I_2 \to 1 \,, \; I_3 \to \cos(\pi \nu) \,,$$

which implies

$$H_{XYZ} \underset{\tau \to +i\infty}{\to} H_{XXZ} \,.$$

Equivalently, we can think about the eight-vertex model with vertices

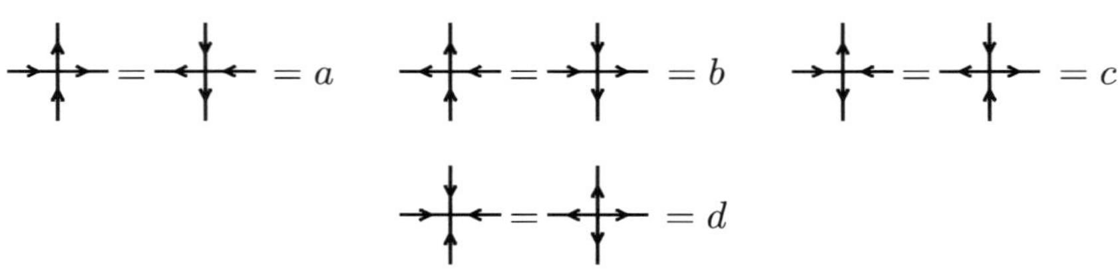

FIG. 14.

Up to an overall multiplier the Boltzmann weights are given in therms of θ-functions of double-modulus. In the homogeneous case they are

$$a : b : c : d = 1 : 1 : \frac{\theta_4(\frac{\nu}{2}, 2\tau)\, \theta_1(\nu, 2\tau)}{\theta_1(\frac{\nu}{2}, 2\tau)\, \theta_4(\nu, 2\tau)} : \frac{\theta_1(\frac{\nu}{2}, 2\tau)\, \theta_1(\nu, 2\tau)}{\theta_4(\frac{\nu}{2}, 2\tau)\, \theta_4(\nu, 2\tau)} \,.$$

From now on all θ-functions will have modulus τ, which will be omitted for the sake of conciseness. The main mathematical object emerging from the study of the

[2] In order to simplify comparison with the critical XXZ case we change in the conventional formulas [**Sk1**] the indices of the theta functions $a+1$ to $5-a$. It corresponds essentially to a modular transformation.

XYZ model is the Sklyanin algebra [**Sk1**] which we denote by $\mathcal{A}$. It is generated by four generators S^a $(a = 0, 1, 2, 3)$ and quadratic commutation relations

$$[S_0, S_a] = iJ_{bc}(S_bS_c + S_cS_b), \tag{5.146}$$

$$[S_b, S_c] = i(S_0S_a + S_aS_0), \tag{5.147}$$

where (a, b, c) runs over cyclic permutations of $(1, 2, 3)$. The J_{bc} are the structure constants given by

$$J_{bc} = -\frac{J_b - J_c}{J_a} = -\varepsilon_a \frac{\theta_1(\frac{\nu}{2})^2\theta_{5-a}(\frac{\nu}{2})^2}{\theta_{5-b}(\frac{\nu}{2})^2\theta_{5-c}(\frac{\nu}{2})^2}, \tag{5.148}$$

$$J_a = \frac{\theta_{5-a}(\nu)\theta_{5-a}}{\theta_{5-a}(\frac{\nu}{2})^2}, \tag{5.149}$$

where

$$\varepsilon_2 = -1, \quad \varepsilon_a = 1 \ (a \neq 2).$$

Here and after, theta functions without arguments stand for the theta zero values, $\theta_a = \theta_a(0)$ and $\theta_1' = \theta_1'(0)$. The J_{bc}'s obey an identity

$$J_{1,2} + J_{2,3} + J_{3,1} + J_{1,2}J_{2,3}J_{3,1} = 0\,. \tag{5.150}$$

Using the Sklyanin algebra one defines the R-matrix which should be considered as an analogue of the R-matrix in the tensor product of $ev_\zeta \otimes \pi^{(1)}$ for the XXZ case:

$$R_{\star,j}(\mu) = \frac{1}{2}\sum_{a=0}^{3} \frac{\theta_{5-a}(2\mu + \frac{\nu}{2})}{\theta_{5-a}(\frac{\nu}{2})} S_a \otimes \sigma_j^a \ \in \ \mathcal{A} \otimes \mathrm{End}(\mathcal{V}^{(1)})\,. \tag{5.151}$$

Here we use an additive spectral parameter μ.

Since the defining relations (5.146), (5.147) are homogeneous, $\mathcal{A}$ is a $\mathbb{Z}_{\geq 0}$-graded algebra, $\mathcal{A} = \oplus_{n\geq 0}\mathcal{A}_n$, where the generators S_a all belong to $\mathcal{A}_1$. We have also a $\mathbb{Z}_2 \times \mathbb{Z}_2$-grading, $\mathcal{A} = \oplus_{(m,n)\in\mathbb{Z}_2\times\mathbb{Z}_2}\mathcal{A}^{(m,n)}$, defined by the assignment $S_a \in \mathcal{A}^{\bar{a}}$, where

$$\bar{0} = (0, 0),\ \bar{1} = (1, 0),\ \bar{2} = (1, 1),\ \bar{3} = (0, 1) \quad \in \mathbb{Z}_2 \times \mathbb{Z}_2. \tag{5.152}$$

To make distinction, the $\mathbb{Z}_{\geq 0}$-grading and the $\mathbb{Z}_2 \times \mathbb{Z}_2$-grading will be referred to as 'degree' and 'colour', respectively. Thus S_a has degree 1 and colour $\bar{a}$.

The Sklyanin algebra has two central elements of degree 2 and colour $\bar{0}$:

$$K_0 = \sum_{a=0}^{3} S_a^2, \quad K_2 = \sum_{a=1}^{3} J_aS_a^2. \tag{5.153}$$

We call them Casimir elements.

We will be concerned with representations in series (a) of [**Sk2**], which are analogs of finite-dimensional irreducible representations of $\mathfrak{sl}_2$. For each non-negative integer k, let $\mathcal{V}^{(k)}$ denote the vector space of entire functions $f(u)$ with the properties

$$f(u + 1) = f(u) = f(-u), \quad f(u + \tau) = e^{-2\pi ik(2u+\tau)}f(u).$$

We have $\dim \mathcal{V}^{(k)} = k+1$. The following formula defines a representation $\pi^{(k)} : \mathcal{A} \to \mathrm{End}(\mathcal{V}^{(k)})$ [**Sk2**]: [3]

$$(5.154)\left(\pi^{(k)}(S_a)f\right)(u) = \frac{\sqrt{\varepsilon_a}\theta_{5-a}(\frac{\nu}{2})}{\theta_1(\nu)\theta_1(2u)}\left(\theta_{5-a}(2u-\frac{\nu}{2}k)f(u+\frac{\nu}{2}) - \theta_{5-a}(-2u-\frac{\nu}{2}k)f(u-\frac{\nu}{2})\right).$$

Here $\sqrt{\varepsilon_2}=i$, $\sqrt{\varepsilon_a}=1$ $(a\neq 2)$.

The Casimir elements act on $\mathcal{V}^{(k)}$ as scalars $K_0(k+1)$, $K_2(k+1)$ with

$$K_0(d) = 4\left(\frac{\theta_1(\frac{1}{2}\nu d)}{\theta_1(\nu)}\right)^2, \quad K_2(d) = 4\frac{\theta_1(\frac{1}{2}\nu(d-1))\theta_1(\frac{1}{2}\nu(d+1))}{\theta_1(\nu)^2},$$

5.5.3. Trace. As in the XXZ case, the main tool is the analytical continuation with respect to the dimension of the trace over the finite-dimensional representations $\pi^{(k)}$. So we need a linear functional $\mathrm{Tr}^{\{d\}}$ on $\mathcal{A}$ satisfying the properties

(1) $\mathrm{Tr}^{\{d\}}A\big|_{d=k+1} = \mathrm{tr}_{\mathcal{V}^{(k)}}\pi^{(k)}(A) \quad (k\in\mathbb{Z}_{\geq 0})$,

(2) Cyclicity
$$\mathrm{Tr}^{\{d\}}(AB) = \mathrm{Tr}^{\{d\}}(BA),$$

(3) Linearity over the ring generated by the Casimir elements
$$\mathrm{Tr}^{\{d\}}(K_iA) = K_i(d)\mathrm{Tr}^{\{d\}}(A) \quad (i=0,2),$$

(4) Colourless property
$$\mathrm{Tr}^{\{d\}}A = 0 \qquad (A\in\mathcal{A}^{(m,n)}, (m,n)\neq(0,0)).$$

We would like to understand for how many really different elements of $\mathcal{A}$ we should define the trace. Denote by $\mathcal{K}$ the polynomial ring generated by the central elements K_0, K_1. In order to eliminate the elements obtained by the cyclicity we consider the quotient of $\mathcal{A}$ by

$$\mathcal{A}' = \sum_{a=0}^{3}[S_a,\mathcal{A}].$$

We want to understand the structure of $\mathcal{A}/\mathcal{A}'$ as a $\mathcal{K}$-module. In [**BJMST3**] the following theorem is proved.

THEOREM 5.11. *The quotient space $\mathcal{A}/\mathcal{A}'$ is a free $\mathcal{K}$-module of rank seven. The following can be taken as a basis:*

$$\{1, S_0, S_1, S_2, S_3, S_0^2, S_1^2\}.$$

The rigorous proof of this theorem is rather long, so, we would like to explain briefly only the main ideas. In addition we shall explain the intuition behind the main formulas. The algebras $\mathcal{A}$, $\mathcal{A}'$ and the ring $\mathcal{K}$ are graded, all of them depend on two parameters encoded in constants $J_{a,b}$ (5.150). Usual argument applicable to finite-dimensional and graded spaces is that the space $\mathcal{A}'$ being an image of linear operators $[S_a,\bullet]$ can turn only smaller at a fixed value of parameter. Hence at a fixed value of parameters the quotient space can be only bigger than is generically. So one has to find a convenient fixed value of parameters in order to prove the theorem. We consider the classical limit $\nu\to 0$ without touching the elliptic modulus τ. We

[3] We have modified eq.(6) of [**Sk2**] by a factor $\theta_1(\nu)$.

have $J_a \to 1-\nu^2 j_a$. Suppose that the generators S_a $(a=1,2,3)$ have finite limits s_a while S_0 behaves as νs_0 with s_a $(a=0,1,2,3)$ being commutative. Then it is easy to see on the resulting commutative algebra we have the Poisson structure

$$\{s_0, s_a\} = 2j_{bc}s_b s_c\,, \qquad \{s_b, s_c\} = -2s_0 s_a\,.$$

The Casimir elements give rise to two elements of the centre of the Poisson algebra $k_0 = \lim K_0$ and $k_1 = \lim \frac{K_0-K_2}{\nu^2}$. Explicitly:

$$\sum_{a=1}^{3} s_a^2 = k_0\,, \quad s_0^2 + \sum_{a=1}^{3} j_a s_a^2 = k_1\,. \tag{5.155}$$

Define vector fields

$$D_a f = \{s_a, f\}\,.$$

The surface S defined by (5.155) inherits the de Rham complex from the one for the affine space $\mathbb{C}^4$. Let us define the algebraic de Rham complex. The differential forms are restricted using

$$\sum_{a=1}^{3} s_a ds_a = 0\,, \quad s_0 ds_0 + \sum_{a=1}^{3} j_a s_a ds_a = 0\,.$$

In particular, two-forms can be written as

$$P(s_0, s_1, s_2, s_3)\frac{ds_1\wedge ds_2}{s_0 s_3}\,, \tag{5.156}$$

with $P(s_0,s_1,s_2,s_3)$ being an arbitrary polynomial. These are closed forms for a trivial reason. The exact two-forms are

$$d\Big(\sum_{a=0}^{3} P_a(s_0,s_1,s_2,s_3)ds_a\Big) = \Big(\sum_{a=0}^{3} D_a P_a(s_0,s_1,s_2,s_3)\Big)\frac{ds_1\wedge ds_2}{s_0 s_3}\,. \tag{5.157}$$

From (5.156) and (5.157) we conclude that the problem of computing the quotient $\mathcal{A}/\mathcal{A}'$ is equivalent to the computation of the top de Rham cohomology $H^2(S,\mathrm{alg})$. By algebraic de Rham theorem, this cohomology group is isomorphic to the singular cohomology group which in turn is isomorphic to the homology group

$$H^2(S,\mathrm{alg}) \simeq H^2(S;\mathbb{C}) \simeq H_2(S;\mathbb{C})\,. \tag{5.158}$$

In order to study the topology of S we use the classical limit of (5.154) which describes just a convenient parametrisation

$$s_a(u,v) = c_a \cdot \frac{\theta_{5-a}(2u-\mu)e^{-2\pi i v} - \theta_{5-a}(-2u-\mu)e^{2\pi i v}}{\theta_1(2u)}\,, \tag{5.159}$$

where $\mu = k\nu/2$ is fixed in the limit $\nu\to 0$, $k\to\infty$

$$c_0 = \frac{1}{2}\,; \qquad c_a = \sqrt{\varepsilon_a}\,\frac{\theta_{5-a}}{\theta_1'}\,,\ a=1,2,3\,.$$

In this parametrisation

$$k_0 = 4\left(\frac{\theta_1(\mu)}{\theta_1'}\right)^2, \quad k_1 = -\left(\frac{\theta_1(\mu)}{\theta_1'}\right)^2 \partial^2 \log\theta_1(\mu)\,. \tag{5.160}$$

For definiteness, we consider the case $\tau\in i\mathbb{R}_{>0}$, $0<\mu<1/2$. The functions s_α have common periods

$$e_1 = (1,0), \quad e_2 = (0,1), \quad e_3 = (\tau, 2\mu),$$

so (u,v) should be regarded as variables on $\mathbb{C}^2/(\sum_{i=1}^3 \mathbb{Z}e_i)$. We shall consider the fundamental domain:

$$-\frac{1}{4} \le \mathrm{Re}(u) < \frac{3}{4}, \quad -\frac{\tau}{4i} \le \mathrm{Im}(u) < \frac{3\tau}{4i}, \quad 0 \le \mathrm{Re}(v) < 1. \tag{5.161}$$

The denominators of $s_a(u,v)$ vanish at $u = p_a$,

$$p_0 = 0,\ p_1 = \frac{1}{2},\ p_2 = \frac{1+\tau}{2},\ p_3 = \frac{\tau}{2}, \tag{5.162}$$

the domain (5.161) is such that the points p_a are not on its boundary. There are eight points where these singularities are cancelled by zeros of the numerators:

$$p_0^S = (0,0), \quad p_1^S = (\frac{1}{2},0), \quad p_0^N = (0,\frac{1}{2}), \quad p_1^N = (\frac{1}{2},\frac{1}{2}),$$
$$p_2^S = (\frac{1}{2}(1+\tau),\mu), \quad p_3^S = (\frac{1}{2}\tau,\mu), \quad p_2^N = (\frac{1}{2}(1+\tau),\mu+\frac{1}{2}), \quad p_3^N = (\frac{1}{2}\tau,\mu+\frac{1}{2}).$$

Taking this into account we conclude that the divisor is the open subset of the fundamental domain consisting of two components $\mathcal{D} = \mathcal{D}' \cup \mathcal{D}''$,

$$\begin{aligned} \mathcal{D}' &= \cup_{a=0,1}\{(p_a,v) \mid 0 \le \Re(v) < 1,\ v \ne 0, 1/2\}, \\ \mathcal{D}'' &= \cup_{a=2,3}\{(p_a,v) \mid 0 \le \Re(v) < 1,\ v \ne \mu, \mu+1/2\}. \end{aligned} \tag{5.163}$$

Carefully analysing this geometry we find seven non-trivial two-cycles: four spheres: $\delta_0, \delta_1, \delta_2, \delta_3$ and three tori $\gamma_0, \gamma_1, \gamma_2$. The spheres δ_a are drown on Fig. 15a, their north (south) poles are regular points p_a^N (p_a^S). The tori $\gamma_0, \gamma_1, \gamma_2$ have generators (e_1,e_3), (e_1,e_2), (e_2,e_3). One should take care that the cycle γ_0 does not hit the divisor, for example, on Fig. 15b it passes through the points p_a^S.

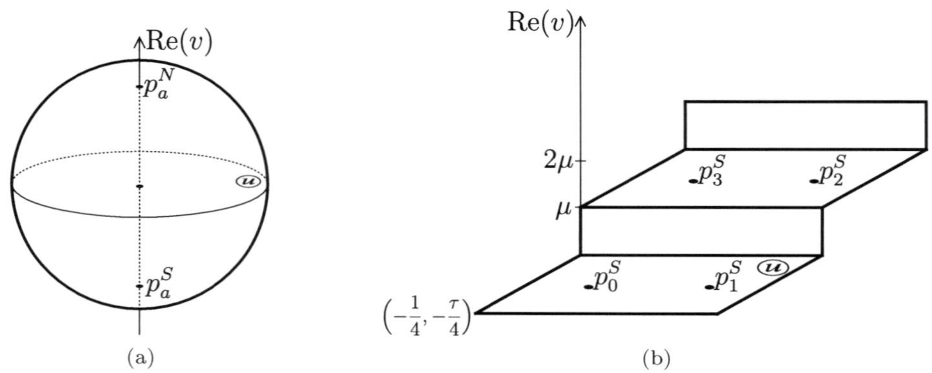

FIG. 15.

Using our seven monomials we construct 2-forms

$$\omega = -\frac{1}{4\pi}\frac{ds_1 \wedge ds_2}{s_0 s_3}, \quad \omega_0 = s_0^2\,\omega, \quad \omega_3 = s_3^2\,\omega,$$
$$\varsigma_a = s_a\,\omega, \quad a = 0,1,2,3.$$

The periods can be evaluated explicitly.

The integrals over δ_a are

$$\int_{\delta_a}\omega = \int_{\delta_a}\omega_b = 0, \quad \int_{\delta_a}\varsigma_b = 2\epsilon_{ab}c_b\frac{\theta_{5-b}(-\mu)}{\theta_1'},$$

where ϵ_{ab} are elements of the matrix:

$$\epsilon = \begin{pmatrix} 1 & 1 & 1 & 1 \\ 1 & 1 & -1 & -1 \\ 1 & -1 & 1 & -1 \\ 1 & -1 & -1 & 1 \end{pmatrix} .$$

The integrals over γ_k $(k = 1, 2)$ are

$$\int_{\gamma_k} \varsigma_a = 0, \quad \int_{\gamma_1} \omega = 1, \quad \int_{\gamma_2} \omega = \tau,$$

$$\int_{\gamma_1} \omega_a = -2\left(\frac{c_a\theta_{5-a}(\mu)}{\theta_1'}\right)^2 \partial^2 \log \theta_{5-a}(\mu),$$

$$\int_{\gamma_2} \omega_a = -2\left(\frac{c_a\theta_{5-a}(\mu)}{\theta_1'}\right)^2 \left(\tau\partial^2 \log \theta_{5-a}(\mu) + 2\pi i\right),$$

These and further formulas hold for all $a = 0, 1, 2, 3$, not only for the linearly independent ω_0, ω_1.

Finally for γ_0 we have

$$\int_{\gamma_0} \varsigma_a = 2\delta_{a,0}\frac{\theta_1(\mu)}{\theta_1'}, \tag{5.164}$$

$$\int_{\gamma_0} \omega = 2\mu, \qquad \int_{\gamma_0} \omega = 2\left(\frac{\theta_{5-a}(\mu)}{\theta_1'}\right)^2 (\partial_\mu - \mu\partial_\mu^2) \log \theta_{5-a}(\mu).$$

Using the above formulas we can calculate the determinant of the period-matrix as

$$\det(\mathcal{P}) = \text{Const} \cdot \theta_1(2\mu)\left(\theta_1(\mu)\theta_0(\mu)\right)^2 \frac{\partial}{\partial\mu} \log\left(\frac{\theta_1(\mu)}{\theta_0(\mu)}\right),$$

whence we conclude that for generic μ the period matrix is non-degenerate.

The classical limit of the trace functional should coincide with integral of some two-cycle. It is easy to see that only the integral over γ_0 satisfies the forth of the properties of trace above (colourless). The classical formulas (5.164) provide an important piece of intuition for guessing exact quantum ones:

$$\mathrm{Tr}^{\{d\}} 1 = d, \tag{5.165}$$

$$\mathrm{Tr}^{\{d\}} S_a = 2\delta_{a,0}\frac{\theta_1(\frac{1}{2}\nu d)}{\theta_1(\nu)},$$

$$\mathrm{Tr}^{\{d\}} S_a^2 = \epsilon_a \frac{2\theta_{5-a}(\frac{\nu}{2})^2}{\theta_1'\theta_1(\nu)}\Big(\partial_\mu - d\cdot\partial_\nu\Big)\left(\theta_{5-a}(\mu + \frac{1}{2}\nu)\theta_{5-a}(\mu - \frac{1}{2}\nu)\right).$$

These formulas can be guessed from the classical expressions (5.164). Once the formulas are guessed one checks them using the fusion relations. Let us elaborate a little on this point. Define the transfer-matrices

$$t_n^{(k)}(\mu) = \mathrm{Tr}^{\{k+1\}}\{R_{\star,1}(\mu - \lambda_1)\cdots R_{\star,n}(\mu - \lambda_n)\}.$$

They can be shown to satisfy the fusion relation

$$t^{(1)}(\mu - \frac{1}{4}k\nu)t^{(k)}(\mu + \frac{1}{4}\nu) = \phi(\mu + \frac{1}{2}\nu)t^{(k+1)}(\mu) + \phi(\mu)t^{(k-1)}(\mu + \frac{1}{2}\nu),$$

where

$$\phi(\mu) = \prod_{j=1}^{n} \frac{\theta_1(2\mu - 2\lambda - \frac{1}{2}k\nu)}{\theta_1(\nu)} .$$

For $n = 1, 2$ the transfer-matrix $t^{(1)}(\mu - \frac{1}{4}k\nu)$ is easy to compute explicitly since $\pi^{(1)}(S_a) = \sigma^a$. So, we have a linear equation for $t^{(l)}$, $l = k-1, k, k+1$ which for $n = 1, 2$ provides a sufficient check for (5.165).

From the exact formulas (5.165) together with the explicit expressions for $K_0(d)$, $K_2(d)$ we come to the important conclusion concerning the analytical structure of the trace. If $A \in \mathcal{A}_n$, its trace has the functional form

$$\mathrm{Tr}^{\{\frac{2\mu}{\nu}\}} A = \theta_1(\mu)^n \times \begin{cases} g_{A,0}(\mu) & (\ n\text{: odd }), \\ g_{A,1}(\mu) - \mu g_{A,2}(\mu) & (\ n\text{: even }), \end{cases} \tag{5.166}$$

where $g_{A,0}(\mu)$, $g_{A,2}(\mu)$ and $g_{A,3}(\mu) := g_{A,1}(\mu + \tau) - g_{A,1}(t)$ are elliptic functions with periods $1, \tau$. In addition, $g_{A,1}(\mu + 1) = g_{A,1}(\mu)$.

5.5.4. Formula for Correlation Functions. In this section we consider the case of homogeneous chain only, generalisation to the inhomogeneous case is straightforward.

The adjoint monodromy-matrix is defined as usual, for $J = [k, l]$

$$\begin{aligned} &\mathbb{T}_{\star,J}(\mu)(X_J) \\ &= R_{\star,l}(\mu) R_{\star,l-1}(\mu) \cdots R_{\star,k}(\mu)\ X_J\ R_{\star,k}(\mu)^{-1} \cdots R_{\star,l-1}(\mu)^{-1} R_{\star,l}(\mu)^{-1} , \end{aligned}$$

Using the adjoint monodromy matrices we construct the operator

$$\begin{aligned} &Y(\mu_1, \mu_2)(X_J) \\ &= \mathrm{Tr}_{a,b}\Big\{ B_{a,b}(\mu_{1,2}) \mathbb{T}_{a,J}(\mu_1) \mathbb{T}_{b,J}(\mu_2) \Big\} \mathrm{Tr}^{\{\frac{2}{\nu}\mu_{1,2}\}} \Big\{ \mathbb{T}_{\star,J}\Big(\frac{\mu_1 + \mu_2}{2}\Big)(X_J) \Big\} , \end{aligned}$$

where

$$B(\mu) = -\frac{1}{4} \frac{\theta_1(2\mu)\theta_1(2\nu)}{\theta_1(2\mu - \nu)\theta_1(2\mu + \nu)} \sum_{a=1}^{3} \frac{\theta_{5-a}(2\mu)}{\theta_{5-a}(\nu)} \sigma^a \otimes \sigma^a . \tag{5.167}$$

Let us discuss the analytical properties of $Y(\mu_1, \mu_2)$. We have the decomposition following from (5.166)

$$Y(\mu_1, \mu_2) = Y_1(\mu_1, \mu_2) - \frac{\mu_{1,2}}{\nu} Y_2(\mu_1, \mu_2) .$$

Define further

$$Y_3(\mu_1, \mu_2) = Y_1(\mu_1 + \tau, \mu_2) - Y_1(\mu_1, \mu_2) .$$

Then

$$\begin{aligned} Y_p(\mu_1 + 1, \mu_2) &= Y_p(\mu_1, \mu_2) , \quad (p = 1, 2, 3) , \\ Y_p(\mu_1 + \tau, \mu_2) &= Y_p(\mu_1, \mu_2) , \quad (p = 2, 3) . \end{aligned}$$

The singular parts (in both variables) of the operators $Y_p(\mu_1, \mu_2)$ satisfy the reduction property usual for annihilation operators, so one can compute the adjoint transfer-matrices on any interval covering the support of the local operator X_J.

As usual, the last ingredient of our construction is the set of functions ω. We consider only the zero-temperature case (infinite homogeneous *Matsubara* direction). In the XXX case we had one function ω, in the XXZ case we had two functions ω_0, ω_1. It should not be very surprising that in the XYZ case we have three functions ω_1, ω_2, ω_3. These functions are constructed form the partition function $\rho(\mu,\nu,\tau)$ which depends on three parameters: spectral parameter μ, coupling constant ν and elliptic modulus τ (temperature for the 8 vertex model):

$$\frac{1}{4\pi i} d\log\varphi = \omega_1 d\mu + \omega_2 d\nu + \omega_3 d\tau\,, \tag{5.168}$$

where

$$\varphi(\mu,\nu,\tau) = \rho(\mu,\nu,\tau)\left(\frac{\theta_1(\nu-2\mu)}{\theta_1(\nu+2\mu)}\right)^{\frac{1}{4}}. \tag{5.169}$$

$$\log\rho(\mu,\nu,\tau) = \frac{1-\nu}{2}a\mu + \sum_{n=1}^{\infty}\sinh(4an\mu)\,\frac{\sinh(an(1-\nu))}{n\cosh(an\nu)\sinh(an)}\,,$$

with $a = 2\pi/|\tau|$. To take the XXZ limit one sets

$$\zeta = e^{2\pi i\mu}\,,$$

and sends $\tau\to+i\infty$ which is the same as $a\to 0$. Obviously (5.169) turns into (5.144).

Conjecturally, the expectation value of a local operator $\mathcal{O}$ is given by the formula

$$\frac{\langle\mathrm{vac}|\mathcal{O}|\mathrm{vac}\rangle}{\langle\mathrm{vac}|\mathrm{vac}\rangle} = \mathrm{Tr}\Big\{e^{\Omega}(\mathcal{O})\Big\}\,,$$

where

$$\Omega = \frac{1}{2\pi i}\oint\oint\sum_{p=1}^{3}\omega_p(\mu_{1,2})Y_p(\mu_1,\mu_2)d\mu_1 d\mu_2\,. \tag{5.170}$$

5.5.5. Discussion. The Sklyanin algebra is of Poincaré-Birkhoff-Witt (PBW) type, which means that as a module over the ring $\mathbb{C}[K_0,K_1]$ it has the basis

$$S_3^m S_2^n S_1^{\epsilon} S_0^{\epsilon'} \qquad m,n\in\mathbb{Z}_{\geq 0}\,,\quad \epsilon,\epsilon'=0,1\,.$$

This is proved in [**FO**, **Od**] in a rather indirect way using the representation (5.154). To the best of our knowledge an algorithmic method of reducing to PWB basis is not known. Similarly, we do not know an algorithmic way of computing traces by virtue of Theorem 5.11. The latter remains an existence theorem, it does not allow a constructive procedure. So, finding an algorithmic procedure looks very important. However, there may be another way.

In Sections of this Chapter the fermionic formulas available for arbitrary *Matsubara* data were used in order to find the coefficients of the decomposition of local operators. Then these coefficients were used for computation of the expectation values for physically relevant *Matsubara* data. Is it possible that a similar procedure is applicable to the eight-vertex case? This would require the formula (5.170) holds for the eight-vertex model on a cylinder with arbitrary *Matsubara* chain if $\omega_p(\mu_{1,2})$ are replaced by appropriate $\omega_p(\mu_1,\mu_2)$ depending on the *Matsubara* data. For the moment we cannot assert this. However, if such a generalised formula exists, it can be used as an existence theorem allowing computation of coefficients, similar to the one discussed above. One important ingredient, which is the formula for scalar product of on-shell and off-shell Bethe vectors, is available now for the XYZ case [**SZZ**]. This is a very tempting possibility which is worth investigation.

APPENDIX A

Quasi-classical Limit and Algebraic Geometry

In this Appendix we explain the relation of our construction to hyper-elliptic Riemann surfaces and their Jacobi varieties. This is a topic closely related to the theory of classical integrable models and their finite-zone solutions. We do not pretend to give an exhaustive list of references on this extensive subject, mentioning only papers which were important for our understanding [**Du**, **IM**, **It**, **DMN**, **FMcL**, **Mu**]. For our goals in this section, it will be convenient to use the genus

$$g = \mathbf{n} - 1$$

in place of the length $\mathbf{n}$ of the *Matsubara* chain.

A.1. Algebraic Interpretation of Quantum Results

In the present volume we considered mostly the homogeneous case, developing operators around the point $\zeta^2 = 1$. In this section it will be convenient to switch to $\zeta^2 = 0$. Recall the *Matsubara* monodromy matrix (2.1). We apply a similarity transformation by ζ^{σ^3} to it and develop its entries in the variable ζ^2:

$$A(\zeta) = \sum_{j=0}^{g+1} \zeta^{2(g+1-j)} A_j\,, \quad D(\zeta) = \sum_{j=0}^{g+1} \zeta^{2(g+1-j)} D_j\,, \tag{A.1}$$

$$B(\zeta) = \sum_{j=0}^{g} \zeta^{2(g-j)} B_j\,, \qquad C(\zeta) = \sum_{j=0}^{g} \zeta^{2(g+1-j)} C_{j+1}\,.$$

Consider the algebra created by A_j, B_j, C_j, D_j. The centre of this algebra is generated by the coefficients F_j of the quantum determinant

$$F(\zeta) = A(\zeta q) D(\zeta) - B(\zeta q) C(\zeta) = \sum_{j=0}^{2(g+1)} \zeta^{2(2g+2-j)} F_j\,.$$

On the *Matsubara* spin chains considered in the present volume, it takes the value

$$F(\zeta) = a(\zeta) d(\zeta q)\,.$$

Let us use the fermionic basis created by the rational operators $\mathbf{b}^*_{\mathrm{rat}}(\zeta)$, $\mathbf{c}^*_{\mathrm{rat}}(\zeta)$. The corresponding function $\omega(\zeta,\xi)$ has the form (4.35) with

$$\omega_{\mathrm{sing}}(\zeta,\xi) = \frac{4}{T(\zeta,\kappa)T(\xi,\kappa)} \Big(a(\zeta) d(\xi) \psi(q\zeta/\xi,\alpha) - a(\xi) d(\zeta) \psi(q^{-1}\zeta/\xi,\alpha) \Big)\,.$$

The function $\omega(\zeta,\xi)$ is uniquely defined, but its splitting into the part containing the determinant and $\omega_{\mathrm{sing}}(\zeta,\xi)$ is not: one can add to $\omega_{\mathrm{sing}}(\zeta,\xi)$ any function of the form $4x(\zeta,\xi|\alpha)/(T(\zeta,\kappa)T(\xi,\kappa))$, where $x(\zeta,\xi|\alpha)$ satisfies

$$x(\zeta,\xi|\alpha) = x(\xi,\zeta|-\alpha)\,,$$

and $x(\zeta,\xi|\alpha)=(\zeta/\xi)^{\alpha}Pol(\zeta^2,\xi^2)$ with $Pol(\zeta^2,\xi^2)$ a polynomial of degree $g+1$ in both variables. This should be compensated by redefinition of $\widetilde{\Omega}_j(z)$, which corresponds to subtracting $x(\zeta,\xi|\alpha)$ from $r^+(\zeta,\xi)$ (4.22). Informally, we change the definition of differentials of the second kind by adding a symmetric linear combination of those of the first kind, without changing the function $r(\zeta,\xi)$.

The coefficients of the function $\omega_{\rm sing}(\zeta,\xi)$ entering the final formula of this subsection depend on those of $a(\zeta)$ and $d(\zeta)$. In what follows it is more logical to have dependence on the elements F_j of the centre. This is achieved by choosing

$$\begin{aligned}x(\zeta,\xi|\alpha)=&-\Big(a(\zeta)d(\xi)\psi(q\zeta/\xi,\alpha)-a(\xi)d(\zeta)\psi(q^{-1}\zeta/\xi,\alpha)\Big)\\&+\Big(H(\zeta,\xi)\psi(q\zeta/\xi,\alpha)-H(\xi,\zeta)\psi(q^{-1}\zeta/\xi,\alpha)\Big)\,,\end{aligned}$$

where the polynomial

$$H(\zeta,\xi)=\sum_{j=0}^{g+1}F_{2j}\zeta^{2j}\xi^{2j}q^{-2j}+\sum_{j=0}^{g}F_{2j+1}\zeta^{2(j+1)}\xi^{2j}q^{-2j}\,,$$

satisfies

$$H(\zeta,\zeta q)=F(\zeta)\,.$$

For simplicity we work only in the spin zero subspace, defining as usual

$$T(\zeta,\kappa)=q^{\kappa}A(\zeta)+q^{-\kappa}D(\zeta)\,.$$

We shall not distinguish notationally $T(\zeta,\kappa)$ and its eigenvalue; the difference should be clear from the context.

Explicitly, the changes are

$$\begin{aligned}&r^{\pm}(\zeta,\xi)=T(\zeta,\kappa)\Delta_{\zeta}^{-1}\left(\psi(\zeta/\xi,\pm\alpha)(T(\zeta,\kappa)-T(\xi,\kappa))\right)\\&+T(\zeta,\kappa+\alpha)\Delta_{\zeta}^{-1}\left(\psi(\zeta/\xi,\pm\alpha)(T(\zeta,\kappa+\alpha)-T(\xi,\kappa+\alpha))\right)\\&-T(\zeta,\kappa)\Delta_{\zeta}^{-1}\left(\psi(q\zeta/\xi,\pm\alpha)(T(\zeta q,\kappa+\alpha)-T(\xi,\kappa+\alpha))\right)\\&-T(\zeta,\kappa+\alpha)\Delta_{\zeta}^{-1}\left(\psi(q^{-1}\zeta/\xi,\pm\alpha)(T(\zeta q^{-1},\kappa)-T(\xi,\kappa))\right)\\&+\big(F(\zeta)-H(\zeta,\xi)\big)\psi(q\zeta/\xi,\pm\alpha)-\big(F(\zeta q^{-1})-H(\xi,\zeta)\big)\psi(q^{-1}\zeta/\xi,\pm\alpha)\,,\end{aligned}$$

and

$$\omega_{\rm sing}(\zeta,\xi|\alpha)=\frac{4}{T(\zeta,\kappa)T(\xi,\kappa)}\Big(H(\zeta,\xi)\psi(q\zeta/\xi,\alpha)-H(\xi,\zeta)\psi(q^{-1}\zeta/\xi,\alpha)\Big)\,.$$

Now everything is written in terms of $T(\zeta,\kappa)$, $T(\zeta,\kappa+\alpha)$ and $F(\zeta)$.

We consider a generalised *Matsubara* chain with twists κ and $\kappa+\alpha$. The main theorem and completeness in the inhomogeneous case imply that for any product $T_{i_1,j_1}(\xi_1)\cdots T_{i_k,j_k}(\xi_k)$ of elements of the *Matsubara* monodromy matrix (where $T_{1,1}(\xi)=A(\xi)$ etc.) we have

(A.2)

$$\begin{aligned}&\frac{\langle\kappa+\alpha|T_{i_1,j_1}(\xi_1)\cdots T_{i_k,j_k}(\xi_k)|\kappa\rangle}{\langle\kappa+\alpha|\kappa\rangle}\\&=\sum_{\substack{\{1,\cdots,k\}\\=I_0\sqcup I_{\bar 0}\sqcup I_+\sqcup I_-}}c_{I_0,I_{\bar 0},I_+,I_-}(\xi_1,\cdots\xi_k)\prod_{j\in I_0}T(\xi_j,\kappa+\alpha)\det\left(\omega(\xi_i,\xi_j|\alpha)\right)_{\substack{i\in I_+\\ j\in I_-}}\prod_{j\in I_{\bar 0}}T(\xi_j,\kappa)\,,\end{aligned}$$

with certain coefficients $c_{I_0,I_{\bar 0},I_+,I_-}(\xi_1,\cdots\xi_k)$ which are independent of the *Matsubara* data. These coefficients may have singularities: practice teaches that they are located only on the diagonals $\xi_i^2 = \xi_j^2$. In addition, in the right hand side there are singularities at $\xi_i^2 = q^{\pm 2}\xi_j^2$ present in $\omega_{\text{sing}}(\xi_i, \xi_j)$. All these singularities must cancel since the left hand side is a polynomial in $\xi_1^2, \cdots, \xi_k^2$.

We have the following simple identity from linear algebra. Quite generally, let A, B be $m\times m$ matrices. For two subsets I, J of $\{1, \cdots, m\}$ of the same cardinality, denote by $A^{(I,J)}$ the matrix obtained from A by replacing the i-th row by the j-th row of B for all $i \in I$, $j \in J$. Considering the case $I = \{i\}$ and $J = \{j\}$ we set $\Delta_{i,j} = \det A^{(\{i\},\{j\})}$. Then we have

$$\det\left(\Delta_{i,j}\right)_{i\in I, j\in J} = (\det A)^{\# I-1} \det A^{(I,J)} .$$

Denote by $T_{i,j;m}$ the coefficients in (A.1). From the above identity, it is easy to conclude that

$$\begin{aligned}(\text{A.3})\qquad &\frac{\langle \kappa+\alpha | T_{i_1,j_1;m_1} \cdots T_{i_k,j_k;m_k} | \kappa \rangle}{\langle \kappa + \alpha | \kappa \rangle} \\ &= \sum_{\substack{\#(K)=\#(\widetilde{K}) \\ K,\widetilde{K}\subset\in\{1,\cdots g\}}} P_{\text{left}}^{(K,\widetilde{K})}(\{T_p(\kappa+\alpha)\}_{p=0}^{g+1})\, \mathcal{D}_{K,\widetilde{K}}\; P_{\text{right}}^{(K,\widetilde{K})}(\{T_p(\kappa)\}_{p=0}^{g+1}) ,\end{aligned}$$

where $P_{\text{left}}^{(K,\widetilde{K})}$ (resp. $P_{\text{right}}^{(K,\widetilde{K})}$) is a polynomial of $T_p(\kappa+\alpha)$ (resp. $T_p(\kappa)$) with coefficients depending on the central elements F_j, $\mathcal{D}_{K,\widetilde{K}}$ is the determinant of a $g\times g$ matrix consisting of periods over $\Gamma_1, \ldots, \Gamma_g$ of Ω_i ($i \in \{1, \cdots, g\}\backslash K$) and $\widetilde{\Omega}_j$ ($j \in \widetilde{K}$). The original determinants are of the size $(g+2)\times(g+2)$, however, they vanish unless the differentials Ω_0, Ω_{g+1} are present because

$$\int_{\Gamma_0} \Omega_j = 2\pi i \delta_{j,0} , \qquad \int_{\Gamma_\infty} \Omega_j = 2\pi i \delta_{j,g+1} .$$

Due to these identitites the size of the determinant reduces automatically.

The determinant $\mathcal{D}_{K,\widetilde{K}}$ can be interpreted as follows. Define the fermions $\mathbf{b}_k^*$, $\mathbf{c}_k^*$ as Taylor coefficients of $\mathbf{b}^*(\zeta)$, $\mathbf{c}^*(\zeta)$ at $\zeta^2 = 0$ (instead of $\zeta^2 = 1$ considered previously), then the determinant gives the expectation value of $\mathbf{b}^*_{\widetilde{K}}\mathbf{c}^*_K$ applied to the primary field.

In the rest of this section we shall explain the meaning of the decomposition (A.3) in the quasi-classical limit.

A.2. Canonical Differential in the Classical Case

There are several possible ways for taking the quasi-classical limit. Our goal is to illustrate the fermionic construction by algebraic geometry, so we consider the one which is most convenient for this goal:

$$\begin{aligned}(\text{A.4})\qquad &\nu \to 0 , \quad s_m \to \infty , \quad \kappa \to \infty , \\ &\sigma_m = s_m \nu , \ \varkappa = \kappa\nu \quad \text{are finite} .\end{aligned}$$

In this limit q^κ and $a(\zeta), d(\zeta)$ stay finite, and we use the same letters to denote their classical limit.

For the transfer-matrix we assume

$$(\text{A.5})\qquad T(\zeta,\kappa) \simeq T_{\text{cl}}(\zeta,\varkappa) + O(\nu) .$$

Then from the Baxter equation

$$a(\zeta)Q^\pm(\zeta q, \kappa) + d(\zeta)Q^\pm(\zeta q^{-1}, \kappa) = T(\zeta,\kappa)Q^\pm(\zeta,\kappa) ,$$

we conclude that in the quasi-classical limit the functions $Q^{\pm}(\zeta,\kappa)$ are given by different branches of the same quasi-classical expression

$$Q(\zeta,\kappa)=S(\zeta,\varkappa)\exp\Big\{\frac{1}{2\pi i\nu}\int_{\zeta_0^2}^{\zeta^2}\log\eta(\xi,\varkappa)\frac{d\xi^2}{\xi^2}\Big\}, \tag{A.6}$$

where $S(\zeta,\varkappa)$ is a power series in ν,

$$\eta(\zeta,\varkappa)=\frac{T_{\rm cl}(\zeta,\varkappa)+\sqrt{T_{\rm cl}(\zeta,\varkappa)^2-4a(\zeta)d(\zeta)}}{2a(\zeta)},$$

and it is convenient to choose ζ_0^2 as one of branch points of the square root. The functions $Q^{\pm}(\zeta,\kappa)$ correspond respectively to the branches of the square root such that $\sqrt{(q^{\kappa}-q^{-\kappa})^2}=\pm(q^{\kappa}-q^{-\kappa})$. It will be technically simpler to take $s_m\in\mathbb{Z}+1/2$. Let us require that the inhomogeneities τ_m are sufficiently separated. We assume that for the ground state the roots of $Q^{+}(\zeta,\kappa)$ in the ζ^2 plane come in $g+1$ groups of size $s_m+1/2$ each, and that the m-th group of roots is not far from the zeros of $\varphi_{s_m}(\zeta/\tau_m)$. Similarly for the roots of $Q^{-}(\zeta,\kappa+\alpha)$. Altogether this means that the contours Γ_m in the ζ^2-plane can be drawn in such a way that inside each of them the measure of integration

$$d\mu^{+}(\zeta)=Q^{-}(\zeta,\kappa+\alpha)Q^{+}(\zeta,\kappa)\varphi(\zeta)\frac{d\zeta^2}{\zeta^2}$$

has exactly $2s_m+1$ poles and $2s_m+1$ zeros. In the quasi-classical limit, the contours Γ_m remain well separated, and the poles and the zeros turn into cuts c_m which the contours Γ_m encircle. We shall need the quasi-classical limit of the ratio

$$h(\zeta)=\frac{Q(\zeta,\kappa+\alpha)}{Q(\zeta,\kappa)}.$$

First, notice that

$$T(\zeta,\kappa+\alpha)\simeq T(\zeta,\kappa)+\nu\alpha\,\partial_{\varkappa}T_{\rm cl}(\zeta,\varkappa)+O(\nu^2),$$

which implies

$$\log\eta(\zeta,\varkappa+\nu\alpha)-\log\eta(\zeta,\varkappa)=\nu\alpha\frac{\partial_{\varkappa}T_{\rm cl}(\zeta,\varkappa)}{\sqrt{T_{\rm cl}(\zeta,\varkappa)^2-4a(\zeta)d(\zeta)}}+O(\nu^2). \tag{A.7}$$

In order to keep close to the usual notation, we shall gradually switch to the parameter $z=\zeta^2$. Also from now on, we shall suppress the dependence on $\varkappa$ as much as possible. The discriminant

$$P(z)=T_{\rm cl}(\zeta)^2-4a(\zeta)d(\zeta)$$

is a polynomial in z.

Obviously

$$\partial_{\varkappa}T_{\rm cl}(\zeta)=U(z) \tag{A.8}$$

is a polynomial of z. From (A.7) we find

$$h(\zeta)=\exp\Big\{\frac{\alpha}{2}\int_{\zeta_0^2}^{\zeta^2}\frac{U(z)}{\sqrt{P(z)}}\frac{dz}{z}\Big\}. \tag{A.9}$$

The differential

$$\varrho=\frac{U(z)}{\sqrt{P(z)}}\frac{dz}{z} \tag{A.10}$$

should satisfy the following requirements.

- ϱ has simple poles at $0, \infty$ with residues

$$\mathrm{res}_0\ \varrho = 1\,, \quad \mathrm{res}_\infty\ \varrho = -1\,. \tag{A.11}$$

- We have

$$\int_{\Gamma_m} \varrho = 0, \quad m = 1, \dots, g+1. \tag{A.12}$$

Condition (A.12) ensures that $h(\zeta)$ is single-valued on the first sheet of the Riemann surface for $\sqrt{P(z)}$ (see Section A.3 below). The integrals in (A.12) sum up to integrals around 0 and ∞, which vanishes by (A.11). Hence it is sufficient to take $m = 1, \dots, g$. The requirements (A.11), (A.12) completely define the differential ϱ. So $U(z)$ is a uniquely defined polynomial with coefficients depending on the coefficients of the polynomial $T_{\mathrm{cl}}(\zeta)$, and (A.8) is a differential equation on these latter.

We shall perform one more quasi-classical computation using the quantum Wronskian

$$\begin{aligned}(q^{-\kappa} - q^{\kappa})W(\zeta) &= Q^+(\zeta,\kappa)Q^-(\zeta q,\kappa) - Q^-(\zeta,\kappa)Q^+(\zeta q,\kappa)\\ &\underset{\nu\to 0}{\to} (\eta^-(\zeta,\varkappa) - \eta^+(\zeta,\varkappa))Q^+(\zeta,\kappa)Q^-(\zeta,\kappa)\,.\end{aligned}$$

This implies

$$Q^+(\zeta,\kappa)Q^-(\zeta,\kappa)\varphi(\zeta) \underset{\nu\to 0}{\to} \frac{q^{\kappa} - q^{-\kappa}}{\sqrt{P(z)}}\,, \tag{A.13}$$

where we used the identity $W(\zeta)a(\zeta)\varphi(\zeta) = 1$.

Altogether, we have the quasi-classical limit of the measures $d\mu^\pm(\zeta)$ (4.14)

$$d\mu^\pm(\zeta) \underset{\nu\to 0}{\to} 2i\sin\pi\varkappa\, dm^\pm(z)\,,$$

where

$$dm^\pm(z) = \exp\Bigl\{\mp\frac{\alpha}{2}\int_{z_0}^{z} \varrho(z)\Bigr\}\frac{1}{\sqrt{P(z)}}\frac{dz}{z}\,. \tag{A.14}$$

Let us consider the quasi-classical limit of important objects considered in Section 4.2. The exact forms are

$$\begin{aligned}&E_z\bigl(g^\pm(z)\bigr)\\ &= \Bigl\{-\sqrt{P(z)}z\frac{d}{dz}\bigl(\sqrt{P(z)}g^\pm(z)\bigr) + \frac{\alpha^2}{4}U(z)d^{-1}\Bigl(U(z)g^\pm(z)\frac{dz}{z}\Bigr)\Bigr\}dm^\pm(z)\end{aligned} \tag{A.15}$$

where $g^\pm(z)$ are $z^{\pm\alpha/2}$ times a polynomial. For the sake of definiteness we fix the primitive d^{-1} by

$$d^{-1}\bigl(z^{\alpha/2+m}\frac{dz}{z}\bigr) = \frac{2}{\alpha+2m}z^{\alpha/2+m}\,.$$

As before we shall need the exact form for $g^\pm(z) = \chi(z/x, \pm\alpha)$ where $\chi(z,\alpha)$ stands for $\psi(\zeta,\alpha)$ written in the variable z,

$$\chi(z,\alpha) = z^{\frac{\alpha}{2}}\frac{z+1}{2(z-1)}\,.$$

In that case the exact form contains a transcendental piece. We find it convenient to separate it from the rest:

$$
\begin{aligned}
\text{(A.16)} \qquad E_z\big(\chi(z/x,\pm\alpha)\big) = \Big\{ & -\sqrt{P(z)}z\frac{d}{dz}\big(\sqrt{P(z)}\chi(z/x,\pm\alpha)\big) \\
& + \frac{\alpha^2}{4}U(z)d^{-1}\Big((U(z)-U(x))\chi(z/x,\pm\alpha)\frac{dz}{z}\Big) \\
& + \frac{\alpha^2}{4}U(z)U(x)X(z/x,\pm\alpha)\Big\}dm^{\pm}(z),
\end{aligned}
$$

where $X(u,\alpha)$ is defined by

$$d(X(u,\alpha)) = \chi(u,\pm\alpha)\frac{du}{u}, \qquad X(u,\alpha) = X(1/u,-\alpha).$$

It is easy to write a Mellin transform for this function which is the quasi-classical limit of (3.72).

Define $r(z,x)$ taking the difference (see (4.24))

(A.17)

$$r(z,x)dm^+(z)dm^-(x) = E_z\big(\chi(z/x,\alpha)\big)dm^-(x) - dm^+(z)E_z\big(\chi(x/z,-\alpha)\big).$$

The function $r(z,x)$ is of the form $r(z,x) = (z/x)^{\alpha/2}Pol(z,x)$. Similarly to Lemma 4.10 we have for $i,j = 0,\cdots,g+1$

$$\text{(A.18)} \qquad \int_{\Gamma_i}\int_{\Gamma_j} r(z,x)dm^+(z)dm^-(x) = 0.$$

Define $\tilde\sigma_i^{\pm}(z)$ by

$$\text{(A.19)} \qquad r(z,x)dm(z)dm(x) = \sum_{i=0}^{g+1}\Big(\sigma_i^+(z)\tilde\sigma_i^-(x) - \tilde\sigma_i^+(z)\sigma_i^-(x)\Big),$$

where

$$\text{(A.20)} \qquad \sigma_j^{\pm}(z) = z^{j\pm\alpha/2}dm^{\pm}(z).$$

The canonical differential is defined like in (4.33) (we combine the differentials $\sigma_i^{\pm}$ and $\tilde\sigma_i^{\pm}$ into vectors $\sigma^{\pm}$ and $\tilde\sigma^{\pm}$):

$$\text{(A.21)} \quad \Omega(z,x|\alpha) = E_z(\chi(z/x,\alpha))dm^-(x) - \tilde\sigma^+(z)^t\sigma^-(x) + \sigma^+(z)^t\,\mathcal{X}^+\,\sigma^-(x),$$

where

$$\text{(A.22)} \qquad \mathcal{X}^+ = (\mathcal{A}^+)^{-1}\mathcal{B}^+, \qquad \mathcal{A}_{ij}^+ = \int_{\Gamma_i}\sigma_j^+, \quad \mathcal{B}_{ij}^+ = \int_{\Gamma_i}\tilde\sigma_j^+.$$

By definition the integrals with respect to the first argument over the contours Γ_j vanish. In addition, we have the symmetry following from (A.18):

$$\Omega(z,x|\alpha) = \Omega(x,z|-\alpha).$$

A.3. Riemann Surfaces

Let us take a more systematic approach having in mind that the quasi-classical limit is related to a hyper-elliptic Riemann surface, denoted hereafter by S. Introduce $w = 2a(\zeta)\eta(\zeta) - T_{\mathrm{cl}}(\zeta)$. This brings the Riemann surface to the canonical form

$$S: \quad w^2 = P(z)\,. \tag{A.23}$$

The genus of S equals g. For a-cycles we choose the cycles $\Gamma_1, \ldots, \Gamma_g$. The surface S is realised as two sheets with cuts which are glued together in the usual way. We denote the points on S by $p = (z, w)$. Topologically it looks as follows:

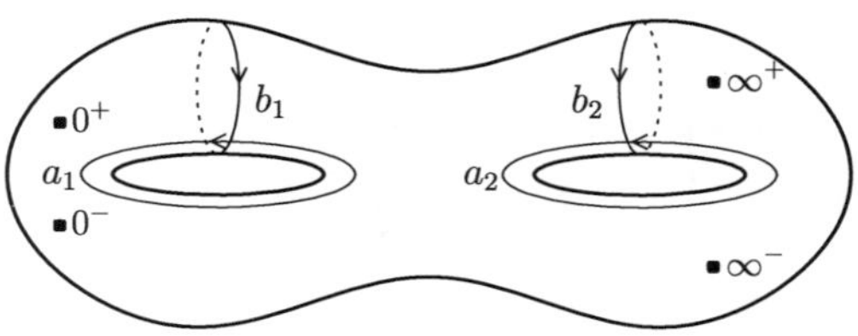

FIG. 16.

The marked points are identified below. We use ς to denote the hyper-elliptic involution which interchanges two sheets. The points with $z = \infty$, 0 lying on the first sheet are denoted by ∞^+, 0^+. Then corresponding points on the second sheet are $\infty^- = \varsigma(\infty^+)$, $0^- = \varsigma(0^+)$.

The b-cycles can be chosen as follows. The cycle b_i starts from the left bank of Γ_i, goes to the left bank of Γ_{g+1} by one sheet of S and returns to the starting point by the other sheet as is indicated on Fig. 17.

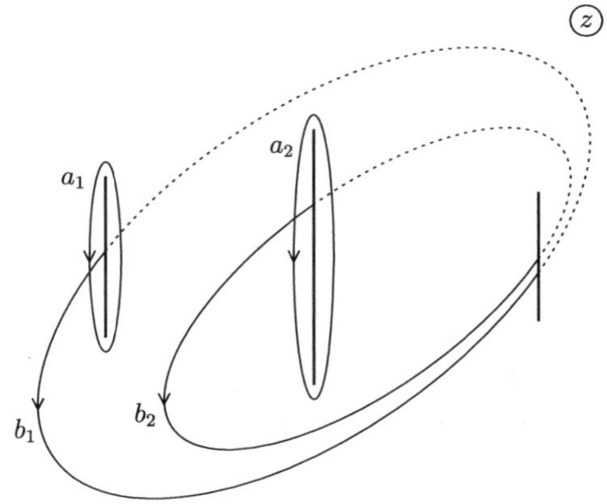

FIG. 17.

Thus we obtain a canonical basis of the first homology

$$a_i \circ b_j = -b_j \circ a_i = \delta_{i,j}\,,$$

where $\circ$ stands for the intersection number.

We shall need some information concerning differential forms on S. First let us consider differentials without residues (of the first and the second kind). Modulo exact forms there are $2g$ of them: g of the first kind and g of the second kind. Introduce the pairing

$$\sigma_1 \circ \sigma_2 = \sum_{\text{poles}} \operatorname{res}(\sigma_1 d^{-1} \sigma_2)\,.$$

The primitive function $d^{-1}\sigma_2$ is well defined on the polygon obtained by cutting the surface along a- and b-cycles.

Consider the two form on $S \times S$

$$\eta(p_1, p_2) = d\Big(\frac{w_1}{(z_1 - z_2)w_2} dz_2 + \frac{w_2}{(z_2 - z_1)w_1} dz_1\Big)\,.$$

This is an exact form, but it is singular, so computing integrals one has to take care of contact terms. This is straightforward, and the result of integration is

$$\int_{\gamma_1} \int_{\gamma_2} \eta = 2\pi i \; \gamma_1 \circ \gamma_2\,. \tag{A.24}$$

In particular,

$$\int_{a_i} \int_{a_j} \eta = 0\,. \tag{A.25}$$

It is easy to see that this identity coincides with (A.18) in the case $\alpha = 0$. In the general case $\alpha \neq 0$ the measures $dm^{\pm}(z)$ are not single-valued on S, and we cannot integrate over the b-cycles. So, the only thing available are the relations (A.18) similar to (A.25), which we would call one quarter of the Riemann bilinear relations. Fortunately, as we have seen, they are sufficient to define the normalised second kind differential $\Omega(z, x|\alpha)$ and to prove its symmetry property.

The form $\eta(p_1, p_2)$ is singular only at $z_{1,2} = \infty$. It can be decomposed as

$$\eta(p_1, p_2) = \sum_{i=1}^{g} \big(\eta_i(p_1)\tilde{\eta}_i(p_2) - \tilde{\eta}_i(p_1)\eta_i(p_2)\big)\,,$$

and the differentials η_i, $\tilde{\eta}_j$ constitute a canonical basis

$$\eta_i \circ \eta_j = \tilde{\eta}_i \circ \tilde{\eta}_j = 0, \qquad \eta_i \circ \tilde{\eta}_j = \delta_{i,j}\,.$$

Inverting symplectic matrix we find from (A.18) a more familiar form of the Riemann bilinear relations

$$\sum_{i=1}^{g} \Big(\int_{a_i} \sigma_1 \int_{b_i} \sigma_2 - \int_{a_i} \sigma_2 \int_{b_i} \sigma_1\Big) = \sigma_1 \circ \sigma_2\,. \tag{A.26}$$

In what follows we shall need several simple facts and definitions.

- The normalised holomorphic differentials $\omega_i(p)$, $i = 1, \cdots, g$, are linear combinations of $z^{j-1}dz/w$, $j = 1, \cdots, g$, satisfying

$$\int_{a_i} \omega_j = \delta_{ij}\,. \tag{A.27}$$

It is always possible to find such differentials, since by a well-known theorem the matrix of a-periods of holomorphic differentials is non-degenerate. The differentials $\omega_i(p)$ will be combined into the vector ω.

- The canonical second kind differential $\omega(p_1,p_2)$ is a differential form on $S\times S$ whose only singularity is situated on the diagonal

(A.28) $$\omega(p_1,p_2)\underset{p_1\sim p_2}{=}\frac{dz_1dz_2}{(z_1-z_2)^2}+O(1)\,.$$

  This differential is normalised:

$$\int_{a_i}\omega(\cdot,p_2)=0\,,\quad\forall p_2\,,$$

  where the integral is taken with respect to the first argument. Due to the Riemann bilinear relation $\omega(p_1,p_2)$ is symmetric, and its b-periods are expressed in terms of the values of the normalised holomorphic differentials

$$\int_{b_i}\omega(\cdot,p_2)=\omega_i(p_2)\,.$$

- The normalised third kind differential $\omega_{p_1,p_2}(p)$ has simple poles at p_1,p_2 and zero a-periods:

(A.29) $$\underset{p=p_1}{\mathrm{res}}\,\omega_{p_1,p_2}(p)=-1\,,\qquad\underset{p=p_2}{\mathrm{res}}\,\omega_{p_1,p_2}(p)=1\,,\qquad\int_{a_i}\omega_{p_1,p_2}=0\,,\ \forall i\,.$$

  Its b-periods are important for our goals,

(A.30) $$\int_{b_i}\omega_{p_1,p_2}=\int_{p_1}^{p_2}\omega_i\,.$$

  The Riemann bilinear identity is applicable to this case because the primitive function of ω_i is well defined on the polygon.

It is easy to see that

$$\omega(p_1,p_2)=\Omega(z_1,z_2|0)\,,$$

for both $p_1=(z_1,w_1)$, $p_2=(z_2,w_2)$ lying on the first sheet of S. This identification explains the terminology used in Chapter 4.

Let us return to the case of generic α. The integral $f(p)=\int^p\varrho$ does not define a single-valued function on S, but rather a section of a bundle. Notice that

$$\varrho=\omega_{\infty^+,\infty^-}+\omega_{0^-,0^+}\,.$$

Up to now we considered the quasi-classical limit of matrix elements taken between two ground states with different twists. Our quantum formulae work for the exited states as well. Let us consider the twists κ, $\kappa+\alpha$ as before. The different states with twist κ are counted by integers $N_1,\dots,N_{g+1}$ which are the numbers of zeros of $Q^+(\zeta,\kappa)$ in the vicinity of poles of $\varphi_{s_1}(\zeta),\dots,\varphi_{s_{g+1}}(\zeta)$. The quasi-classical meaning of the integers is clear: we have $g+1$ actions J_m which satisfy the Bohr-Sommerfeld quantisation condition:

$$J_m=2\pi\nu\left(N_m+1/2\right)\,.$$

The sum of these actions corresponds to the total spin of the state

$$S=\sum_{m=1}^{g+1}\left(N_m-s_m-1/2\right)\,.$$

Similarly, we have the numbers $N_1',\cdots N_{g+1}'$ of zeros of $Q^-(\zeta,\kappa+\alpha)$. Introduce

$$k_m=N_m'+N_m-(2s_m+1)\,,$$

which are finite in the quasi-classical limit. The differential $\Omega(z,x|\alpha)$ does not vanish only if the total spins are balanced:

$$\sum_{m=1}^{g+1} k_m = 0\,.$$

We shall denote $k=(k_1,\cdots,k_g)$. In order to describe the quasi-classical limit we use the normalised differentials $\omega_i(p)$. It is easy to conclude that the formulae for $E_z(\chi(z/x,\alpha))$, $\mathcal{A}^+$, $\mathcal{B}^+$, $\omega(z,x)$, etc. remain the same if we replace everywhere

$$\alpha\varrho\to\alpha\varrho+4\pi i(k,\omega)\,, \tag{A.31}$$

Correspondingly $U(z)$ changes according to (A.10). For future use let us write explicitly the measure in this case:

$$dm^{\pm}(p)=\exp\Bigl\{-\frac{\alpha}{2}\int_{p_0}^{p}\varrho-2\pi i\int_{p_0}^{p}(k,\omega)\Bigr\}\frac{dz}{zw}\,. \tag{A.32}$$

A.4. Affine Jacobi Variety

Following the conventions above, introduce the classical monodromy matrix

$$t(z)=\begin{pmatrix} a(z) & b(z)\\ c(z) & d(z)\end{pmatrix}, \tag{A.33}$$

where

$$a(z)=\sum_{m=0}^{g+1} a_m z^{g+1-m}\,,\qquad d(z)=\sum_{m=0}^{g+1} d_m z^{g+1-m}\,,$$

$$b(z)=\sum_{m=0}^{g} b_m z^{g-m}\,,\qquad c(z)=\sum_{m=0}^{g} c_{m+1} z^{g+1-m}\,.$$

The dependence on $\varkappa$ is implicit: $\log(a_0/d_0)=2\pi i\varkappa$. The geometry of the model is non-trivial due to the constraint

$$a(z)d(z)-b(z)c(z)=f(z)\,, \tag{A.34}$$

where the polynomial in the right hand side is equal to the product $f(\zeta^2)=a(\zeta)d(\zeta)$. Geometry of the intersection of $(2g+3)$ quadrics (A.34) is well studied. Let us sketch the most important results in the spirit of [**Mu**].

The quasi-classical limit of the R-matrix relations are [**Sk3**]

$$\{t(z_1)\overset{\otimes}{,}t(z_2)\}=[r(z_1/z_2),t(z_1)\otimes t(z_2)]\,, \tag{A.35}$$

where

$$r(z)=\frac{1}{z-1}\Bigl((1+z)(\tau_+\otimes\tau^-+\tau_-\otimes\tau^+)+2(\sigma^+\otimes\sigma^-+z\,\sigma^-\otimes\sigma^+)\Bigr)\,.$$

Certainly, the determinant (A.34) is in the centre of the Poisson algebra A defined by (A.35). The integrals of motion are generated by the trace $a(z)+d(z)$, they are in involution. We introduce a convenient parametrisation corresponding, from physicist's point of view, to separation of variables [**FMcL**]:

$$b(z)=b_0\prod_{j=1}^{g}(z-z_j)\,,\quad w_j=a(z_j)-d(z_j)\,. \tag{A.36}$$

We have

$$\{a(z_i), z_j\} = 2\delta_{i,j} z_i a(z_i), \qquad \{d(z_i), z_j\} = -2\delta_{i,j} z_i d(z_i) .$$

In addition

$$a(z_j)d(z_j) = f(z_j) ,$$

which implies that the points (z_j, w_j) belong to the curve S: $w_j^2 = P(z_j)$. So, we can write $a(z_i) = \alpha(z_j)y_j$, $a(z_i) = \delta(z_j)y_j$, where $\alpha(z)\delta(z) = f(z)$ and $\log y_j$, $\log z_j$ are canonically conjugate variables. These variables are not sufficient to parametrise the phase space, since we have the variable $\varkappa$. Let us introduce the canonically conjugate variable ϖ:

$$\{\varpi, \varkappa\} = 1 . \tag{A.37}$$

The functions b_j/b_0 have singularities when one of z_j's equals infinity. The trace $a(z) + d(z)$ is fixed while the difference $a(z) - d(z)$ can be presented as

$$\begin{aligned} &a(x) - d(x) \\ &= x\sum_{j=1}^{g} \frac{w_j}{z_j} \prod_{p\neq j} \frac{x - z_p}{z_j - z_p} + b(x)\Big(\frac{a_0 - d_0}{b_0}x + \frac{a_{g+1} - d_{g+1}}{b_g}\Big) . \end{aligned} \tag{A.38}$$

The coefficients a_0, d_0, a_{g+1}, d_{g+1} are fixed since the trace and the determinant are. Finally, $c(x)$ is found from

$$c(x)b_0 = \big(a(x)d(x) - f(x)\big)\frac{b_0}{b(x)} . \tag{A.39}$$

The numerator in the right hand side is by definition divisible by $b(x)/b_0$, hence the ratio equals the polynomial part of the Laurent series at $x = \infty$ of the right hand side of (A.39) which is a polynomial in a_j, d_j, b_j/b_0. Consider the the symmetric product of S, the quotient of the product by the action of the symmetric group:

$$S(k) = S^k/\mathfrak{S}_k . \tag{A.40}$$

The formulae (A.36), (A.38), (A.39) provide a parametrisation of

$$b(x)/b_0 , \quad a(x) - d(x) , \quad c(x)b_0$$

in terms of $S(g)$. There are singularities situated at

$$D = D_0 \cup D_\infty \cup D_\varsigma ,$$

where

$$\begin{aligned} D_0 &= \{(p_1, \cdots, p_g) \in S(g) |\ p_i = 0^\pm \text{ for some } i\}, \\ D_\infty &= \{(p_1, \cdots, p_g) \in S(g) |\ p_i = \infty^\pm \text{ for some } i\}, \\ D_\varsigma &= \ \{(p_1, \cdots, p_g) \in S(g) |\ p_i = \varsigma(p_j) \text{ for some } i \neq j\ \}. \end{aligned} \tag{A.41}$$

According to (A.39), the classical analogue of the quantum subalgebra of spin zero operators is the algebra A_0 created by b_j/b_0, $j = 1, \cdots, g$, and a_j, d_j, $j = 0, \cdots, g+1$.

For the separated variables we have the equations of motion:

$$\{a(x) + d(x), z_j\} = 2xw_j \prod_{p\neq j} \frac{x - z_p}{z_j - z_p} .$$

To our knowledge, this kind of equations appeared for the first time in the work of Kovalevskaya [**Kv**]. In the modern time they were discussed in wider context by

Dubrovin [**Du**]. These equations imply that the dynamics is linearised by the Abel transformation:

$$\varphi = \sum_{k=1}^{g} A(p_k), \qquad A(p) = \int_{p_0}^{p} \omega. \tag{A.42}$$

It is convenient to choose the initial point p_0 as one of the branch points.

We compute

$$\frac{1}{2x} \cdot \{\mathcal{a}(x) + \mathcal{d}(x), \varphi\} = \omega(x)/(\frac{dx}{\sqrt{P(x)}}). \tag{A.43}$$

The right hand side is a polynomial of degree $g-1$.

The period matrix

$$B_{ij} = \int_{b_i} \omega_j$$

defines a point B in the Siegel upper half space:

$$B^t = B, \qquad \operatorname{Im}(B) > 0.$$

The Jacobi variety of S is a g-dimensional complex torus:

$$\mathcal{J} = \frac{\mathbb{C}^g}{\mathbb{Z}^g + B\mathbb{Z}^g}.$$

The Riemann theta function associated with $\mathcal{J}$ is defined by

$$\theta(\eta) = \sum_{m \in \mathbb{Z}^g} \exp 2\pi i \left(\frac{1}{2}\, {}^t m B m + {}^t m \eta \right),$$

where $\eta \in \mathbb{C}^g$. The theta function satisfies

$$\theta(\eta + m + Bn) = \exp 2\pi i \left(-\frac{1}{2}\, {}^t n B n - {}^t n \eta \right) \theta(\eta), \tag{A.44}$$

for $m, n \in \mathbb{Z}^g$.

The divisor Θ is the $(g-1)$-dimensional subvariety of $\mathbb{J}$ defined by

$$\Theta = \{\eta \mid \theta(\eta) = 0\}. \tag{A.45}$$

Its points are given by Abel transformation to the $(g-1)$-divisors $\{p_1, \cdots, p_{g-1}\}$,

$$\eta = \sum_{j=1}^{g-1} \int_{p_0}^{p_j} \omega + \Delta, \tag{A.46}$$

where Δ is the Riemann characteristic corresponding to the choice of p_0 for the reference point.

For the sake of completeness of our exposition, let us express the observables in terms of the Riemann theta function. We begin with the identity

$$\sum_{j=1}^{g} \int_{p_0}^{p_j} \omega_{q_1,q_2} = \log \frac{\theta\Big(\sum_{j=1}^{g} A(p_j) - A(q_2) + \Delta\Big)}{\theta\Big(\sum_{j=1}^{g} A(p_j) - A(q_1) + \Delta\Big)}. \tag{A.47}$$

The theta functions in the right hand side have zeros at $p_j = \varsigma(p_j)$ which cancel between the numerator and the denominator, in addition the numerator (denominator) has a zero at $p_j = q_2$ ($p_j = q_1$). So, the logarithmic contributions to the left hand side and the right hand side coincide. When p_j goes around a b-cycle, both sides acquire the same addition due to (A.30) and (A.44)

The identity (A.47) for $q_1 = \infty^+$, $q_2 = \infty^-$ allows us to solve the Jacobi inversion problem which consists in expressing the symmetric functions of z_j's in terms of the theta-functions. We have

$$\omega_{\infty^+,\infty^-} = \frac{U_\infty(z)}{w}dz\,,$$

where $U_\infty(z)$ is a polynomial of degree g whose leading coefficient is fixed by residues of $\omega_{\infty^+,\infty^-}(z)$

$$U_\infty(z) = (a_0 - d_0)z^g + \cdots\,.$$

The Poisson bracket similar to (A.43) is computed as follows

$$\text{(A.48)}\qquad \frac{1}{2x}\Big\{a(x)+d(x),\ \sum_{j=1}^{g}\int_{p_0}^{p_j}\omega_{\infty^+,\infty^-}\Big\} = U_\infty(x) - \frac{a_0-d_0}{b_0}\,b(x)\,.$$

Simply rewriting this we obtain the solution to the Jacobi inversion formula

(A.49)

$$\frac{b(x)}{b_0} = \frac{1}{2(d_0-a_0)}\Big[\sum_{i,j=1}^{g} x^{i-1}(\mathcal{A}^{-1})_{i,j}\partial_{\varphi_j}\log\frac{\theta(\varphi+\Delta-A(\infty^-))}{\theta(\varphi+\Delta-A(\infty^+))} - 2U_\infty(x)\Big]\,,$$

where

$$\mathcal{A}_{ij} = \int_{a_i}\frac{x^{j-1}}{\sqrt{P(x)}}dx\,.$$

Similarly to U_∞ we define U_0 by

$$\omega_{0^-,0^+} = \frac{U_0(z)}{zw}dz\,,$$

so that we have $U(z) = zU_\infty(z) + U_0(z)$ for $U(z)$ defined in (A.10). We have

$$\text{(A.50)}\qquad \frac{1}{2}\Big\{a(x)+d(x),\ \sum_{j=1}^{g}\int_{p_0}^{p_j}\omega_{0^-,0^+}\Big\} = U_0(x) - \frac{a_{g+1}-d_{g+1}}{b_g}b(x)\,,$$

from which one readily derives the Jacobi inversion formula for symmetric polynomials of the variables z_j^{-1} in terms of $\log\frac{\theta(\varphi+\Delta-A(0^+))}{\theta(\varphi+\Delta-A(0^-))}$.

Collecting (A.48) and (A.50) we find

$$\text{(A.51)}\quad \frac{1}{2}\Big\{a(x)+d(x),\ \sum_{j=1}^{g}\int_{p_0}^{p_j}\varrho\Big\} = U(x) - \Big(\frac{a_0-d_0}{b_0}x + \frac{a_{g+1}-d_{g+1}}{b_g}\Big)b(x)\,.$$

On the other hand

$$\text{(A.52)}\qquad \frac{1}{2}\Big\{a(x)+d(x),\log\prod_{j=1}^{g}z_j\Big\}$$
$$= a(x) - d(x) - \Big(x\frac{a_0-d_0}{b_0} + \frac{a_{g+1}-d_{g+1}}{b_g}\Big)b(x)\,.$$

Together with (A.51) this gives

$$\text{(A.53)}\qquad \frac{1}{2}\Big\{a(x)+d(x),\ \sum_{j=1}^{g}\Big(\int_{p_0}^{p_j}\varrho - \log z_j\Big)\Big\} = U(x) + d(x) - a(x)\,.$$

The application of this formula is twofold. First, due to the simple equality

$$\prod_{j=1}^{g} z_j = \frac{\theta(\varphi+\Delta-A(0^+))\theta(\varphi+\Delta-A(0^-))}{\theta(\varphi+\Delta-A(\infty^-))\theta(\varphi+\Delta-A(\infty^+))}\,, \tag{A.54}$$

(A.53) allows us to express $a(x)-d(x)$ in terms of theta functions:

$$a(x)-d(x) = \sum_{i,j=1}^{g} x^i(\mathcal{A}^{-1})_{i,j}\partial_{\varphi_j}\log\frac{\theta(\varphi+\Delta-A(0^-))}{\theta(\varphi+\Delta-A(\infty^-))} + U(x)\,. \tag{A.55}$$

Second, it gives rise to an elegant relation which will be important in the next subsection

$$\frac{1}{2}\Big\{a(x)+d(x),\ \sum_{j=1}^{g}\Big(\int_{p_0}^{p_j}\varrho - \log z_j\Big)+\varpi\Big\} = d(x)-a(x)\,, \tag{A.56}$$

where ϖ is given in (A.37), and we use (A.8). Notice that (A.48), (A.50), (A.54) provide three different ways to express $\prod_{j=1}^{g} z_j$ in terms of theta functions, which reflects the fact that there are many theta-function identities in the hyper-elliptic case.

Let us summarise. The relations (A.49), (A.55) allow us to express b_j/b_0, a_j-d_j and, consequently, c_jb_0 as single-valued meromorphic functions on the Jacobi variety $\mathcal{J}$ with singularities on the divisor

$$\widetilde{\Theta} = (\Theta+A(\infty^+))\sqcup(\Theta+A(\infty^-))\sqcup(\Theta+A(0^+))\sqcup(\Theta+A(0^-))\,. \tag{A.57}$$

The compact varieties $\mathcal{J}$ and $S(g)$ are not isomorphic since the Abel map is not one-to-one. However, we have an isomorphism of affine varieties

$$S(g)-D\simeq \mathcal{J}-\widetilde{\Theta}\,.$$

The algebra A_0 is the algebra of functions on $S(g)-D$.

Notice that shifted theta divisors of the kind (A.57) was used in the study of integrable models for the first time by Its [**It**] who investigated periodic solutions to the non-linear Schrödinger equation .

A.5. Classical Interpretation of Fermionic Basis

The main problem considered in Volume I consists in computation of matrix elements

$$\langle\kappa+\alpha|x|\kappa\rangle\,,$$

with x being an element of the algebra $\mathcal{A}$ of matrix elements of the quantum monodromy matrix. This formulation implies that we are actually interested in the quotient of $\mathcal{A}$ by the left (right) action of the transfer-matrix with the twist κ $(\kappa+\alpha)$.

In the quasi-classical limit, the algebra $\mathcal{A}$ becomes a commutative Poisson algebra $\mathcal{A}_{\mathrm{cl}}$. The coefficients of the determinant of $t(z)$ constitute the centre of the Poisson algebra, and we factor them out. Further it is easy to conclude that in the limit (A.4) one takes the quotient by the action of the quantum transfer-matrices to two operations

- Multiplication by $a(z)+d(z)$ which is the quasi-classical limit of the transfer-matrix. We consider the quotient of $\mathcal{A}_{\mathrm{cl}}$ by multiplication by a_j+d_j.

- Starting with the Hamiltonian vector fields
$$D_j(x)=\{\mathscr{a}_j+\mathscr{d}_j,x\}\,,\quad j=1,\cdots,g\,,$$
we define the connection ∇_j by
$$\nabla_j(x)=D_j(x)+\alpha(\mathscr{a}_j-\mathscr{d}_j)x\,. \tag{A.58}$$
It is easy to see that this connection is flat
$$[\nabla_i,\nabla_j]=0\,.$$
We consider the quotient by the action of ∇_j.

The formulae of the previous subsection allow us to trivialise this flat connection. Consider the points $p_1,\cdots,p_g$ lying on the first sheet of S, i.e., on the plane with cuts on which we have the single-valued function
$$\hbar(z_1,\cdots,z_g)=-\frac{1}{2}\Bigl(\sum_{j=1}^{g}\bigl(\int_{p_0}^{p_j}\varrho-\log z_j\bigr)+\varpi\Bigr)\,.$$
According to (A.56) we have
$$e^{\alpha\hbar}\,\nabla_j x=D_j\,e^{\alpha\hbar}x\,,\quad x\in A_0\,. \tag{A.59}$$
Under Abel transformation, the function $e^{\alpha\hbar}x$ become a function of φ with real periods equal to 1 in any φ_j. Such functions allows a Fourier series. Fourier coefficients are integrals with the measure $d^g\varphi$ whose pullback under the Abel transformation is
$$d^g\varphi=\prod_{i<j}(z_i-z_j)\prod_{j=1}^{g}\frac{dz_j}{w_j}\,.$$
So, the Fourier series are

(A.60)
$$x=\sum_{k\in\mathbb{Z}^g}e^{-\alpha\hbar+2\pi i(k,\varphi)}$$
$$\times\int_{a_1}\cdots\int_{a_g}x\prod_{i<j}(z_i-z_j)\exp\Bigl\{-\sum_{m=1}^{g}\frac{1}{2}\int_{p_0}^{p_m}(\alpha\varrho+4\pi i(k,\omega))\Bigr\}\prod_{j=1}^{g}\frac{z_j^{\frac{\alpha}{2}}dz_j}{z_jw_j}\,,$$
where the φ_j are are given by the Abel transformation, x is considered as a function of $p_1,\cdots,p_g$ via the substitution of (A.36), (A.38). Application of ∇_j to x corresponds to multiplication of the Fourier coefficient by k_j.

Define
$$\boldsymbol{\sigma}_j=\sum_{m=1}^{g}\sigma_j(p_m)\,,\quad\tilde{\boldsymbol{\sigma}}_j=\sum_{m=1}^{g}\tilde{\sigma}_j(p_m)\,.$$
It is easy to see that the integrand in (A.60) equals
$$x\cdot\boldsymbol{\sigma}_1\wedge\cdots\wedge\boldsymbol{\sigma}_g\,.$$

For any subsets $I=\{i_1,\dots,i_m\}$, $J=\{j_1,\dots,j_m\}$ of $\{1,\dots,g\}$, there are such observables $x_{I,J}$ that
$$x_{I,J}\cdot\boldsymbol{\sigma}_1\wedge\cdots\wedge\boldsymbol{\sigma}_g=\boldsymbol{\sigma}_{i_1}\wedge\cdots\wedge\boldsymbol{\sigma}_{i_m}\wedge\tilde{\boldsymbol{\sigma}}_{\bar{j}_1}\wedge\cdots\wedge\tilde{\boldsymbol{\sigma}}_{\bar{j}_{g-m}}\,, \tag{A.61}$$
where $\bar{J}=\{1,\cdots,g\}\backslash J$. The definition (A.61) should be taken with some care. Recalling (A.31), (A.10), (A.16) one concludes that $x_{I,J}$ in (A.61) is a symmetric

polynomial in z_j, in other words polynomial of $\mathscr{b}_j/\mathscr{b}_0$ with coefficients polynomial in $\mathscr{a}_j+\mathscr{d}_j$ and k_j. Our convention is that these latter are replaced by ∇_j according to explanation above.

The Fourier coefficients of $\mathscr{x}_{I,J}$ coincide with the quasi-classical limit of the determinants in the quantum formula (A.3). This is not surprising because for an integrable model the Fourier coefficients are generally given by the quasi-classical limit of the quantum matrix elements. The interesting fact is that the quantum formula (A.3) implies in the quasi-classical limit

THEOREM A.1. *Any observable $\mathscr{y}$ can be presented as*

$$\mathscr{y}=\sum_{\#(I)=\#(J)} P_{I,J}(\{\mathscr{a}_p+\mathscr{d}_p\}_{p=1}^{g})Q_{I,J}(\{D_p\}_{p=1}^{g})\mathscr{x}_{I,J}\,, \tag{A.62}$$

with $P_{I,J}$, $Q_{I,J}$ being polynomials whose coefficients depend on $\mathscr{f}_j$ defining the quadrics (A.34).

This statement is quite non-trivial. Generally an observable $\mathscr{y}$ contains in addition to singularities at infinities those at $p_j=\varsigma(p_j)$ which can be poles of arbitrary order. The theorem above asserts that applying the Hamiltonian vector fields D_j and taking linear combinations with $\mathscr{a}_j+\mathscr{d}_j$ one can completely eliminate these singularities. It is possible to give a cohomological interpretation of this statement in the spirit of [**NS**].

Consider the affine variety $\mathcal{M}$ obtained by intersecting the intersection of quadrics (A.34) by the plane with fixed $\mathscr{a}_j+\mathscr{d}_j$. We have seen that this variety is equivalent to the Jacobi variety of S with $\widetilde{\Theta}$ removed. Define the differential forms

$$\Omega^{(p)}=\sum_{i_1<\cdots<i_p}\mathscr{x}_{i_1,\cdots,i_p}\boldsymbol{\sigma}_{i_1}\wedge\cdots\wedge\boldsymbol{\sigma}_{i_p}\,,$$

and the differential

$$d\Omega^{(p)}=\sum_{j=1}^{g}\sum_{i_1<\cdots<i_p}\nabla_j(\mathscr{x}_{i_1,\cdots,i_p})\,\boldsymbol{\sigma}_j\wedge\boldsymbol{\sigma}_{i_1}\wedge\cdots\wedge\boldsymbol{\sigma}_{i_p}\,.$$

Similarly to the de Rham complex the differential d based on the flat connection ∇_j defines a local system. Then the Theorem A.1 implies that the top cohomology of the local system can be represented as elements of

$$\bigwedge^{g}V\,,\qquad V=\bigoplus_{j=1}^{g}\mathbb{C}\boldsymbol{\sigma}_j\oplus\bigoplus_{j=1}^{g}\mathbb{C}\tilde{\boldsymbol{\sigma}}_j\,.$$

Analysing this statement one comes to the conclusion that our fermions have purely classical, cohomological, origin: the operators $\mathbf{c}_j^*$ remove one of the first kind forms $\boldsymbol{\sigma}_j$ and the operator $\mathbf{b}_j^*$ adds one second kind form $\tilde{\boldsymbol{\sigma}}_j$.

Notation

Chapter 1

1.2

q parameter q in the six vertex model, (1.1)
P permutation matrix, (1.3)
$T^{6\text{v}}_{j,\mathbf{M}}$ *Matsubara* monodromy matrix, (1.4)
$T^{6\text{v}}_{\mathbf{S},\mathbf{M}}$ monodromy matrix with *Space* and *Matsubara* directions, (1.5)
$T^{6\text{v},\kappa}_{\mathbf{M}}$ *Matsubara* transfer matrix with twist, (1.8)

1.3

$q^{2\alpha S(0)}\mathcal{O}$ quasi-local operator, (1.9)
$\mathcal{Z}^{6\text{v},\kappa}_{L,\mathbf{n}}$ expectation-value functional (finite chain) (1.10)
$\mathcal{Z}^{6\text{v},\kappa}_{\mathbf{n}}$ expectation-value functional (infinite chain) (1.11)

1.4

$L_{j,\mathbf{m}}$ L operator (1.12)
$R(\zeta)$ R-matrix of 6 vertex model, (1.13)
$T^{\kappa}_{\mathbf{M}}(\xi,\kappa)$ *Matsubara* transfer matrix with spectral parameter and twist, (1.18)

1.5

$\mathbf{A}(\xi_1,\dots,\xi_l,\alpha)$ reduced qKZ operator, (1.21), (1.22)
θ charge conjugation, (1.23)
$\mathbb{R}_{j,j+1}(\xi_j/\xi_{j+1})$ adjoint action of R-matrix ,(1.25)

1.6

$T_{\mathbf{S}}(\zeta)$ *Space* transfer matrix (1.29)
$\widetilde{R}_{j,\mathbf{m}}(\zeta)$ $\tilde{R}$ matrix (1.32)
$\check{R}_{j,\mathbf{m}}(\zeta)$ $\check{R}$ matrix (1.33)
$\mathbf{I}_p$ local Hamiltonians (1.37)

1.7

$\rho(n,T)$ density matrix, (1.49)
$s(n,T)$ entanglement entropy, (1.51)

Chapter 2

2.1

$A(\xi)$, $B(\xi)$, $C(\xi)$, $D(\xi)$ entries of *Matsubara* monodromy matrix (2.1)
$a(\xi)$, $d(\xi)$ vacuum eigenvalues of $A(\xi)$, $D(\xi)$ (2.7)
$|\lambda_1^+,\dots,\lambda_k^+\rangle$ Bethe vector for *Matsubara* transfer matrix (2.8), (2.93)
$\langle\lambda_1^-,\dots\lambda_{\mathbf{n}-k}^-|$ Bethe covector for *Matsubara* transfer matrix, (2.13)
$Q^+(\xi,\kappa)$, $Q^-(\xi,\kappa)$ Q-functions (2.11),(2.15)
$M_{\mathbf{n}}(\lambda_1,\dots\lambda_{\mathbf{n}}|\tau_1,\dots,\tau_{\mathbf{n}})$ domain wall partition function, (2.18), (2.20)

2.2

$\mathcal{R}$ universal R-matrix, (2.21),(2.22),(2.23),(2.24)
E, F, K generators of $U_q(\mathfrak{sl}_2)$, (2.25)
C Casimir element of $U_q(\mathfrak{sl}_2)$, (2.27),(2.30)
V^Λ Verma module of $U_q(\mathfrak{sl}_2)$, (2.28)

2.3

e_i, f_i, t_i generators of $U_q(\widehat{\mathfrak{sl}}_2)$, (2.32),(2.33),(2.34),(2.35)
$U_q\mathfrak{b}^+$, $U_q\mathfrak{b}^-$ Borel subalgebras of $U_q(\widehat{\mathfrak{sl}}_2)$, p.37
ev_ζ evaluation homomorphism, (2.43)
π_ζ^Λ evaluation Verma representation of $U_q(\widehat{\mathfrak{sl}}_2)$, (2.45)
$\pi_\zeta^{(2s)}$ spin s representation of $U_q(\widehat{\mathfrak{sl}}_2)$, (2.45)
$R^{(ev,1)}(\zeta)$ evaluation R-matrix with spin 1/2, (2.46)
Osc q-oscillator algebra, (2.49)
$\mathbf{a}$, $\mathbf{a}^*$, $q^{\pm D}$ generators of Osc, (2.49)
$W^\pm$ representations of Osc, (2.50)
$R_A^+(\zeta)$, $R_A^-(\zeta)$ R-matrix with Osc as auxiliary space ,(2.53), (2.57)
$R_a(\zeta)$ R-matrix with $\pi^{(1)}$ as auxiliary space (2.55)
$\check{R}_{A,B}(\zeta)$ intertwiner for $W_{\zeta_1}^+\otimes W_{\zeta_2}^+$, (2.59),(2.61)
$\mathfrak{A}^{\mathrm{left}}(\zeta)$, $\mathfrak{A}^{\mathrm{right}}(\zeta)$ (2.66), (2.71)
$\mathfrak{A}(\zeta)$ quotient algebra of $U_q(\mathfrak{sl}_2)$, (2.67)
$F_{a,A}$ matrix of conjugation used for fused R-matrix (2.79)

2.4

$\mathfrak{H}_{\mathbf{M}}$ *Matsubara* space with arbitrary spin representations, (2.80), (2.81)
$R^{(ev,2s)}(\zeta)$ evaluation R-matrix with spin s, (2.82)
$R^{(+,2s)}(\zeta)$ R-matrix in oscillator and spin s representations, (2.84)
$\mathfrak{A}_{\mathrm{loc}}^{\mathrm{left}}(\zeta)$, $\mathfrak{A}_{\mathrm{loc}}^{\mathrm{right}}(\zeta)$ localisation of algebras $\mathfrak{A}^{\mathrm{left}}(\zeta)$, $\mathfrak{A}^{\mathrm{right}}(\zeta)$, Definition 2.7
$a(\xi)$,$d(\xi)$ functions $a(\xi)$, $d(\xi)$ for general spin, (2.93)

2.5

$Q_{\mathbf{M}}^\pm(\zeta,\kappa)$ Q-operators, (2.102)
$W(\xi)$ scalar function in the Wronskian identity, (2.109), (2.110)

2.6

$\mathfrak{a}(\zeta|\kappa)$ unknown function in DdV equation, (2.111)
$K_0(\zeta)$ kernel function in DdV equation, (2.113)
$R_0(\zeta)$ resolvent kernel in DdV equation, (2.115)

Chapter 3

$\mathcal{W}$, $\mathcal{W}_{\alpha,s}$ space of quasi-local operators, (3.1)
$\mathcal{W}_\alpha$ block α of $\mathcal{W}$, (3.4)

3.1

$R^+_{\{a,A\}}(\zeta)$ fusion of oscillator and spin 1/2 R-matrix, (3.5)
$R_{\{a,A\},\{b,B\}}(\zeta)$ R-matrix intertwining fused spaces, (3.6)
$\mathcal{F}_{\{a,A\},\{b,B\}}(\zeta)$ quasi R-matrix, (3.8)

3.2

$\mathbb{T}_{\mathrm{aux},J}(\zeta)$ adjoint monodromy matrix, (3.10)
$\mathbb{R}_{\mathrm{aux},j}(\zeta)$ adjoint R-matrix, (3.11)
$T_{\{a,A\},J}(\zeta)$ fused monodromy matrix, (3.13)
$\mathbb{T}_{\{a,A\},J}(\zeta,\alpha)$ fused adjoint monodromy matrix, (3.15)
$\mathbf{k}_J(\zeta,\alpha)$ operator $\mathbf{k}(\zeta,\alpha)$ on interval J, (3.16),(3.17)
Δ_ζ q-difference operator, (3.20)
$\mathbf{v}_J(\zeta,\alpha)$ operator $\mathbf{v}(\zeta,\alpha)$ on interval J, (3.22)
ϕ spin reversal operation on operators, Definition 3.6
$\psi(\zeta,\alpha)$ Cauchy kernel, (3.30)
$\mathbf{q}_J(\zeta,\alpha)$ adjoint Q-operator, (3.35)

3.3

$\bar{\mathbf{c}}_J(\zeta,\alpha)$ annihilation operator $\bar{\mathbf{c}}$ on J, (3.38)
$\mathbf{c}_J(\zeta,\alpha)$ annihilation operator $\mathbf{c}$ on J, (3.39)
$\mathbf{c}(\zeta,\alpha)$ annihilation operator $\mathbf{c}$, (3.42)
$\mathbf{b}(\zeta,\alpha)$ annihilation operator $\mathbf{b}$, (3.43)

3.4

$\mathbf{t}^*_J(\zeta,\alpha)$ creation operator $\mathbf{t}^*$ on J, (3.52)
$\mathbf{r}_{i,j}(\zeta^2)$ 'tail' of adjoint R-matrix, (3.54)
$\mathbf{t}^*(\zeta,\alpha)$ creation operator $\mathbf{t}^*$, (3.59)

3.5

$\mathbf{f}_J(\zeta,\alpha)$ operator $\mathbf{f}^*$ on J, (3.67)
$\mathbf{f}_{\mathrm{rat},J}(\zeta,\alpha)$ rational version of $\mathbf{f}_J$, (3.69)
$\Delta_\zeta^{-1}\psi(\zeta,\alpha)$ q-primitive of $\psi(\zeta,\alpha)$, (3.72)
$\mathbf{b}^*_J(\zeta,\alpha)$ creation operator $\mathbf{b}^*$ on J, (3.74)
$\mathbf{g}_{c,L}(\zeta,\alpha)$ operator $\mathbf{g}$ on L, (3.75)
$\mathbf{b}^*(\zeta,\alpha)$ creation operator $\mathbf{b}^*$, (3.85)
$\mathbf{c}^*(\zeta,\alpha)$ creation operator $\mathbf{c}^*$, (3.87)

3.6

$\mathbf{g}_j^{\epsilon}(\xi_j,\alpha)$ Russian-doll operators, (3.90), (3.92)
$\mathbf{x}^{\epsilon *}(\zeta)$ combined notation for creation operators, (3.95)
$\mathbf{b}^*_{\mathrm{rat}}(\zeta)$ rational version of $\mathbf{b}^*$, (3.109)
$D_\zeta F(\zeta)$ q-Laplacian (operator version), (3.110)

3.8

$\mathbf{b}_p$, $\mathbf{c}_p$ Fourier components of annihilation operators (3.124)
$\mathbf{b}_p^*$, $\mathbf{c}_p^*$, $\mathbf{t}_p^*$ Fourier components of creation operators (3.125), (3.126), (3.127)

Chapter 4

4.2

$\varphi(\zeta)$ logarithmic q-primitive of $a(\zeta)/d(q\zeta)$, (4.12),(4.13)
$d\mu^{\pm}(\zeta)$ measures used for deformed Abelian differentials, (4.14)
$\rho(\zeta|\alpha)$ ratio of transfer matrix eigenvalues, (4.16)
δ_ζ^- q-difference operator, (4.17)
$E_\zeta(g^{\pm}(\zeta))$ exact form, (4.19)
$r(\zeta,\xi)$, $r^+(\zeta,\xi)$, $r^-(\xi,\zeta)$ generating functions of deformed Abelian differentials, (4.22)
$\Omega^{\pm}_{\mathbf{m}}(\zeta)$, $\widetilde{\Omega}^{\pm}_{\mathbf{m}}(\zeta)$ deformed Abelian differentials, (4.27)
$\mathcal{A}^{\pm}_{\mathbf{i},\mathbf{j}}$, $\mathcal{B}^{\pm}_{\mathbf{i},\mathbf{j}}$ deformed period matrices, (4.28)
$\omega(\zeta,\xi|\alpha)$ function ω, (4.35)
$\omega_{\mathrm{sing}}(\zeta,\xi|\alpha)$ singular part of ω, (4.35)

4.3

$\mathbf{g}_{\mathrm{rat}}(\zeta,\alpha)$ rational version of $\mathbf{g}(\zeta,\alpha)$, (4.51)

4.4

$\bar{\mathbf{t}}^*(\zeta)$ dual version of $\mathbf{t}^*(\zeta)$, (4.70), (4.71)
$\bar{\mathbf{b}}^*(\zeta)$, $\bar{\mathbf{c}}^*(\zeta)$ dual version of $\mathbf{b}^*(\zeta)$, $\mathbf{c}^*(\zeta)$, (4.72)
$\mathbf{s}(\zeta,\alpha)$ spin 2 annihilation operator, (4.77), Remark 4.27
$B_n(\alpha)$ basic operator supported on $[1,n]$, (4.78), (4.79)
$\mathbf{s}_p$ Fourier components of $\mathbf{s}(\zeta,\alpha)$, (4.80)
$\mathbf{y}_m$ operators used to describe quasi-local operators of given length, (4.88)

Chapter 5

5.1

$K_\alpha(\zeta)$ generalised kernel function, (5.2)
$dm(\zeta)$ generalised measure, (5.3)
$(F\star G)(\zeta,\xi)$ convolution with respect to dm, (5.4)
R_{dress} dressed resolvent kernel, (5.5)
$f_{\mathrm{left}}(\zeta,\xi)$, $f_{\mathrm{right}}(\zeta,\xi)$ (5.6)
$G_{\mathrm{right}}(\zeta,\xi)$ right auxiliary function,(5.8)

δ_ζ^+ q-difference operator, (5.12)
$G_{\rm left}(\zeta,\xi)$ left auxiliary function, (5.17)
$\psi_+(\zeta,\alpha)$ modified Cauchy kernel, (5.23)
$dm_0(\zeta)$ generalised measure, (5.25)
$(F\circ G)(\zeta,\xi)$ convolution with respect to dm_0, (5.26)
$(F*G)(\zeta,\xi)$ convolution $*$, (5.27)
$F_{\rm left}$, $F_{\rm right}$ (5.30)

5.2

$\omega_{\rm rat}(\zeta,\xi|\alpha)$ rational version of ω, (5.40)
$\overline{D}_\zeta$ q-Laplacian, (5.41)
$(\mathcal{A}f)(x_1,\cdots,x_l)$ operator A on symmetric polynomials, (5.45)
$(\mathcal{D}f)(x_1,\cdots,x_l)$ operator D on symmetric polynomials, (5.46)
$(\mathcal{B}f)(x_1,\cdots,x_l)$ operator B on symmetric polynomials, (5.47)
$(\mathcal{C}f)(x_1,\cdots,x_l)$ operator C on symmetric polynomials, (5.48)
$\mathcal{T}_{++}$ operator A on Young diagrams, (5.53)
$\mathcal{T}_{--}$ operator D on Young diagrams,(5.54)
$\mathcal{T}_{+-}$ operator B on Young diagrams,(5.55)
$\mathcal{T}_{-+}$ operator C on Young diagrams,(5.56)

5.3

Ω annihilation operator Ω, (5.65),(5.66)
$B^\varepsilon(\zeta)$ (5.67)
$\omega_0(\zeta_2,\zeta_1)$, $\omega_1(\zeta_2,\zeta_1)$ Taylor coefficients of $\omega_{\rm rat}$ as α(5.71)
Ω_0, Ω_1 annihilation operator Ω as $\alpha\to 0$, (5.75),(5.76)
$\mathcal{Q}$ charge operator, (5.80)
$\tilde{X}_1(\zeta_1,\zeta_2)$, $\tilde{X}_0(\zeta_1,\zeta_2|\alpha)$ gauge transformed $X_1(\zeta_1,\zeta_2)$, $X_0(\zeta_1,\zeta_2|\alpha)$, (5.110)
$Y^\vee$ vector corresponding to operator Y, (5.116)
$\widetilde{O}(J)$ A-basis of invariant operators, (5.121)
$v_{k,J,m}$ vector corresponding to a Bratteli diagram J, (5.122)
$O(J)$ $3j$-basis of invariant operators, (5.125)

5.4

$\omega(x)$ function ω in the XXX case, (5.137)

5.5

$\varphi(\zeta,\nu)$ generating function for ω_0, ω_1 in XXZ case, (5.144)
J_a, J_{bc} structure constants of Sklyanin algebra, (5.148), (5.149)
K_0, K_2 Casimir elements of Sklyanin algebra, (5.153)
$\varphi(\mu,\nu,\tau)$ generating function for ω_1, ω_2, ω_3 in XYZ case, (5.169)

Appendix

A.1

g genus of the hyper-elliptic curve,

A.2

$\varkappa$ rescaled twist, finite in the quasi-classical limit, (A.4)
$T_{cl}(\zeta,\varkappa)$ quasi-classical limit of transfer matrix, (A.5)
ϱ differential of the third kind with poles at $0,\infty$ (A.10)
$dm^{\pm}(z)$ quasi-classical measure, (A.14), (A.32)

A.3

S hyper-elliptic curve, (A.23)
ω_i normalized differential of the first kind, (A.27)
$\omega(p_1,p_2)$ canonical differential of the second kind, (A.28)
ω_{q_1,q_2} normalized differential of the third kind, (A.29)

A.4

$t(z)$ classical monodromy matrix, (A.33)
$a(z)$, $b(z)$, $c(z)$, $d(z)$ matrix elements of $t(z)$, (A.33)
z_j, w_j separated variables, (A.36)
ϖ variable conjugate to $\varkappa$, (A.37)
$S(k)$ symmetric power of S, (A.40)
$A(p)$ Abel transformation, (A.42)
Θ Theta divisor, (A.45)

A.5

∇_j flat connection, (A.58)

Bibliography

[Be] H. Bethe. *Zur Theorie der Metalle. I. Eigenwerte und Eigenfunktionen der linearen Atomkette. (On the theory of metals. I. Eigenvalues and eigenfunctions of the linear atom chain).* Zeitschrift für Physik, **71** (1931), 205–226.

[O1] L. Onsager. *Two-dimensional model with an order-disorder transition.* Phys. Rev., **65** (1944), 117–149.

[O2] L. Onsager. *The Ising model in two dimensions.* In Critical phenomena in alloys, magnets and superconducters, 3–12. McGraw-Hill, New York, 1971.

[YY] C. N. Yang and C. P. Yang. *Thermodynamics of a one-dimensional system of bosons with repulsive delta-function interaction.* J. Math. Phys., **10** (1969), 1115–1122.

[Li] E. Lieb. *Exact solution of the problem of the entropy of two-dimensional ice.* Phys. Rev. Lett., **18** (1967), 692–694.

[Su] B. Sutherland. *Exact solution of a two-dimensional model for hydrogen-bonded crystals.* Phys. Rev. Lett., **19** (1967), 103–104.

[Ba1] R. Baxter. *Partition function of the eight-vertex lattice model.* Ann. Physics, **70** (1972), 193–228.

[Ba2] R. Baxter. *Eight-vertex model in lattice statistics and one-dimensional anisotropic Heisenberg chain. I, II, III.* Ann. Physics, **76** (1973), 1–24, 25–47, 48–71.

[Ba3] R. Baxter. *Exactly Solved Models in Statistical Mechanics.* Academic Press, London, 1982.

[GGKM] C. Gardner, J. Greene, M. Kruskal, and R. Miura. *Method for solving the Korteweg-de Vries equation.* Phys. Rev. Lett., **19** (1967), 1095–1097.

[La] P. Lax. *Integrals of nonlinear equations of evolution and solitary waves.* Comm. Pure Appl. Math., **21** (1968), 467–490.

[ZF] V. Zakharov and L. Faddeev. *Korteweg-de Vries equation: A completely integrable Hamiltonian system.* Funct. Anal. Appl., **5** (1971), 18–27.

[FT] L. Faddeev and L. Takhtajan. *Essentially nonlinear one-dimensional model of classical field theory.* Teor. Mat. Fiz., **21** (1974), 160–174.

[FK] L. Faddeev and V. Korepin. *Quantum theory of solitons.* Physics Reports, **42** (1978), 1–87.

[C] S. Coleman. *Quantum sine-Gordon equation as the massive Thirring model.* Phys. Rev. D, **11** (1975), 2088–2097.

[Ma] S. Mandelstam. *Soliton operators for the quantized sine-Gordon equation.* Phys. Rev. D, **11** (1975), 3026–3030.

[ZaS] A. Zamolodchikov. *Exact two-particle S-matrix of quantum sine-Gordon solitons.* Commun. Math. Phys., **55** (1977), 183–186.

[ZZ] A. Zamolodchikov and Al. Zamolodchikov. *Factorized S-matrices in two dimensions as the exact solutions of certain relativistic quantum field theory models.* Ann. Physics, **120** (1979), 253–291.

[STF] E. Sklyanin, L. Takhtajan, and L. Faddeev. *Quantum inverse problem method.* Theoret. and Math. Phys., **40** (1979), 194–220.

[KIB] V. E. Korepin, N. M. Bogoliubov, and A. G. Izergin. *Quantum Inverse Scattering Method and Correlation Functions.* Cambridge University Press, Cambridge, 1993.

[KR] P. Kulish and N. Reshetikhin. *Quantum linear problem for the sine-Gordon equation and higher representations.* Zapiski Nauch. Sem. LOMI, **101** (1980), 101–110.

[Dr] V. Drinfeld. *Quantum Groups.* In Proceedings of the International Congress of Mathematicians, 798–820, Berkeley, 1990.

[Ji1] M. Jimbo. *A q-difference analogue of $U(\mathfrak{g})$ and the Yang-Baxter equation.* Lett. Math. Phys., **10** (1985), 63–69.

[Sm] F. Smirnov. *Form Factors in Completely Integrable Models of Quantum Field Theory.* World Scientific, Singapore, 1992.

[JM] M. Jimbo and T. Miwa. *Algebraic Analysis of Solvable Lattice Models.* **85**. Amer. Math. Soc., 1995.

[KMT] N. Kitanine, J.-M. Maillet, and V. Terras. *Correlation functions of the XXZ Heisenberg spin-$\frac{1}{2}$-chain in a magnetic field.* Nucl. Phys. B, **567** (2000), 554–582.

[GKS] F. Göhmann, A. Klümper, and A. Seel. *Integral representations for correlation functions of the XXZ chain at finite temperature.* J. Phys. A, **37** (2004), 7625–7656.

[Sl] N. Slavnov. *Calculation of scalar products of wave functions and form-factors in the framework of the algebraic Bethe ansatz.* Theoret. and Math. Phys., **79** (1989), 502–508.

[BLZ1] V. Bazhanov, S. Lukyanov, and A. Zamolodchikov. *Integrable structure of conformal field theory, quantum KdV theory and thermodynamic Bethe ansatz.* Commun. Math. Phys., **177** (1996), 381–398.

[BLZ2] V. Bazhanov, S. Lukyanov, and A. Zamolodchikov. *Integrable structure of conformal field theory II. Q-operator and DDV equation.* Commun. Math. Phys., **190** (1997), 247–278.

[BLZ3] V. Bazhanov, S. Lukyanov, and A. Zamolodchikov. *Integrable structure of conformal field theory III. the Yang-Baxter relation.* Commun. Math. Phys., **200** (1999), 297–324.

[BJMST1] H. Boos, M. Jimbo, T. Miwa, F. Smirnov, and Y. Takeyama. *Hidden Grassmann structure in the XXZ model.* Commun. Math. Phys., **272** (2007), 263–281.

[BJMST2] H. Boos, M. Jimbo, T. Miwa, F. Smirnov, and Y. Takeyama. *Hidden Grassmann structure in the XXZ model II. Creation operators.* Commun. Math. Phys., **286** (2009), 875–932.

[JMS1] M. Jimbo, T. Miwa, and F. Smirnov. *Hidden Grassmann structure in the XXZ model III: Introducing Matsubara direction.* J. Phys. A, **42** (2009), 304018.

[BJMS] H. Boos, M. Jimbo, T. Miwa, and F. Smirnov. *Hidden Grassmann structure in the XXZ model IV: CFT limit.* Commun. Math. Phys., **299** (2010), 825–866.

[JMS2] M. Jimbo, T. Miwa, and F. Smirnov. *Hidden Grassmann structure in the XXZ model V: sine-Gordon model.* Lett. Math. Phys., **96** (2011), 325–365.

[T] M. Takahashi. *Half-filled Hubbard model at low temperature.* J. Phys. C, **10** (1977), 1289–1301.

[BK] H. Boos and V. Korepin. *Quantum spin chains and Riemann zeta functions with odd arguments.* J. Phys. A, **34** (2001), 5311–5316.

[SST] J. Sato, M. Shiroishi, and M. Takahashi. *Correlation functions of the spin-1/2 antiferromagnetic Heisenberg chain: exact calculation via the generating function.* Nucl. Phys. B, **729** (2005), 441–466.

[Lk1] S. Lukyanov. *Low energy effective hamiltonian for the XXZ spin chain.* Nucl. Phys. B, **522** (1998), 533–549.

[LT] S. Lukyanov and V. Terras. *Long-distance asymptotics of spin-spin correlation functions for the XXZ spin chain.* Nucl. Phys. B, **654** (2003), 323–356.

[BGKS] H. Boos, F. Göhmann, A. Klümper, and J. Suzuki. *Factorization of the finite temperature correlation functions of the XXZ chain in a magnetic field.* J. Phys. A, **40** (2007), 10699–10728.

[LZ] S. Lukyanov and A. Zamolodchikov. *Exact expectation values of local fields in quantum sine-Gordon model.* Nucl. Phys. B, **493** (1997), 571–587.

[FFLZZ] V. Fateev, D. Fradkin, S. Lukyanov, A. Zamolodchikov, and Al. Zamolodchikov. *Expectation values of descendent fields in the sine-Gordon model.* Nucl. Phys. B, **540** (1999), 587–609.

[ZaA] Al. Zamolodchikov. *Mass scale in sine-Gordon model and its reductions.* Int. J. Mod. Phys. A, **10** (1995), 1125–1150.

[JM2] M. Jimbo and T. Miwa. *Quantum Knizhnik-Zamolodchikov equation at $|q| = 1$ and correlation functions of the XXZ model in the gapless regime.* J. Phys. A, **29** (1996), 2923–2958.

[Kl] A. Klümper. *Free energy and correlation length of quantum chains related to restricted solid-on-solid lattice models.* Annalen der Physik, **1** (1992), 540–553.

[Su] M. Suzuki. *Transfer matrix method and Monte Carlo simulation in quantum spin systems.* Phys. Rev. B, **31** (1985), 2957–2965.

[HLW] C. Holzhey, F. Larsen, and F. Wilczek. *Geometric and renormalized entropy in conformal field theory.* Nucl. Phys. B, **4**24 (1994), 443–467.

[Ko] V. Korepin. *Universality of entropy scaling in one dimensional gapless models.* Phys. Rev. Lett., **9**2 (2004), 096402.

[CC] P. Calabrese and J. Cardy. *Entanglement entropy and conformal field theory.* J. Phys. A, **42** (2009), 504005–504036.

[G] M. Gaudin. *La Fonction d'Onde de Bethe.* Masson, Paris, 1983.

[I] A. Izergin. *Partition function of a six-vertex model in a finite volume.* Sov. Phys. Dokl., **32** (1987), 878–879.

[Ja] C. Jantzen. *Lectures on Quantum Groups.* Amer. Math. Soc., 1996.

[Lu] G. Lusztig. *Introduction to Quantum Groups.* Birkhäuser, Boston, 1993.

[KT] S. Khoroshkin and V. Tolstoy. *Universal R matrix for quantized (super) algebras.* Commun. Math. Phys., **141** (1991), 599–617.

[FH] E. Frenkel and D. Hernandez. *Baxter's relations and spectra of quantum integrable models.* Duke Math. J., **164** (2015), 2407–2460.

[FV] L. Faddeev and A. Volkov. *The quantum method of the inverse problem on a discrete space time.* Theoret. and Math. Phys., **92** (1992), 837–842.

[KRS] P. Kulish, N. Reshetikhin, and E. K. Sklyanin. *Yang-Baxter equation and representation theory I.* Lett. Math. Phys., **5** (1981), 393–403.

[Ji2] M. Jimbo. *Introduction to the Yang-Baxter equation.* Int. J. Mod. Phys. A, **4** (1989), 3759–3777.

[KBP] A. Klümper, M. Batchelor, and P. Pearce. *Central charges of the 6- and 19-vertex models with twisted boundary conditions.* J. Phys. A, **2**4 (1991), 3111–3133.

[DDV] C. Destri and H. de Vega. *Unified approach to thermodynamic Bethe Ansatz and finite size corrections for lattice models and field theories.* Nucl. Phys. B, **438** (1995), 413–454.

[BJMS] H. Boos, M. Jimbo, T. Miwa, and F. Smirnov. *Completeness of a fermionic basis in the homogeneous XXZ model.* J. Math. Phys., **50** (2009), 095206.

[BG] H. Boos and F. Göhmann. *On the physical part of the factorized correlation functions of the XXZ chain.* J. Phys. A, **42** (2009), 1–27.

[MS] M. Mori and M. Sugihara. *The double-exponential transformation in numerical analysis.* J. Comput. Appl. Math., **127** (2001), 287–296.

[Oo] T. Ooura. *Double exponential quadratures for various kinds of integral.* Talk given at Second International ACCA-JP/UK Workshop, January 19, 2016 Kyoto University.

[KMST] N. Kitanine, J.-M. Maillet, N. Slavnov, and V. Terras. *Emptiness formation probability of the XXZ Heisenberg spin-$\frac{1}{2}$-chain at $\Delta = 1/2$.* J. Phys. A, **35** (2002), L385–L391.

[RS] N. Reshetikhin and F. Smirnov. *Hidden quantum group symmetry and integrable perturbations of conformal field theories.* Commun. Math. Phys., **131** (1990), 157–177.

[JMT] M. Jimbo, T. Miwa, and Y. Takeyama. *Counting minimal form factors of the restricted sine-Gordon model.* Moscow J. Math., **4** (2004), 787–846.

[Sk1] E. Sklyanin. *Some algebraic structures connected with the Yang-Baxter equation.* Funct. Anal. Appl., **16** (1982), 27–34.

[Sk2] E. Sklyanin. *Some algebraic structures connected with the Yang-Baxter equation.* Representations of quantum algebras. Funct. Anal. Appl., **17** (1983), 34–48.

[BJMST3] H. Boos, M. Jimbo, T. Miwa, F. Smirnov, and Y. Takeyama. *Traces on the Sklyanin algebra and correlation functions of the eight-vertex model.* J. Phys. A, **38** (2005), 7629–7659.

[FO] A. Odesskii and B. Feigin. *Sklyanin elliptic algebra.* Funct. Anal. Appl., **23** (1990), 207–214.

[Od] A. Odesskii. *Elliptic algebras.* Russian Math. Surveys, **57** (2002), 1–50.

[SZZ] N. Slavnov, A. Zabrodin, and A. Zotov. *Scalar products of Bethe vectors in the 8-vertex model.* JHEP, **06** (2020), 123.

[Du] B. Dubrovin. *Inverse problem for periodic finite-zoned potentials in the theory of scattering.* Funct. Anal. Appl., **9** (1975), 61–62.

[IM] A. Its and V. Matveev. *Hill's operator with finitely many gaps.* Funct. Anal. Appl. (1975), pages 65–66.

[It] A. Its. *Canonical systems with finite-gap spectrum and periodical solutions of non-linear Schrödinger equation.* Vestn. LGU. Ser. matem., **7** (1976), 121–129.

[DMN] B. Dubrovin, V. Matveev, and S. Novikov. *Non-linear equations of Korteweg-de Vries type, finite-zone linear operators, and Abelian varieties.* Russ. Math. Surveys, **31** (1976), 59–146.

[FMcL] H. Flashka and D. McLaughlin. *Canonically conjugate variables for the Korteweg-de Vries equation and the Toda lattice with periodic boundary conditions.* Progr. Theoret. Phys., **55** (1976), 438–456.

[Mu] D. Mumford. *Tata Lectures on Theta II.* Modern Birkhäuser Classics book series. Birkhäuser, Boston, 2007.

[Sk3] E. Sklyanin. *Quantum version of the method of inverse scattering problem.* J. Soviet Math., **19** (1982), 1546–1596.

[Kv] S. Kovalevskaya. *Sur le problème de la rotation d'un corps solide autour d'un point fixe.* Acta Math., **12** (1889), 177–232.

[NS] A. Nakayashiki and F. Smirnov. *Cohomologies of affine hyperelliptic Jacobi varieties and integrable systems.* Commun. Math. Phys., **217** (2001), 623–652.

Index

Q-operator, 41
$U_q(\widehat{\mathfrak{sl}}_2)$, 25
$U_q\mathfrak{b}^+$, 25
$U_q\mathfrak{b}^-$, 25
φ-divisor, 85
q-exact 2-form, 54
q-exact form, 52
q-oscillator algebra, 27
3j-basis, 144

$U_q(\mathfrak{sl}_2)$, 23

Abel transformation, 176
adjoint action, 11
adjoint monodromy matrix, 51
algebraic Bethe ansatz, 19
annihilation operators, 60
antipode, 22
auxiliary space, 27

Bazhanov-Lukyanov-Zamolodchikov construction, 27
Bethe equations, 20
Bratteli diagram, 143

canonical second kind differential, 173
Casimir element, 23
Casimir elements (Sklyanin algebra), 157
classical monodromy matrix, 174
completeness (homogeneous case), 106
completeness (inhomogeneous case), 79
coproduct, 22
counit, 22
crossing symmetry relation, 11

deformed Abelian differential, 85
deformed Abelian integral, 86
deformed normalised differential of the second kind, 90
deformed second kind Abelian differential, 86
density matrix, 16
Destri-de Vega equation, 45
disordered phase, 2
divisor, 51
domain-wall partition function, 21
dressed resolvent, 111
dressed Russian doll, 73
Drinfeld generators, 25
dual reduced qKZ equation, 11

entanglement entropy, 16
evaluation homomorphism, 26

fusion procedure, 37

Hopf algebra, 22

inhomogeneous lattice, 9
invariant operators, 141
inverse problem, 118

Jacobi inversion formula, 177
Jacobi variety, 176

left (right) admissible operator, 41
left reduction, 52
length, 47

Matsubara direction, 1
Matsubara monodromy matrix, 4
Matsubara transfer matrix, 4

normalised holomorphic differential, 172

normalised third kind differential, 173

ordered phase, 2

quantum space, 27
quantum Wronskian relation, 43
quasi-intertwiner, 50
quasi-local, 6

R-matrix, 8
reduced qKZ equation, 10
Riemann bilinear relations, 172
right reduction, 52
Russian doll construction, 72

Schur polynomial, 123
separation of variables, 174
six vertex model, 1
Sklyanin algebra, 157
Space direction, 1
spectral parameter, 8
spin, 3
spin 2 annihilation operator, 105
support, 47

TQ equation, 20

universal R-matrix, 23

Verma module, 24

weight, 3

Yang-Baxter equation, 8
Young diagram, 123

Zamolodchikov algebra, 101

Selected Published Titles in This Series

256 **Michio Jimbo, Tetsuji Miwa, and Fedor Smirnov,** Local Operators in Integrable Models I, 2021

252 **Julie Déserti,** The Cremona Group and Its Subgroups, 2021

251 **David Hoff,** Linear and Quasilinear Parabolic Systems, 2020

250 **Bachir Bekka and Pierre de la Harpe,** Unitary Representations of Groups, Duals, and Characters, 2020

249 **Nikolai M. Adrianov, Fedor Pakovich, and Alexander K. Zvonkin,** Davenport–Zannier Polynomials and Dessins d'Enfants, 2020

248 **Paul B. Larson and Jindrich Zapletal,** Geometric Set Theory, 2020

247 **István Heckenberger and Hans-Jürgen Schneider,** Hopf Algebras and Root Systems, 2020

246 **Matheus C. Bortolan, Alexandre N. Carvalho, and José A. Langa,** Attractors Under Autonomous and Non-autonomous Perturbations, 2020

245 **Aiping Wang and Anton Zettl,** Ordinary Differential Operators, 2019

244 **Nabile Boussaïd and Andrew Comech,** Nonlinear Dirac Equation, 2019

243 **José M. Isidro,** Jordan Triple Systems in Complex and Functional Analysis, 2019

242 **Bhargav Bhatt, Ana Caraiani, Kiran S. Kedlaya, Peter Scholze, and Jared Weinstein,** Perfectoid Spaces, 2019

241 **Dana P. Williams,** A Tool Kit for Groupoid C^*-Algebras, 2019

240 **Antonio Fernández López,** Jordan Structures in Lie Algebras, 2019

239 **Nicola Arcozzi, Richard Rochberg, Eric T. Sawyer, and Brett D. Wick,** The Dirichlet Space and Related Function Spaces, 2019

238 **Michael Tsfasman, Serge Vlăduţ, and Dmitry Nogin,** Algebraic Geometry Codes: Advanced Chapters, 2019

237 **Dusa McDuff, Mohammad Tehrani, Kenji Fukaya, and Dominic Joyce,** Virtual Fundamental Cycles in Symplectic Topology, 2019

236 **Bernard Host and Bryna Kra,** Nilpotent Structures in Ergodic Theory, 2018

235 **Habib Ammari, Brian Fitzpatrick, Hyeonbae Kang, Matias Ruiz, Sanghyeon Yu, and Hai Zhang,** Mathematical and Computational Methods in Photonics and Phononics, 2018

234 **Vladimir I. Bogachev,** Weak Convergence of Measures, 2018

233 **N. V. Krylov,** Sobolev and Viscosity Solutions for Fully Nonlinear Elliptic and Parabolic Equations, 2018

232 **Dmitry Khavinson and Erik Lundberg,** Linear Holomorphic Partial Differential Equations and Classical Potential Theory, 2018

231 **Eberhard Kaniuth and Anthony To-Ming Lau,** Fourier and Fourier-Stieltjes Algebras on Locally Compact Groups, 2018

230 **Stephen D. Smith,** Applying the Classification of Finite Simple Groups, 2018

229 **Alexander Molev,** Sugawara Operators for Classical Lie Algebras, 2018

228 **Zhenbo Qin,** Hilbert Schemes of Points and Infinite Dimensional Lie Algebras, 2018

227 **Roberto Frigerio,** Bounded Cohomology of Discrete Groups, 2017

226 **Marcelo Aguiar and Swapneel Mahajan,** Topics in Hyperplane Arrangements, 2017

225 **Mario Bonk and Daniel Meyer,** Expanding Thurston Maps, 2017

224 **Ruy Exel,** Partial Dynamical Systems, Fell Bundles and Applications, 2017

223 **Guillaume Aubrun and Stanisław J. Szarek,** Alice and Bob Meet Banach, 2017

222 **Alexandru Buium,** Foundations of Arithmetic Differential Geometry, 2017

For a complete list of titles in this series, visit the
AMS Bookstore at **www.ams.org/bookstore/survseries/**.